Advances in Environmental Microbiology

Volume 5

Series Editor
Christon J. Hurst
Cincinnati, Ohio
USA

and

Universidad del Valle
Santiago de Cali, Valle
Colombia

More information about this series at http://www.springer.com/series/11961

Christon J. Hurst
Editor

The Connections Between Ecology and Infectious Disease

Springer

Editor
Christon J. Hurst
Cincinnati, Ohio, USA

Universidad del Valle
Santiago de Cali, Valle, Colombia

ISSN 2366-3324 ISSN 2366-3332 (electronic)
Advances in Environmental Microbiology
ISBN 978-3-030-06419-8 ISBN 978-3-319-92373-4 (eBook)
https://doi.org/10.1007/978-3-319-92373-4

Softcover re-print of the Hardcover 1st edition 2018

Cover illustration: "Volvox reimagined" is a montage image created in 2015 and is being used with permission of the artist, Christon J. Hurst. Those images incorporated into this montage were: center image "Aspergillus flavus" (author: Hossein Mirhendi, image supplied by and used with author's permission); clockwise from upper right "Micrasterias" (author: Rogelio Moreno Gill, image supplied by and used with author's permission), "BrownGiantKelp3600ppx" (author: FASTILY, Creative Commons Attribution Share Alike 3.0 Unported license), "Stumpfungus" (author: Ecornerdropshop, public domain image), "Letharia vulpina JHollinger crop" (author: Jason Hollinger, Creative Commons Attribution-Share Alike 3.0 Unported license), "Phytoplankton - the foundation of the oceanic food chain", (author: NOAA MESA Project, public domain image), "Spirogyra cell" (author: Jasper Nance, Creative Commons Attribution-Share Alike 3.0 Unported license.

This Springer imprint is published by the registered company Springer Nature Switzerland AG
The registered company address is: Gewerbestrasse 11, 6330 Cham, Switzerland

I am very fortunate by having been allowed two set of parents in this lifetime. My parents by birth were Hubert and Joy Hurst. They met when my father, Hubert, was a college student and would eat his noontime meal at the luncheon counter of a pharmacy in Cincinnati, Ohio near where he was studying. My mother, Joy, worked there at the pharmacy and perhaps the best part of my father's day was a smile that she served to him along with that daily lunch as the two of them fell into romance. He graduated, they married, and then they had three children which were in order of birth my brother Preston, myself, and our sister Embeth. Hubert and Joy loved their children and they generally tolerated our childhood misdeeds. That tolerance required patience because some of my misdeeds must have seemed nearly intolerable! But, most importantly for me, my parents encouraged my efforts to understand science and lovingly they showed pride of my achievements.

Joy and Hubert Hurst on vacation in 1975

When I still was a young child, my family moved into a large housing complex in Cincinnati, Ohio where I soon met Rebecca Wadlin and also met Becky's older sister Caroline, plus their parents Herbert and Renko. It often must have seemed as though Renko and Herbert had five children instead of just their two, Carrie and Becky, because my brother, sister, and I spent a great many hours in their apartment and Herb invited us along on numerous local adventures. Herb was from New York state and fortunately met Renko in Tokyo, Japan while he was on assignment there with the United States Air Force after the Second World War had ended. When they met, Renko was studying English and her instructor had invited Herb to speak with their class. Herb and Renko corresponded when he was assigned to Okinawa. Eventually, they reunited in Tokyo, where they married and had two children. Herbert then brought their family to Cincinnati so that he could study here for a Master's Degree which he succcessfully earned. It was a sad day for me when their family moved onward to California. Decades later, Herb and Renko's family informally adopted me and that has brought me great pleasure. I happily have considered their family as also being my mother, father, and sisters.

Renko and Herbert Wadlin with Christon Hurst in 2003

My personal beliefs being what they are, I also wish to thank Johann Meyer and

Eleonora Diebold Meyer who were the parents of Guilelmina Eleonora "Mina" Meyer. John and Ellen, as they came to be known, were immigrants to the United States from the Kingdom of Württemberg. Mina was born in Sedamsville, which was then just slightly west of Cincinnati's city boundary in Hamilton County, Ohio. She was baptized on November 7th, 1868 in Saint Michael the Archangel church, located in the neighboring Lower Price Hill section of Cincinnati. Unfortunately, Mina passed away in Lower Price Hill on May 28th, 1882 and she was laid to rest at a pretty location on top of the hill, at Saint Joseph Cemetery in the Price Hill neighborhood of Cincinnati. I do as well thank Dr. Max Cohn and Anna Weiland Cohn who were the parents of Herbert Max Cohn. Herbert Cohn was born in Berlin, Germany on August 23rd, 1900. Eventually Herbert lived as a family with his parents, along with his wife Ursula Sara Gerling Cohn and their own son Charlie Max Cohn, in Bad Harzburg, Germany where his family operated their home as a pension called Haus Frohsinn and Villa Frohsinn.

Postcard image of Haus Frohsinn in Bad Harzburg

Sadly, Herbert Cohn was arrested from their Bad Harzburg home in November, 1938 as part of the Reichspogromnacht raids and taken to Buchenwald Concentration Camp, Germany where he became ill. Herbert Cohn died in Buchenwald Camp on November 30th, 1938 and the historical records suggest that he was shot, listing even the minute of his death. Herbert's father Max was attacked by a mob in Bad Harzburg and he thus died on December 26th, 1938. Despite the tragedy of those two months, things ended better for the rest of Herbert's family. Herbert's wife Ursula and their son Charles fortunately were sponsored into England as refugees during April, 1939 where they survived the Second World War and remained living near London. Ursula eventually had a second chance at romance and in September, 1945 remarried which gave Herbert's son a new father. Herbert's son remains alive and in good health. Anna eventually returned to living in the familys Bad Harzburg home in 1940 and there she survived the war. Their home in Bad Harzburg still stands and has been well kept. Bad Harzburg thoughtfully memorialized both Herbert and his father with their names on a plaque in the local cemetery. I can attest that Bad Harzburg again is a cheerful place and that the people now residing there are very kind.

Christon Hurst at memorial in Bad Harzburg Cemetary July 25th 2013

Series Preface

The light of natural philosophy illuminates many subject areas including an understanding that microorganisms represent the foundation stone of our biosphere by having been the origin of life on Earth. Microbes therefore comprise the basis of our biological legacy. Comprehending the role of microbes in this world which together all species must share, studying not only the survival of microorganisms but as well their involvement in environmental processes, and defining their role in the ecology of other species, does represent for many of us the Mount Everest of science. Research in this area of biology dates to the original discovery of microorganisms by Antonie van Leeuwenhoek, when in 1675 and 1676 he used a microscope of his own creation to view what he termed "animalcula," or the "little animals" which lived and replicated in environmental samples of rainwater, well water, seawater, and water from snow melt. Van Leeuwenhoek maintained those environmental samples in his house and observed that the types and relative concentrations of organisms present in his samples changed and fluctuated with respect to time. During the intervening centuries we have expanded our collective knowledge of these subjects which we now term to be environmental microbiology, but easily still recognize that many of the individual topics we have come to better understand and characterize initially were described by van Leeuwenhoek. van Leeuwenhoek was a draper by profession and fortunately for us his academic interests as a hobbyist went far beyond his professional challenges.

It is the goal of this series to present a broadly encompassing perspective regarding the principles of environmental microbiology and general microbial ecology. I am not sure whether Antonie van Leeuwenhoek could have foreseen where his discoveries have led, to the diversity of environmental microbiology subjects that we now study and the wealth of knowledge that we have accumulated. However, just as I always have enjoyed reading his account of environmental microorganisms, I feel that he would enjoy our efforts through this series to summarize what we have learned. I wonder, too, what the microbiologists of still future centuries would think of our efforts in comparison with those now unimaginable discoveries which they will have achieved. While we study the many wonders of microbiology, we also further our recognition that the microbes are our biological critics, and in the end they undoubtedly will have the final word regarding life on this planet.

Christon J. Hurst in Heidelberg

Indebted with gratitude, I wish to thank the numerous scientists whose collaborative efforts will be creating this series and those giants in microbiology upon whose shoulders we have stood, for we could not accomplish this goal without the advantage that those giants have afforded us. The confidence and very positive encouragement of the editorial staff at Springer DE has been appreciated tremendously and it is through their help that my colleagues and I are able to present this book series to you, our audience.

Cincinnati, OH Christon J. Hurst

Volume Preface

Infectious disease is part of an interconnected ecology that involves a pathogenic microorganism and the host species in which that microbe causes illness. The ecology of that microorganism may include a more extended set of connections which could involve a natural environmental presence, its possible carriage by vehicles such as air, water, and food, and interactions with other host species including vectors for which the microbe either may or may not be pathogenic. This book explains these connections. The contents of this volume are divided into three sections, the first of which presents an introduction to the field of science currently titled disease ecology and explains both the role of biological community interactions plus the impact of biodiversity. The second section considers diseases directly affecting humans with a focus on waterborne and foodborne illnesses and importantly examines the critical aspect of microbial biofilms. The third section presents ecology of infectious diseases from the perspective of their impact upon mammalian livestock and wildlife plus includes understanding of the fact that those same diseases often affect humans.

Decades ago I presented two specialty courses on disease ecology at Universidad del Valle in Cali, Colombia for their graduate students in the School of Public Health and professional members of the community. The first of those courses was titled "La Ecología de la Transmisión de Infecciones y la Aplicación de Modelos en Salud Pública" [The Ecology of Infection Transmission and the Application of Models in Public Health] during August and September of 1998, which was sponsored by the Fulbright program. The second course was titled "La Ecología de Enfermedades Infecciosas Transmitidas por el Agua" [The Ecology of Infectious Diseases Transmitted by Water] during March of 2000, for which I served as an International Professor for Latin America and was sponsored by the American Society for Microbiology. Including those courses among my previous experience, I presumed that I would be qualified to write the introductory chapter for this book. However, while reading the chapters written by this books other contributing authors, the realization came to me that those others have a level of expertise which now outweighs mine and so the banner representing his subject area clearly has been passed onward. Many of these authors and also our readers for this book represent the next generation of scientists to be exploring this field of endeavor and I feel reassurance from knowing how well they and you will proudly carry the banner forward.

I am tremendously grateful to Hanna Hensler-Fritton, Andrea Schlitzberger, and Isabel Ullmann at Springer DE, for their help and constant encouragement which has enabled myself and the other authors to achieve publication of this collaborative project.

Cincinnati, OH Christon J. Hurst

Contents

Part I
Introduction to Disease Ecology

Chapter 1
Interkingdom Community Interactions in Disease Ecology

M. Eric Benbow, Jennifer L. Pechal, Jeffery K. Tomberlin, and Heather R. Jordan

Abstract A key tenet of community ecology is the interactions of individual organisms contribute to the ecological structure and function of ecosystems. Within these networks of interacting organisms are those taxa important for human and animal health: disease systems defined by combinations of host, pathogen, reservoir, and vector or a subset of these components. While the simplest disease system is that of the host and pathogen, more complex systems include the direct interactions of a pathogen with other hosts and the microbial communities of those hosts, reservoirs, and sometimes vectors. Each of these disease system components is made up of species that directly and indirectly interact with other species in ways that affect their individual fitness, population biology, and role in communities of the ecosystem. This chapter recognizes the direct interactions of those species that make up the primary components of disease systems; however, the focus and examples provided relate to the more indirect interkingdom (or domain) interactions that impact disease system components. The examples provided include how microbial communities mediate invertebrate and vertebrate fitness and behavior, often in systems where the

M. E. Benbow (✉)
Department of Entomology, Michigan State University, East Lansing, MI, USA

Department of Osteopathic Medical Specialties, Michigan State University, East Lansing, MI, USA

Ecology, Evolutionary Biology and Behavior Program, Michigan State University, East Lansing, MI, USA
e-mail: benbow@msu.edu

J. L. Pechal
Department of Entomology, Michigan State University, East Lansing, MI, USA
e-mail: pechalje@msu.edu

J. K. Tomberlin
Department of Entomology, Texas A&M University, College Station, TX, USA
e-mail: jktomberlin@tamu.edu

H. R. Jordan
Department of Biological Sciences, Mississippi State University, Mississippi State, MS, USA
e-mail: jordan@biology.msstate.edu

C. J. Hurst (ed.), *The Connections Between Ecology and Infectious Disease*, Advances in Environmental Microbiology 5,
https://doi.org/10.1007/978-3-319-92373-4_1

hosts play important roles in pathogen transmission and disease emergence. The potential mechanisms of these interkingdom interactions are also developed in detail, as the mechanisms of such interactions are likely the target of future studies that could directly inform disease management strategies. Based on these examples and mechanisms, the existing literature suggests there are likely undiscovered and complex interactions of species within communities that affect disease systems.

1.1 Introduction

A key tenet of community ecology is the interactions of individual organisms contribute to the biological structure and function of ecosystems. Within ecosystems, infectious disease systems (i.e., the network of species involved in the existence and emergence of disease caused by pathogenic microorganisms) are defined by at least two species interacting in a way that leads to a pathogenic outcome affecting at least one of those species. This minimum system is composed of the pathogenic microorganism species (e.g., bacteria, virus, fungi, or eukaryotic parasite; hereafter referred to as pathogen) and a host species, and inherently implies a minimal interaction between the two with the host negatively affected by the pathogen (Fig. 1.1). This minimum infectious disease system is the simplest scenario in disease ecology defined as an interkingdom interaction or, as we define here, the ecological relationship where two or more species from different domains or kingdoms of life interact in a way that has a demonstrated or hypothesized biological, ecological, or evolutionary importance to one or all of the species in the disease system.

Disease ecology involves at least two different species that define a disease system within a larger community of organisms in nature. However, many disease systems of humans, other animals, and plants involve multiple intermediate species with the pathogen transmitted to one or more host or reservoir species (Fig. 1.1). These multi-species pathogen transmission pathways are also interkingdom interactions that lead to the movement of the pathogen in the environment, and ultimately to the host, resulting in pathogenesis and disease symptoms. Therefore, in its simplest form, understanding the ecology of a disease rests on defining the components, constraints, and dynamics of interacting organisms from different domains and kingdoms. The discipline of disease ecology also includes asking questions about coevolutionary relationships and the broader impact of species interactions on the community that are not a direct component of the disease system but might indeed be influential (Burdon and Thrall 2008; Ricklefs 2010). The more traditional interkingdom interactions defining many infectious disease systems will be covered in this chapter; however, this will be done with a focus on how the interactions mediate transmission and pathogenesis. This chapter will also cover how interkingdom interactions may influence infectious disease systems without the interacting organisms being any of the disease system component species (i.e., pathogen, host, reservoir, vector; Fig. 1.1). Further, because an entire book could

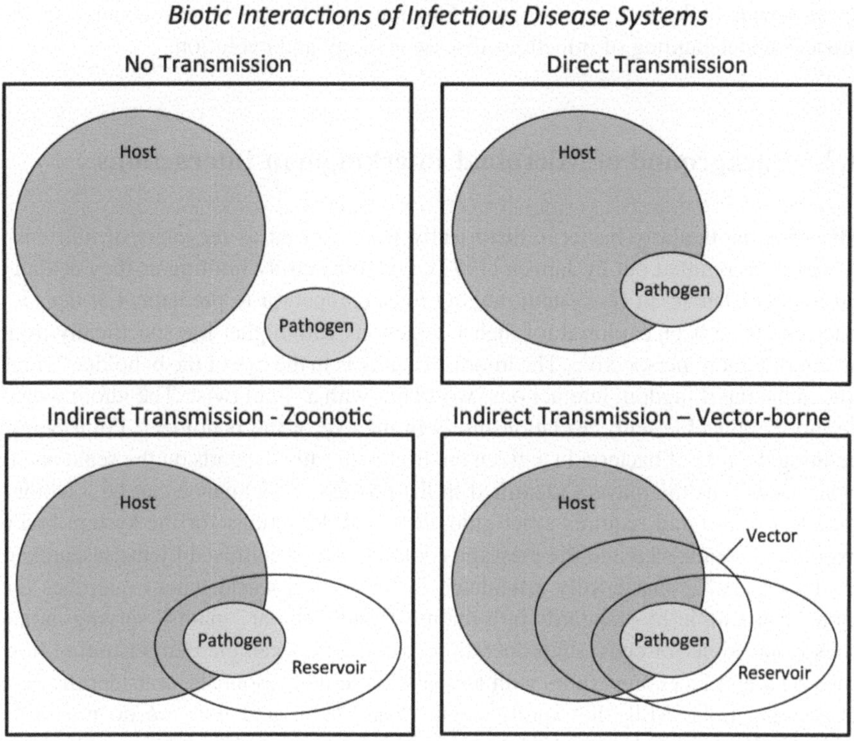

Fig. 1.1 Conceptual depiction of the interkingdom interactions involved in an infectious disease system. When there is no transmission (top left), there are no interactions among organisms involved in the disease system. When there is direct transmission (top right), the only interaction is with the host and pathogen—the minimal disease system. Indirect zoonotic transmission involves the interaction of the host, a reservoir (often animal), and pathogen (bottom left), whereas indirect, vector-borne transmission includes these interactions but also a fourth, vector species (bottom right)

be written on all interkingdom interactions important to disease ecology, this chapter will highlight recent advancements in understanding how microbial taxa influence species of other domains and kingdoms. By introducing aspects and examples of interkingdom interactions that have been investigated in the primary ecological literature, the chapter will synthesize how this understanding can influence the scientific approach to studying disease ecology.

Lastly, disease ecology is a discipline involving the study of how hosts, reservoirs, vectors, and pathogens interact within and among populations and communities in a way that ultimately defines ecosystem structure. Since populations are inherent components of communities, we focus here on the interactions among different species within larger communities of organisms where disease systems are nested. While it is acknowledged that specific examples of interkingdom interactions occur within a broader and more complex ecological network, we focus on examples and provide discussion of interactions that often involve two or three species with biological or ecological function, which may have relevance to the broader understanding of infectious disease ecology and evolution.

1.2 Background of Microbial Interkingdom Interactions

Microbes, particularly bacteria, historically were viewed as recyclers of nutrients; however, as pointed out by Janzen (1977), this role is truly limiting as they actually fill many niches in an ecosystem ranging from competitor to predator. For decades microbes have been explored for their interactions with higher life specifically from an interkingdom perspective. The idiom, “Beauty is in the eye of the beholder,” rings true in the interkingdom interactions’ world but with a slight twist. The idiom would be more applicable with “Relationship is in the eye of the beholder.” Point being, defining the role of bacteria in a given environment truly depends on the scale one is using as well as the players identified in the process. This nuance can be a tedious road to explore and requires strict guidelines and definitions for the system being explored. One misstep and the message could be lost or confused by the researcher.

Janzen (1977) masterfully articulated in his view a world where microbes are more than simple by-standards in a realm of multicellular “giants” serving as the cogs responsible for ecosystem operations. As a great example demonstrating how bacteria are in fact competitors with animals for similar resources, consider the case of spoiling milk. Milk that spoils smells “bad” to us and thus we do not drink it. Janzen’s presumed interpretation would be bacteria modified the resource in a manner resulting in reduced competition from vertebrates that would otherwise drink this resource that is common to the bacteria. While this example is anecdotal, others have tested his hypothesis and empirically demonstrated this competition does occur in nature. Burkepile et al. (2006) demonstrated reducing microbes associated with carrion in shoreline estuaries allows for crustacean scavengers to feed for longer periods of time on the decaying fish resource; however, like with the red queen (Van Valen 1973), such an arms race opens opportunities for other organisms to take advantage of these resources. Thus, as the resource decomposes and the interest of consumers of fresh carrion decreases, other animals are attracted to and consume the degraded resource.

This interaction is prominent in the vertebrate carrion literature, specifically when one considers vultures and their affinity for decomposing vertebrate remains (Kruuk 1967). These species have developed an ability to tolerate bacteria-laden resources. Even brown tree snakes are known to use volatile organic compounds (VOCs) resulting from microbial activity in mouse carcasses as a means to locate these resources (Shivik and Clark 1997).

While insects, such as blow fly (Diptera: Calliphoridae) larvae, competing for these resources can release antimicrobial products that kill competing or pathogenic bacteria (Greenberg 1968) on the resource, the bacteria may also affect the behavior (Ma et al. 2012) and ecology (e.g., oviposition) (Tomberlin et al. 2012) of flies on a resource. For a comprehensive review of these interkingdom interactions involving bacteria and insects, see Tomberlin et al. (2017a). Briefly, bacterial communities are known to go through community succession on decomposing vertebrate remains (Pechal et al. 2013). During this process, blow fly attraction to carrion can be related to the sex and physiological state-specific aspects of the blow flies (gravid vs non-gravid) (Mohr and Tomberlin 2014, 2015). Subsequent research has determined that the potential mechanisms regulating these blow fly behavioral responses (e.g., attraction and colonization) are in part regulated by VOCs associated with microbial degradation of essential amino acids associated with the resource (Liu et al. 2016). Further, Flores et al. (2017) demonstrated that bacteria associated with blow fly larvae impact the development of larvae of a competing species. These data by Flores et al. (2017), as well as several other studies of similar interkingdom interactions (Tomberlin et al. 2012; Pechal et al. 2013; Mohr and Tomberlin 2014, 2015; Liu et al. 2016), provide new data suggesting that microbes could be serving a mechanism regulating arthropod succession patterns during colonization of vertebrate carrion. These interactions could result in pathogenic bacteria associated with a decomposing resource to either proliferate or be suppressed. Thus, complex interkingdom interaction may mediate pathogen emergence and dispersal in natural environments.

Additional studies have demonstrated that interkingdom interactions may manifest within trophic networks as well. Previous research has shown that insects, such as blow fly (Greenberg 1971a; Greenberg and Klowden 1972) and house fly [*Musca domestica* L. (Diptera: Muscidae)] (Zurek et al. 2001) larvae, consume microbes associated with decomposing organic matter. Some microbes are thought to be digested and assimilated, while others are thought to pass through the alimentary canal (Mumcuoglu et al. 2001), thus being potentially dispersed to resources in other habitats, as reviewed in Nayduch (2017). Microbial pathogen dispersal in the environment via flies is a key aspect of the transmission of some foodborne illnesses (Greenberg 1971b). For instance, Weatherbee et al. (2017) described interkingdom interactions of microbial communities with blow fly larvae on swine carcasses where taxa from the carcass became integrated into the internal microbiome of feeding and developing larvae, with distinct shifts in microbial phyla composition that showed convergence of the carcass and larval microbiomes later in decomposition.

Further Pechal and Benbow (2016) demonstrated that the internal microbial communities of aquatic insects in streams with decomposing salmon were structurally different from the same insect taxa from streams without carcasses or historic salmon spawning. Similar to Weatherbee et al. (2017), they also reported the internal microbial communities of larvae collected from masses of the salmon carcasses were similar to the carcass microbial communities. These field studies confirmed previous findings from culture-based laboratory research of bacteria that showed differential acquisition of microbial taxa from decomposing organic matter by blow flies and other filth flies (see review of Nayduch 2017). Lastly, Pechal and Benbow (2016) showed in the same field studies that the internal microbial communities of adult blow flies captured above salmon carcasses varied in time, and adults captured nearer the time of larval collections had microbiomes more similar to the larvae than adults captured 10 days later. These results suggest the microbial ecology of carcass microbial communities in an ecosystem plays important roles in determining the internal microbial communities of both larvae that colonize and consume the resource but likely also influence adults that emerge from those larvae and then disperse into the landscape.

While descriptive, the previous examples provide evidence that the dispersal of microbes into surrounding communities and ecosystems may involve aspects of trophic ecology that provided the foundation for more deeply rooted evolutionary relationships, an area of inquiry that requires additional investigation. The current mechanisms of these potentially evolved interactions are only recently being uncovered, with recent findings showing a future for exciting avenues of research. This new area of research is especially true for those interkingdom interactions among prokaryotic microbes and eukaryotes.

1.3 Mechanisms of Interkingdom Interactions in Disease Ecology

Eukaryotes have variable relationships with prokaryotes, and these interactions can range from beneficial to detrimental. These associations are clearly facilitated by complex bidirectional communication taking place between them and that information exchange, physical interactions, and chemical signaling likely resulted from coevolutionary processes. Furthermore, though bacterial growth and virulence are influenced by local environmental parameters such as temperature, pH, and nutrient availability, the influence of host chemical stimuli on bacterial behavior has only recently become apparent (Rumbaugh 2007; Hughes and Sperandio 2008). Microbial receptors are able to recognize that the particular microbe is within the locality of a suitable host and, for commensals, that it is the appropriate time to initiate expression of genes involved in host colonization. Pathogens can then hijack these signals, leading to activation of their virulence genes (Rumbaugh 2007; Hughes and Sperandio 2008). Moreover, hosts and microbes have developed multiple mechanisms to protect themselves from each other and have, in some cases, also evolved mechanisms that allow a mutualistic coexistence. In this section, we discuss several mechanisms that are used for communication interaction and exchange between bacteria and their hosts. We focus particularly on methods of information transfer, as well as mechanisms for detection of and response to host signals. Deciphering such communication is needed for understanding the evolutionary biology of signal development and information exchange. Also, in terms of practical applications, greater understanding of the mechanisms mediating these processes could lead to strategies which disrupt the more damaging aspects of the information exchange and exploitation of the more beneficial and efficient segments within each process (Fig. 1.2).

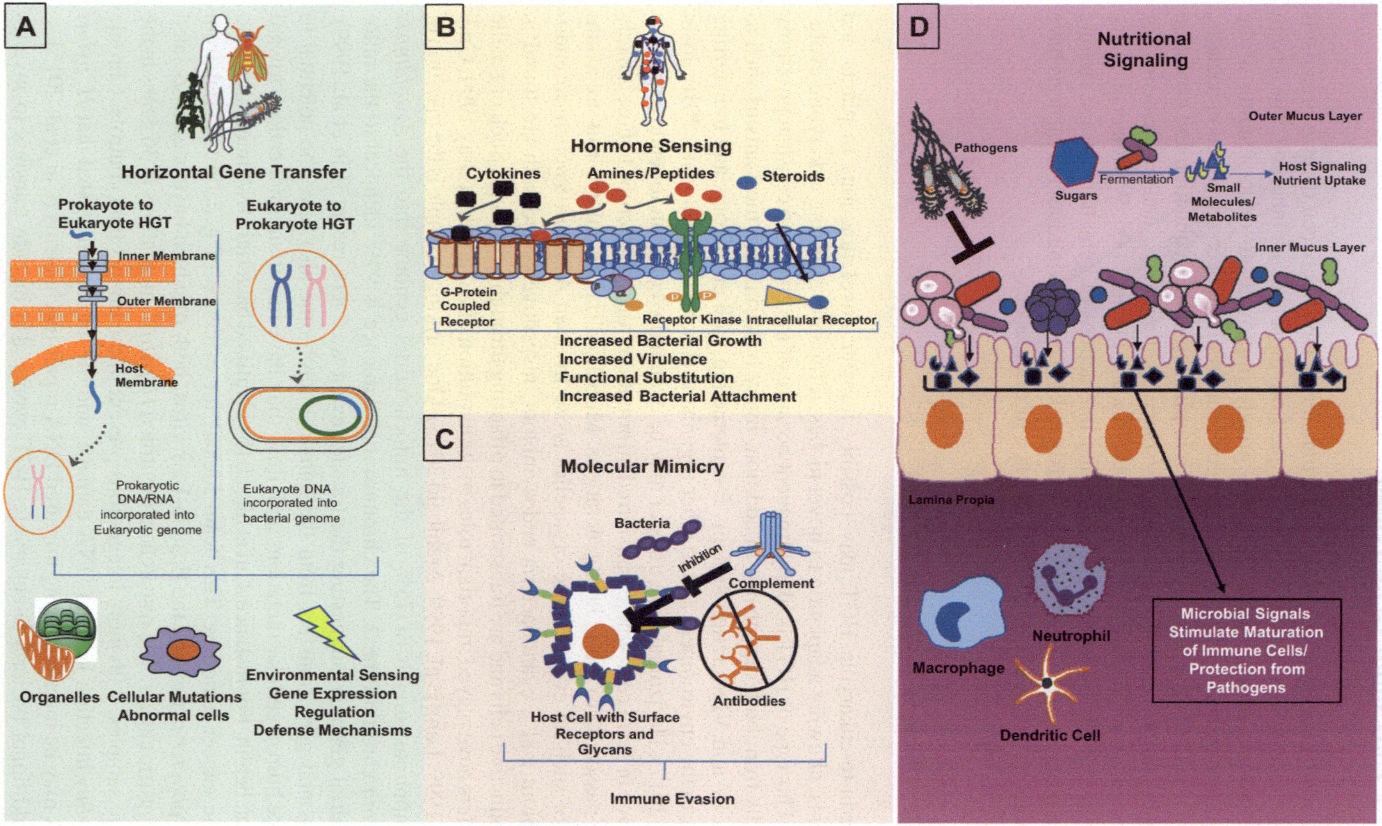

Fig. 1.2 Conceptual mechanisms and consequences of interkingdom communication between hosts and microbes through (**a**) horizontal gene transfer, (**b**) hormone signaling, (**c**) molecular mimicry, and (**d**) nutritional signaling

1.3.1 *Horizontal Gene Transfer*

Horizontal gene transfer (HGT) is the transfer of DNA between diverse organisms leading to acquisition of novel traits unique from those that are inherited. The advent of large-scale genome sequencing has greatly improved our understanding of the importance of HGT. Most cases of genetic transfers have been described in bacteria, where transfer between bacteria has been implicated in the acquisition and evolution of many traits including antibiotic resistance, pathogenesis, and mechanisms for bioremediation (Hooper et al. 1999; de la Cruz and Davies 2000; Werren et al. 2010; Ruzzini and Clardy 2016). The molecular interactions of gene transfer and sharing are an intimate example of how previously unknown interkingdom interactions have the potential to greatly expand the science of disease.

1.3.1.1 Bacteria-to-Eukaryote HGT

Bacteria-to-eukaryote HGT may possibly transfer novel functions to plants and animals, allowing adaptation to novel niches, ultimately affecting eukaryotic evolution. This type of HGT is likely frequent because of the close and constant proximity of cells from both organisms. In addition, many bacteria have mechanisms such as types IV and VI secretion systems for interacting with eukaryotes and transferring DNA to the environment and to other organisms (Lacroix and Citovsky 2016; Joshi et al. 2017). Horizontal gene transfer may be even more common for endosymbionts that live in germ cells where transferred information from the microbe to host is likely to be passed on to future host generations. Mitochondria and chloroplasts are examples of such an event. These organelles arose from an *Alphaproteobacteria* endosymbiont and a cyanobacteria endosymbiont, respectively, and both are found in reproductive cells and are transmitted to offspring through germ cells. Such organelle transfers have played important roles in the evolution of eukaryotes (Gray 1999; Gray et al. 1999; Lang et al. 1999; Archibald 2015). One striking example is the insertion of a continuous stretch of 75% of the mitochondrial genome into the *Arabidopsis thaliana* chromosome 2. The chromosomal sequence is 99% identical to the mitochondrial genome, suggesting the transfer event was very recent (Lin et al. 1999). Horizontal gene transfers from the human microbiome also have the potential to induce human cellular mutations and, thus, may have important implications toward human health. For instance, Cui et al. (2015) identified bacterial DNA integration into gastric cancer genomes. This team looked for evidence of genetic material from *Helicobacter pylori* and the Epstein-Barr virus, both of which have been associated with gastric cancer. The researchers identified *H. pylori* integrations in 36 genes in the gastric samples, with more integrations present in the tumors than controls. Chronic *H. pylori* infection leads to DNA damage, and the authors suggested that *H. pylori* DNA may be inserted during impaired DNA repair mechanisms (Cui et al. 2015).

Additional bacteria-to-eukaryote HGT events also include transfers to invertebrates and are particularly prevalent from endosymbiotic bacteria to their host

(Lacroix and Citovsky 2016). The first *Wolbachia*-host transfer described was a variant of the bean beetle *Callosobruchus chinensis* that contained *Wolbachia* genes in its chromosome, and a later study found that approximately 30% of a *Wolbachia* genome was found on the X chromosome (Kondo et al. 2002; Nikoh et al. 2008). *Wolbachia*-host transfer has also been described in the filarial nematode *Onchocerca* spp. (Fenn and Blaxter 2006; Fenn et al. 2006). The HGT to *Drosophila ananassae* Doleschall 1858 (Diptera: Drosophilidae) from its *Wolbachia* endosymbiont is the largest as the entire *Wolbachia* genome has been integrated into the *D. ananassae* chromosome (Dunning Hotopp et al. 2007). *Wolbachia* nuclear inserts were detected in four lines of geographically diverse *D. ananassae* indicating that the insert may be widely distributed (Dunning Hotopp et al. 2007). The authors also found that some of these inserted *Wolbachia* genes were transcribed within eukaryotic cells lacking endosymbionts, suggesting lateral gene transfer occurred into eukaryotic hosts from their prokaryote symbionts, potentially providing a mechanism for acquisition of new genes and functions (Dunning Hotopp et al. 2007).

Another example involves transfers among parasitoid wasps (Hymenoptera), poxviruses, and the endosymbionts, *Wolbachia* and *Orientia tsutsugamushi* (Werren et al. 2010). The authors examined protein domain arrangements in the wasp *Nasonia vitripennis* (Walter 1836) (Hymenoptera: Pteromalidae) and found HGT involving poxviruses, *Wolbachia*, and *Nasonia*. They found 13 poxvirus protein repeats of ankyrin C-terminal domains. Phylogenetic analysis of the repeat sequences suggested that the *Nasonia* lineage acquired one or more of the proteins from *Wolbachia*, with subsequent amplification and divergence. The authors speculate that such lateral gene transfers between bacteria and eukaryotes could be an important source of evolutionary innovation (Werren et al. 2010).

1.3.1.2 HGT of Small RNAs

Small RNAs (sRNAs), including small interfering RNAs (siRNAs) and microRNAs (miRNAs), are conventionally regarded as critical molecular regulators of various intracellular processes. However, recent accumulating evidence indicates that sRNAs can be transferred within cells and tissues and even across species (Skippington and Ragan 2012; Cui et al. 2015; Frohlich and Papenfort 2016). In plants, nematodes, and microbes, these mobile sRNAs can mediate interkingdom communication, environmental sensing, gene expression regulation, host-parasite defense, and many other biological functions. Zhang et al. (2012) used sensitive high-throughput sequencing methods to detect high levels of plant miRNAs in serum and tissues of humans, mice, and calves with plant-based diets and found "cross-kingdom" functional RNA transfer. However, subsequent studies further explored the emerging field of sRNA-mediated cross talk between species; some groups reported negative results and questioned its general applicability (Witwer and Hirschi 2014). Thus, careful considerations are required in future studies.

1.3.1.3 Eukaryote-to-Bacteria HGT

Eukaryote-to-bacteria HGT should be expected to be widespread, as bacteria readily participate as both donors and recipients in HGT and lack the barriers to HGT possessed by eukaryotes, such as presence of a nucleus and segregated germ cells (Koraimann and Wagner 2014). Moreover, acquisition of eukaryotic genes by bacteria is of particular interest because of the possible role of such transferred genes potentially influencing bacterial pathogenicity. However, only a few eukaryote-to-bacteria transfers have been described, though spanning transfers to several bacterial phyla. For instance, *Legionella pneumophila* has been found to encode more than 100 eukaryotic-like proteins with phylogenetic analysis revealing that 29 are of eukaryotic ancestry (Lurie-Weinberger et al. 2010). Furthermore, there seems to be a modest but consistent excess of acquired eukaryotic genes in at least some parasites, such as *Mycobacterium tuberculosis*, *Pseudomonas aeruginosa*, *Xylella fastidiosa*, and *Chlamydia pneumoniae* (Koonin et al. 2001). Cyanobacterium *Synechocystis* spp. encodes a variety of proteins associated with different forms of signaling that have been considered eukaryotic (Kaneko and Tabata 1997; Ponting et al. 1999). The ancient transfer of two tubulin genes and a kinesin light chain gene to *Prosthecobacter dejongeii* has also been described (Jenkins et al. 2002). Some *Wolbachia* strains and numerous mosquito and lepidopteran species share phylogenetically related genes (Ahmed et al. 2016). Klasson et al. (2009) examined the sequenced genome of *Aedes aegypti* L. (Diptera: Culicidae) for any genes that could have originated in *Wolbachia* and found that one of the two *A. aegypti* genes identified showed female salivary gland-specific (SGS) expression. These gene and homologs in *Anopheles* are candidate *Plasmodium* sporozoite receptors, and the authors and others suggest the SGS-type mosquito genes might have arisen from an ancient transfer between *Wolbachia* and mosquitoes with HGT quite possibly happening in both directions, though this has been shown with only weak support (Korochkina et al. 2006; Klasson et al. 2009).

1.3.2 Hormonal Signaling

There are three broad categories of mammalian hormones: peptides (or proteins), steroids, and amines. The structure of a hormone dictates the location of its receptor. For instance, amine and peptide hormones cannot cross the cell membrane and are instead recognized and bound to cell-surface receptors such as receptor kinases and G-protein-coupled receptors, whereas steroid hormones can cross plasma membranes and are primarily recognized and bound to intracellular receptors. Peptide hormones include the epidermal growth factor (EGF), insulin, and glucagons. Steroid hormones are derived from cholesterol, and amines are synthesized from tyrosine. Amine hormones include the catecholamines epinephrine, norepinephrine (NE), and dopamine (Freestone 2013; Kendall and Sperandio 2016). All of these hormones are engaged in interkingdom interactions with microorganisms, where bacteria sense several of these hormones (Sperandio et al. 2003; Freestone 2013; Kendall and Sperandio 2016).

1.3.2.1 Amine Hormones

Epinephrine and NE were the first hormones recognized as influencing bacterial growth and virulence gene expression. Notably, these studies were conducted with bacteria that colonize the human GI tract such as enterohemorrhagic *Escherichia coli* (EHEC) (Kendall and Sperandio 2016). For instance, epinephrine and NE have been reported to enhance the growth of EHEC, possibly by aiding in iron acquisition. More recently, NE was shown to induce enterobactin expression and iron uptake in *E. coli*. Furthermore, epinephrine and NE influence virulence gene expression in EHEC, independent of their influence on EHEC growth. Epinephrine and NE do not cross the cellular membrane; therefore, EHEC relies on the two-component systems QseBC and QseEF, to relay adrenergic signals to the intracellular environment, leading to coordinated expression of genes encoding flagella and motility, attaching-and-effacing lesion formation on enterocytes, stress responses, potassium uptake, and osmolarity (Lyte 1992; Freestone et al. 2002, 2003; Chen et al. 2003; Sperandio et al. 2003; Lyte et al. 2011; Freestone 2013; Kendall and Sperandio 2016).

1.3.2.2 Peptide Hormones

Peptides constitute an extremely diverse class of hormones processed from a precursor and exported out of the cell (Freestone 2013). Gastrin, a peptide hormone secreted by stomach cells, stimulates the release of gastric acid and has been shown to increase the growth rate of *H. pylori*. This gastrin-induced phenotype was mediated by a peptide hormone, cholecystokinin, by the four C-terminal amino acids which are shared by gastrin, and a synthetic analogue, pentagastrin, showing the interaction between gastrin and *H. pylori* was specific and dependent on a defined gastrin sequence (Chowers et al. 1999, 2002). It is also noteworthy that this response is restricted to human gastrin, and this may explain the specificity of *H. pylori* to humans. Conversely, another gastric peptide hormone, somatostatin, which inhibits the secretion of gastric acid, suppresses *H. pylori* growth. Somatostatin has also been shown to directly bind to *H. pylori*, suppressing its proliferation (Yamashita et al. 1998). Furthermore, somatostatin antibody immunoglobulin G fraction blocked the antiproliferative effect of somatostatin, suggesting that *H. pylori* harbors an as-yet-unidentified receptor for this peptide (Yamashita et al. 1998; Kendall and Sperandio 2016). Moreover, *H. pylori* infection also changes the levels of these two hormones, increasing and decreasing gastrin and somatostatin levels, respectively (Joseph and Kirschner 2004; Maciorkowska et al. 2006).

Another example of interkingdom interaction involving peptide hormones is the sensing of EGFs, whose processing from an inactive membrane-associated form into a secreted and active form requires proteolytic activation by the conserved eukaryotic membrane serine protease rhomboid (Werren et al. 2010). Although well studied in *D. melanogaster*, RHO proteins are ubiquitous, with homologies and functional conservation found in bacteria, archaea, yeast, and plants (Gallio et al. 2002). Though they show low-level sequence similarity, they possess similarity at the structural level and with conserved amino acid residues (Gallio et al. 2002). The first prokaryotic member of the RHO family to be genetically characterized was the product of the gene *aarA* from the Gram-negative pathogen *Providencia stuartii* that represses the expression of an acetyltransferase that modifies peptidoglycan (Payie et al. 1995). Gallio et al. (2002) constructed a mutant of *D. melanogaster* defective for the production of RHO and an *aarA* mutant of *P. stuartii*. They showed that *Drosophila* rhomboid-1 and AarA are able to functionally substitute for each other, indicating evolutionary convergence of the RHO proteins, providing a link between eukaryotic signaling and host-microorganism communication. However, despite these similarities, it is currently unknown whether *P. stuartii* responds to the EGF eukaryotic signal or whether the *P. stuartii* Aar-dependent signal influences eukaryotic cells (Gallio et al. 2002).

1.3.2.3 Interplay Between Stress Response and Interkingdom Interactions

Adaptations to environmental, psychosocial, or physical insults in mammals are referred to as stress responses. It is well established that in an organism undergoing stress, several hormones are released to restore homeostasis. The short-term activation of the stress response involves the synchronized interaction of the neuroendocrine system to ensure that energy substrates are available to meet increasing energy demands of the body. Epinephrine and NE are classic stress hormones that have a pivotal role in the stress response (Karavolos and Khan 2014; Clark and Mach 2016; Herman et al. 2016; Myers et al. 2017). Although they increase mammal's ability to respond to the environment, there is some evidence that they can be sensed by bacteria which allows bacteria, especially pathogens, to gauge the metabolic state of the host and exploit a weakened immune system (Karavolos and Khan 2014).

In addition to epinephrine and NE, hosts may release another class of peptide hormones during stress, known as opioids, such as dynorphin (Karavolos and Khan 2014). A recent study suggested the opportunistic pathogen *P. aeruginosa* uses host dynorphin to enhance the expression of its virulence traits; it was also shown that this mammalian stress hormone cross-signals with *P. aeruginosa* quorum sensing signaling systems (a bacterial response system correlated to population density, whose influence on bacterial-eukaryotic interactions is described in more detail in later sections) (Zaborina et al. 2007; Lesouhaitier et al. 2009). Taken together, these studies have shown that there is an intricate connection between host stress signaling, bacterial response sensing, and pathogenesis, suggesting that response to stress, some of the most basic physiological functions in prokaryotic and eukaryotic cells, is central to interkingdom communication (Lesouhaitier et al. 2009).

Finally, the hypothalamic-pituitary-adrenal axis also induces glucocorticoid stress hormone release by the adrenal gland (Liu et al. 1997; Karavolos and Khan 2014). It is therefore interesting that exposure to adrenocorticotropic hormone (ACTH) increased attachment of EHEC to colonic mucosa by acting through melanocortin receptors located on enteric nerves to enhance mucosal adherence of EHEC (Schreiber and Brown 2005). Physically stressing mice by surgery or catabolic stress significantly increased the numbers of *E. coli* adherence to the cecal mucosa of the stressed mice compared to control animals. Another study showed that elevated NE and dopamine release, such as through stress exposure, modulate EHEC adherence to cecal mucosa and may influence EHEC infection susceptibility (Hendrickson et al. 1999; Chen et al. 2003). Verbrugghe et al. (2011, 2016) have shown, through murine and porcine models, that social stress results in elevated serum cortisol levels and that cortisol increased intracellular proliferation of *Salmonella* inside host macrophages. Collectively, these studies show psychological and physical stress of a host is being sensed by its microbiota, and, in the case of pathogens, host stress is leading to increased virulence.

1.3.3 Molecular Mimicry

Microbial pathogens have evolved exquisite mechanisms to interfere and intercept host biological processes, sometimes through molecular mimicry of host proteins. For instance, some bacterial pathogens have capsular polysaccharides surrounding their outer membrane that are composed of linear polysaccharides known as glycosaminoglycans (GAGs) that are identical to the backbones of GAGs found in animals (Cress et al. 2014). Furthermore, a number of pathogens also have outer glycans, such as sialic acid on their cell surface. These self-associated molecular patterns (Sampson and Mazmanian 2015) allow pathogens to evade complement-mediated responses (Langford-Smith et al. 2015). Varki (2006) hypothesized that these glycans did not arise from HGT from vertebrates but, rather, from the recruitment of pre-existing homologous genes or possibly convergent evolution and that these and other forces such as faster evolution through HGT, high mutation rates, fast growth, and short life spans are likely to have further driven the diversification of species-specific glycosylation patterns during evolution (Varki 2006). Additionally, glycosyltransferase genes responsible for biosynthesizing the glycans in pathogen and host are typically not homologous (Gagneux and Varki 1999). Varki (2006) also argued that sexual reproduction-enabled mutations in host glycosyltransferases and subsequent change in glycan profile allow eukaryotic organisms to adapt to pathogenic pressure.

This ability of pathogens to exploit the similarities of their gene products with their host homologs to elude the defense response may have also contributed to drive the rapid diversification of host-defense genes in mammals. For instance, Murphy (1993) conducted a search of protein orthologs and found receptors and ligands involved in antimicrobial host-defense functions have undergone structural diversification that has resulted in species restricted function. Additionally, the authors suggested that pathogenic microbes may have played a role through exerting selective pressure on the host (Murphy 1993).

Another example of molecular mimicry involves ubiquitination, a highly conserved eukaryotic posttranslational modification essential in determining protein fate that is often hijacked by pathogenic bacteria. The conserved SKP1/CUL1/F-box (SCF) E3 ubiquitin ligase complex plays a key role in ubiquitination of proteins in eukaryotic cells. The F-box protein component of the SCF complex provides specificity to ubiquitination by binding to specific cellular proteins, targeting them to be ubiquitinated by the SCF complex. The *L. pneumophila* protein AnkB is essential for establishing intracellular growth within macrophages and protozoa by promoting the decoration of the *Legionella* containing vacuole with polyubiquitinated proteins. The AnkB effector achieves this by mimicking the action of host cell F-box proteins, a highly conserved component of the SCF ubiquitin ligase complex found in both unicellular organisms and mammalian cells (Price et al. 2010; Price and Kwaik 2010). Importantly, *L. pneumophila*, as well as other pathogens such as *Agrobacterium tumefaciens* and *Ralstonia solanacearum*, utilizes either type III or IV secretion systems to inject into the host cell eukaryotic-like F-box effectors (AnkB in the case of *L. pneumophila*) that interact with the host SKP1 component of the SCF complex to trigger ubiquitination of specific host cell targets, a mechanism essential to promote pathogen proliferation. Price et al. (2010) identified at least 74 genes encoding putative F-box proteins belonging to 22 other bacterial species, including human pathogens, plant pathogens, and amoebal endosymbionts, suggesting that subversion of the host ubiquitination machinery by bacterial F-box proteins may be a widespread strategy among pathogenic bacteria. They also found that bacterial F-box proteins harbor ankyrin repeats that are present in F-box proteins of primitive, but not higher, eukaryotes, suggesting that acquisition of many bacterial F-box proteins may have been from primitive eukaryotic hosts rather than the mammalian host (Price et al. 2010).

1.3.4 Sensing the Host Immune System

The host immune system can use bacterial signals to gauge, and respond to, bacterial infections. In kind, bacteria have also evolved the ability to sense and manipulate host immune systems. Cytokines are the chief signaling molecules employed by the immune system, which makes them a prime target to be recognized as cues by pathogens. Several cytokines, such as interleukin-1β (IL-1β), IL-2, and granulocyte macrophage colony-stimulating factor, increase the growth of virulent, but not avirulent, strains of *E. coli*, suggesting that bacteria sense these cytokines and change their behavior using a mechanism that is not due to nutritional factors (Porat et al. 1991).

High-affinity binding of IL-1β to the F1 antigen of the *Yersinia pestis* (causative agent of plague) capsule through the capsule antigen F1 assembly protein (Caf1A) has also been reported (Zav'yalov et al. 1995). The Caf1A protein is present on outer

Table 1.1 Immune signals recognized by some pathogens and consequences of recognition

Pathogen	Signal	Bacterial receptor	Consequence for pathogen	References
Escherichia coli	IL-1β	Unknown	Increase in growth	Porat et al. (1991)
Escherichia coli	IL-2	Unknown	Increase in growth	Denis et al. (1991), Porat et al. (1991)
Escherichia coli	Granulocyte macrophage	Unknown	Increase in growth	Denis et al. (1991), Porat et al. (1991)
Escherichia coli	TNF-α	Unknown	Increase in growth	Luo et al. (1993)
Salmonella enterica	TNF-α	Unknown	Increased virulence	Arnold et al. (1993b), Luo et al. (1993)
Salmonella enterica and *Salmonella typhimurium*	Cationic antimicrobial peptides	PhoPQ	Modification of LPS; increased virulence	Bader et al. (2005), Dalebroux and Miller (2014)
Pseudomonas aeruginosa	IFN-γ	OprF	Quorum sensing; increased expression of PA-I lectin	Wu et al. (2005)
Shigella flexneri	TNF-α	Unknown	Increased virulence	Luo et al. (1993)
Yersinia pestis	IL-1β	Caf1A	Capsular binding	Zav'yalov et al. (1995)
Candida albicans	IL-2	Unknown	Increased growth	Treseler et al. (1992)

membranes and shows significant homology to the human IL-1β receptor, suggesting that the gene encoding Caf1A may have been acquired through HGT from mammals (Zav'yalov et al. 1995). Other pathogenic bacteria such as pathogenic *E. coli*, *Shigella*, and *Salmonella enterica* bind to and show increased virulence in response to tumor necrosis factor alpha (TNF-α) (Arnold et al. 1993a, b; Luo et al. 1993). More recently, TNF-α-dependent increased growth of *E. coli* was shown using both in vitro and in vivo models, but no underlying mechanism was proposed (Lee et al. 2003).

Another cytokine reportedly sensed by bacteria is gamma interferon (IFN-γ). The *P. aeruginosa* OprF surface protein binds host IFN-γ and initiates a signaling cascade in the bacterial cell that reprograms gene expression, resulting in the expression of a quorum sensing-dependent virulence determinant, PA-I lectin (Wu et al. 2005). This IFN-γ enhancement of *P. aeruginosa* pathogenesis is thought to be especially intriguing in light of the observation that IFN-γ treatment exacerbates *P. aeruginosa* infection in several mouse models and from clinical studies (Luo et al. 1993; Kendall and Sperandio 2016) (Table 1.1).

1.3.4.1 Bacterial Sensing of Mammalian Antimicrobial Peptides

One of the front lines of defense of the mammalian innate immune system is the production of cationic antimicrobial peptides (CAMPs) that interact with and disrupt bacterial membranes. Several Gram-negative bacterial pathogens modify the lipid A of their LPS to prevent CAMPs from binding to their outer membrane. For instance, salmonellae sense host cues to regulate properties important for bacterial survival and replication within host tissues. The PhoPQ two-component regulatory system senses phagosome acidification and CAMPs and then regulates the protein and lipid contents of their bacterial envelope that comprises an inner and outer membrane. In *S. typhimurium*, the PhoQ sensor kinase directly binds, and is activated by, host CAMPS. In turn, PhoQ then promotes the expression of virulence genes through a phosphorelay cascade (Bader et al. 2005). The PhoPQ regulation of genes encoding lipid A modification enzymes is vital for *S. enterica* and *S. typhimurium* virulence (Dalebroux and Miller 2014). The CAMPs alter conformation of the PhoQ, activating its kinase activity and resulting in a phosphotransfer to PhoP with subsequent activation of downstream genes necessary for invasion of epithelial cells and intramacrophage replication (Bader et al. 2005).

1.3.5 Nutritional Signaling

The mammalian GI tract provides a complex, competitive, and spatially diverse environment for resident host microbiota (Hooper et al. 1998, 2001, 2002; Groussin et al. 2017). The GI tract is heavily colonized by a complex and highly adapted microbiota. Estimates of the number of bacterial species present in the human gut vary widely among studies, but it is generally accepted that individuals harbor more than 1000 microbial, species-level phylotypes (Bull and Plummer 2014). The GI microbiota has an important symbiotic relationship with its host, providing and gaining nutrients in the form of carbon and nitrogen sources. Additionally, the microbiota provides critical signals that promote maturation of immune cells and tissues, leading to protection from infections by pathogens. Gut microbiota have been shown to become perturbed during several dysbiotic disease states (Chow et al. 2010). Particularly, imbalance of the normal gut microbiota has been linked with inflammatory bowel disease, irritable bowel syndrome, and wider systemic manifestations of disease such as obesity, type 2 diabetes, and allergic dermatitis (Bull and Plummer 2014). Furthermore, once the microbiota is under immune attack, a more virulent or pathogenic profile may provide certain microbial species with a greater chance of success (Bull and Plummer 2014). Successful intestinal colonization by pathogenic bacteria is thought to depend on the microbe's ability for scavenging nutrients, sensing chemical signals present in the intestine, competing with the resident bacteria for space and nutrients, and precisely regulating the expression of virulence genes. The authors remind us that the most fundamental way of bacterial communication is the importation of substrates for metabolism and exporting its end products and that "If your neighbor eats starch and produces 5 mM succinate, you will probably take notice" (Fischbach and Sonnenburg 2011).

One example of nutritional sensing involves sugars. The mammalian intestinal epithelium is protected from direct contact with bacteria by the mucus layer whose major component is a major intestinal secretory mucin, Muc2 (Bergstrom et al. 2010). Sugars bound to mucin are made available to the microbiota through the polysaccharide-degrading activity of glycolytic commensal anaerobes. Subsequently, the mucus layer is an important source of carbohydrates for bacterial communities colonizing mucosal surfaces (Kim and Ho 2010; Kendall and Sperandio 2016).

Fucose is also a major component of mucin glycoproteins abundant in the intestine (Robbe et al. 2004). Hooper et al. (1999) used a gnotobiotic mouse model to show *Bacteroides thetaiotaomicron*, a component of the intestinal microbiota of mice and humans, uses a repressor, FucR, as a molecular sensor of fucose availability. Importantly, FucR coordinates expression of an operon encoding enzymes in the fucose metabolic pathway with expression of another locus that regulates production of fucosylated glycans in intestinal enterocytes. Genetic and biochemical studies suggest that FucR uses fucose as an inducer at one locus and as a corepressor at the other locus. Coordinating this commensal's immediate nutritional requirements with production of a host-derived energy source underscores the bacterium's need to enter and persist within a competitive ecosystem (Hooper et al. 1999). Furthermore, fucose utilization has been found to be important for EHEC intestinal colonization (Pacheco et al. 2012).

The EHEC also regulates virulence gene expression through recognition of host glycolytic and gluconeogenic environments. The lumen is more glycolytic because of predominant glycophagic microbiota degrading complex polysaccharides into monosaccharides readily utilized by nonglycophagic bacterial species such as *E. coli* and *Citrobacter rodentium* (Yanez et al. 2003; Bertin et al. 2014). Furthermore, it has been shown that genes classified in the gluconeogenesis gene ontology category were significantly up-regulated by EHEC incubated in bovine small intestine contents, relative to EHEC grown in minimal medium (Bertin et al. 2014).

1.3.6 Interkingdom Signaling with Nonmammalian Hosts

Just as signaling and recognition systems for host hormones, metabolites, and host defenses used by mammalian pathogens have been discovered, research also indicates that these mechanisms for interkingdom interactions are conserved in bacteria that infect nonmammalian hosts. Some examples have been discussed in earlier sections. However, some additional examples are highlighted here. For instance, *Edwardsiella tarda*, a pathogen leading to septicemia in fish, uses the QseBC two-component system to regulate gene expression controlling flagellar motility, fimbrial hemagglutination, and intracellular virulence (Wang et al. 2011). Moreover, *E. tarda* responds to epinephrine to promote motility, and this effect was lost by a *qseB* or *qseC* deletion strain, suggesting that this mechanism of gene regulation is conserved in diverse pathogens (Wang et al. 2011; Kendall and Sperandio 2016).

Additionally, the *eut* locus encoding ethanolamine metabolism was described as important to infection mechanisms of the insect pathogen *Photorhabdus luminescens* and the plant pathogen *Erwinia chrysanthemi* (Munch et al. 2008). Further studies show that a disruption of *eutR* in *E. chrysanthemi* results in impaired systemic infection of African violet plants. However, it is currently unknown whether the effect was due to defects in ethanolamine metabolism or signaling through EutR-dependent regulation of virulence traits (Munch et al. 2008). Lastly, *Vibrio coralliilyticus*, a coral pathogen of *Pocillopora damicornis*, senses dimethylsulfoniopropionate (DMSP) to target heat-stressed corals, rich in DMSP. Sensing results in chemotaxis, and this signal was used solely for finding the host location, rather than for pathogen metabolism (Garren et al. 2014).

In arthropods, the presence of endosymbiotic prokaryotes plays an important role in metabolism. In some cases, genome integration has occurred from endosymbiotic-host relationships, proving that intracellular symbiosis is not simply a nutritional supplement (Fenn and Blaxter 2006; Dunning Hotopp et al. 2007; Ahmed et al. 2016). Intracellular symbiotic bacteria are also described in nematodes (Fenn et al. 2006). In particular, the presence of intracellular *Wolbachia* in filariae positively influences reproductive biology and host survival, as proved by antibiotic treatment against this bacterium (Fenn et al. 2006).

Another classic example includes the interkingdom interaction of rhizobia and host legume plants. Rhizobia are nitrogen-fixing bacteria that colonize plant roots through an evolutionarily developed association to form root nodules under nitrogen-limiting conditions. In the nodule, these bacteria fix nitrogen that can then be used by the host, and in return, the plant provides carbon useful for bacterial growth. For this mutually beneficial relationship to be established, signals are exchanged between the bacteria and the host. Specifically, legumes produce flavonoids sensed by the bacterial NodD protein. The perception of specific Nod factors by rhizobia triggers a host signaling cascade that leads, in most legumes, to formation of specialized intracellular structures called infection threads that act as a conduit for rhizobial access to the inner root tissues where they are endocytosed into nodule cells to begin nitrogen fixation (Liu and Murray 2016). Furthermore, the bacteria also produce chemicals influencing host response to infection. For example, the bacterial Nod factor may play a role in suppressing the plant immune response, thereby inducing nodule development (Toth and Stacey 2015). Studies on how rhizobia influence plant innate immunity to establish symbiotic associations highlight that interkingdom interactions are not limited to host and pathogen. Indeed, it is likely that additional studies examining interkingdom interactions between host and associated microbiota will demonstrate conserved mechanisms shared by pathogens as well as commensal bacteria. Such conserved mechanisms highlight the need for new and innovative studies in complex disease systems. Understanding such novel mechanisms may yield important advances in disease management.

1.3.7 Quorum Sensing Systems and Disease Ecology

One final system for prokaryotic-eukaryotic interactions involves bacterial responses to eukaryote chemical stimuli that are dependent upon threshold bacterial concentrations, a phenomenon known as quorum sensing, which was previously mentioned. In this case, both pathogenic and nonpathogenic bacteria, and some yeast, may recognize chemical stimuli individually but will not adjust gene expression until threshold microbial densities are reached that then allow for coordinated phenotypes (Sprague and Winans 2006; Plancak et al. 2015; Hawver et al. 2016). Quorum sensing has been identified in eukaryote and prokaryote systems and has been studied extensively. However, much needed research still remains. The next section will focus on this intriguing bacterial communication pathway as a key example of complex mechanisms of microbes influencing eukaryote behavior, lending insight into the potential of similar interactions that may be important to infectious disease systems. Indeed, the precedent (see below) that bacteria influence the behavior of higher level organisms (e.g., hosts or reservoirs) suggests that these interkingdom interactions may be more important to pathogen environmental transmission than previously considered.

1.3.7.1 Quorum Sensing Regulates Behavior of Eukaryotes

Limited information is available on the interkingdom interactions and microbial quorum sensing roles in inducing behavioral responses of hosts or other animals competing for similar resources as the microbes. Several reviews have been published on this topic; however, the first known review was not published until 2008 (Lowery et al. 2008). More recently, the molecular mechanisms of indole as a quorum sensing molecule and regulator of physiological responses across kingdoms or domains were published (Lee et al. 2015), while a subsequent review examined its role as a modulator of behavior and distinguished between its role as a signal and cue (Tomberlin et al. 2017b).

Specific studies exploring the role of bacteria regulating arthropod vector behavior have been published, with much research focused on the role bacteria play in regulating adult mosquito responses to oviposition (Ponnusamy et al. 2008) sites or hosts (Verhulst et al. 2009, 2010a, b, 2011). More recently, efforts have begun exploring the role of quorum sensing as a mechanism regulating mosquito behavior. Cui et al. (2015) determined that isolates of *Staphylococcus epidermidis* that were able to quorum sense attracted 74% of adult *A. aegypti* versus the mutants unable to quorum sense. This interaction has significant ramifications with regard to pathogen transmission from vector (i.e., mosquito) to hosts as the vectors utilize microbial quorum sensing as a means to locate the said hosts.

This relationship between bacterial quorum sensing and arthropod vector behavior has been demonstrated for other species as well, including filth flies that are known to transmit human pathogens (Greenberg 1971b). Research with a blow fly model [i.e., *Lucilia sericata* (Diptera: Calliphoridae)] determined the level of attraction to food, and oviposition substrate inoculated with *Proteus mirabilis*, which is a commensal with this fly species, was partially regulated by the ability of this bacteria to swarm (quorum sensing response) (Ma et al. 2012). The response of the insect was partially regulated by its age, sex, and nutrition history (Tomberlin et al. 2012). Interestingly, the volatiles that regulated these responses, such as indole, are derivatives of essential amino acids (Liu et al. 2016). The reason this information is relevant is that it ties together microbial ecology (e.g., resource utilization), population dynamics (e.g., microbial population structure as related to resource utilization), and either potential competition or mutualistic interactions between the bacteria and the insect. If the insect were to colonize the resource at a suboptimal time point (e.g., when the bacteria are in exponential growth phase and utilizing resources quickly), survivorship of the insect offspring could be threatened. High mortality in vector offspring affects population density and potential vectorial capacity, having direct impact in models of foodborne disease transmission from filth flies.

This aspect of interkingdom interactions was described well with the house fly, *Musca domestica* L. (Diptera: Muscidae) model. Bacterial populations associated with house fly eggs shift as the eggs mature (Lam et al. 2007). Researchers determined that a regulatory threshold existed within the bacterial population that affected house fly attraction and oviposition. Attractant levels below the threshold resulted in attraction and oviposition, while levels above the threshold resulted in reduced attraction (Lam et al. 2007). They determined this was critical since colonization above the threshold might correspond to hatching of eggs associated with the microbes and the resulting larvae would cannibalize the eggs deposited by the adults after the microbial threshold was reached. Furthermore, they determined these bacteria associated with the eggs also served as nutrients for the resulting larvae (Lam et al. 2009a). They also reported that bacteria possessed antifungal properties, thereby protecting the eggs during maturation (Lam et al. 2009b). The above examples of quorum sensing interactions with invertebrates continue to identify new microbe-animal relationships that have potential to reveal new ways to manage pathogen transmission, insect vector population biology and other ecological (e.g., microbe competition) forces of disease systems. There are similar studies of interkingdom interactions that include vertebrates that may be especially important in more comprehensively understanding zoonotic diseases.

1.4 Microbiomes and Vertebrates and Disease Ecology

Understanding the complex dynamics and function of the microbiome in vertebrate disease systems has undergone a surge of research interest and new findings with the advent of next-generation sequencing (Cui et al. 2015). Estimates are that up to 90% of the vertebrate host genetic information is microbial (Sender et al. 2016a, b) and that this genetic material likely plays a significant role in the human condition (McFall-Ngai et al. 2013). Thus, characterization of these microorganisms, primarily bacterial communities, in host populations has led to a better understanding of how microbes can influence behavior (Ezenwa et al. 2012; Wong et al. 2015), physiology (Nicholson et al. 2012), immunity (Eleftherianos et al. 2013), reproduction (Reid et al. 2015), nutrition (Nicholson et al. 2012), and development (Hooper 2004).

The relationships between host phylogeny and microbiomes have led to investigations into the potential mechanisms describing their coevolutionary history (Moeller and Ochman 2014; Colston and Jackson 2016). Indeed, many researchers no longer see a clear delineation between a host and its microbial constituents. Rather the host and all of its symbiotic microbes are deemed a "holobiont" (Bordenstein and Theis 2015) and should be holistically explored as a single biological entity. The complexities and integration of organisms within a holobiont have been documented for multi-host pathogen introgressions and hybridizations. For example, in West Africa there is evidence of an introgression that has been detected between a human (*Schistosoma haematobium*) and two ruminant (*S. bovis* and *S. curassoni*) schistosomes (Webster et al. 2013). Zoonotic pathogens are the cause of one billion cases of human illnesses annually (Han et al. 2015), yet the true host range of vertebrates capable of transmitting disease-causing parasites and pathogens remains underexplored.

As shown in Fig. 1.1 and mentioned in the introduction to this chapter, there are several pathways for infectious disease occurrence, but at a minimum, there must be a host species and a pathogen. An example of this minimal disease system involving a single vertebrate host species would be a human and a virus such as influenza virus. More often there is a combination of vertebrate and invertebrate host species directly and indirectly interacting with pathogens in a disease system. Pathogens can achieve host switching via rapid mutations as is seen in viruses, such as influenza A and avian flu. Thus, the breadth of potential host species for pathogen dispersal increases. Estimates of zoonotic host diversity have been attempted using predictive models based on vertebrate traits (e.g., activity cycle, dispersal age, habitat breadth, sexual maturity age, body mass) (Han et al. 2015). Several studies have determined rodents to be the most abundant zoonotic reservoir as compared to either other mammals or birds (Han et al. 2015; Johnson et al. 2015; Ostfeld and Keesing 2017). Of the extant rodent species, approximately 11% were predicted to be a zoonotic reservoir (Han et al. 2015). Many of these host species involved in disease systems are important players in an array of ecosystem services (Keesing et al. 2010; Ostfeld et al. 2010; Johnson et al. 2015; Ostfeld and Keesing 2017). However, it remains unknown as to the relationship between true pathogen prevalence and associations with vertebrate community diversity. Biodiversity within an ecosystem may play an important role in disease dynamics (Keesing et al. 2010). Increased host biodiversity may lead to a decreased pathogen transmission by reducing the probability of transmission among conspecifics (Keesing et al. 2010; Ostfeld et al. 2010; Johnson et al. 2015; Ostfeld and Keesing 2017).

This premise of how biodiversity affects transmission has occurred from suggestions in multiple disease systems. For instance, there is a documented negative correlation of *Hantavirus* in rodents when small mammal diversity increases within a habitat (Clay et al. 2009), and increased heterospecific snail populations lead to greater probability of schistosomiasis being transmitted to a dead-end host (Laracuente et al. 1979). Continued research is needed to identify how microbiome community composition and assembly potentially impacts disease emergence and dynamics in vertebrate hosts. Furthermore, determining the response of pathogens and their hosts to dynamic selective pressures, such as shifts in host range due to climate and anthropogenic changes or disease control strategies, will be imperative for mitigating infectious disease outbreaks with a changing climate. It is of great importance to develop a better understanding for infectious disease systems with regard to the basic ecological interactions and potential robust mechanisms between microbiomes and their hosts. Below are several examples of these ecological interactions that either directly or indirectly influence disease systems.

1.4.1 Influence of Ontogeny and Life History on Microbiomes

The development of an animal is linked to a multifactor suit of characters selected to optimize its fitness. Primarily the goal of the organism, whether prokaryotic or eukaryotic, is to pass along its genes through successful reproduction. In vertebrate hosts, there are a variety of reproductive strategies including oviparous, oviparity, and viviparity, and one vertebrate is known to have a mixed-mating strategy of hermaphrodites that reproduce by self-fertilization or crossbreeding—the mangrove killifish (Molloy and Gage 2006). Despite how a vertebrate is produced, a number of physiological transformations occur during development that can alter the microbiome. In the earliest stages of development, organisms can be protected against potential deleterious microbes by physical and chemical barriers (McFall-Ngai et al. 2013). These barriers include mucous and antimicrobial peptides that regulate the developing microbiome. Also, host traits can increase predictive capability of acting as a zoonotic reservoir related to an increased reproductive output, such as early age at maturity, large litters, short life span, and host geographic range, as exampled by rodents (Han et al. 2015; Ostfeld and Keesing 2017).

As a vertebrate host grows to maturity and beyond, diet and reproductive status can further regulate their microbial communities. Some research has focused on the horizontal acquisition of microbiota through feeding behaviors. Green iguana (*Iguana iguana*) hatchlings initially increase their hindgut microbiota from consuming soil within the nest chamber; shortly after hatching (*ca.* 1 week), they begin ingesting plants and soil from outside of the nest chamber; and finally, the animals associate with other conspecifics and eat the feces of older individuals, gaining access to more complex microbial communities (Troyer 1984). Other research in humans has demonstrated that the "healthy" gut microbiota is typically a stable community of *Bacteroidetes* and *Firmicutes* (Lloyd-Price et al. 2016), but there is a shift in communities as an individual ages (Backhed et al. 2004; Koenig et al. 2011; Lozupone et al. 2012; Oh et al. 2012). However, there have been significant shifts in the core gut microbiome in as little as 5 days due to changes in dietary intake (David et al. 2014). It has also been well documented that obesity in humans is positively correlated to an increase in the *Firmicutes* to *Bacteroidetes* ratio (Turnbaugh et al. 2006, 2009; Boulangé et al. 2016). Much research is needed to further investigate the impact of other vertebrate hosts and their life history strategies on their microbiome; however, the growing body of literature suggests that the microbiome and its interaction at both molecular and cellular level have a significant role in health and disease. Some of these interactions play out through host manipulation of the microbiome, with potential influence on host and reservoir population dynamics.

1.4.2 Host Manipulation by the Microbiome

Recent research on the interactions between the microbiome and host has elucidated a large number of mechanisms by which the microbiota affects animal behavior (Ezenwa et al. 2012). For example, pathogens can shift animal behavior to optimize transmission and therefore affect the entire disease system network. In this section, we will present key examples of how pathogens and other microbial groups affect vertebrate biology. These interactions likely have broad population biology and natural history effects that are important to disease ecology. A classic example of host behavior modification by a pathogen has been thoroughly described in a feline and rodent model, whereby *Toxoplasma gondii* mediates the behavior of its intermediate host (rodents) by reducing predator risk assessment (Webster 2001; Webster et al. 2017). This pathogen affects the cognitive perception of the mouse thus leading to an increased feeding rate by felines. This modification of risk assessment, however, differs between wild and domesticated feline species. Rats demonstrated an attraction to wild cats over domesticated cats when infected with *T. gondii* (Kaushik et al. 2014). There is also documented increased aggression in vertebrate populations caused by rabies, Borna disease, and Hantaan or Seoul viruses, which increased the rate of blood/saliva transmission and thus pathogen proliferation (Klein 2003).

The residing microbiota of a vertebrate host population is also known in some instances to impact sociality. In hyenas, there is a strong correlation between fermentative bacteria detected in the scent glands of different hyena species (spotted versus striped), sex, and reproductive state (Theis et al. 2013). *Firmicutes* was the predominate phylum in each species representing over 95.59–99.96% relative abundance, but constituents of *Bacteroidetes* (0.04–1.16%) and *Actinobacteria* (3.73%) were the next predominate phyla in striped and spotted hyenas, respectively (Theis et al. 2013). Furthermore there were significant differences in microbiome structure comparisons between males and females, with additional differences between lactating versus pregnant females. These differences in bacterial communities have direct implications in the chemical communication among individuals within a group and thus the overall social structure. If the same is true in other social animals, these interkingdom interactions likely play important roles in pathogen transmission in many disease systems.

For instance, in humans the gut microbiome has been linked to various mood disorders, obesity, autism, and other physical diseases, such as allergies and asthma (Cho and Blaser 2012; Han et al. 2012; Liu et al. 2012a; Hsiao et al. 2013; Christian et al. 2015). Furthermore, many studies are demonstrating associations between brain development and behavior with the microbiome (Sampson and Mazmanian 2015). For instance, Butterworth (2015) found that changes in the gut microbiome trigger changes in the central nervous system and the brain and that these changes are related to anxiety, depression, and other forms of mental illnesses in human. Christian et al. (2015) reported that the diversity of toddler's microbiome was related to temperament traits such as extroversion. In a related way, another study found that the microbiome may act to stimulate the production of serotonin and important neurotransmitter that affects mood and is related to several mood disorders (Hsiao et al. 2013). It is evident that both pathogens and commensal microbes are key in modulating host behavior, which is of importance for further exploring transmission dynamics of infectious diseases. The microbiome also affects other aspects of the human condition, with potentially important effects on disease presentation and pathogenesis.

1.4.2.1 Human Hosts

The human body is a dynamic vessel for microorganisms that together are a population constantly in flux. The composition and role of microbial consortia that endogenously and exogenously comprise the human microbiome have been extensively studied for human health (The Human Microbiome Jumpstart Reference Strains Consortium 2010). The human microbiome can be extremely variable within an individual as distinct taxa of bacteria reside in different parts of the body (Grice et al. 2009). Put simply, research suggests that the human body is an ecosystem, which consists of multiple distinct natural environments (e.g., the gut, skin, eyes, mouth), with each natural environment containing a unique microbiome community. A healthy microbial community can be characterized in terms of diversity, stability, and resilience (Lozupone et al. 2012). However, a healthy microbial community that becomes disturbed will often result in a state of dysbiosis or an abnormal microbial community (structure and function) (Levy et al. 2017). Diversity is typically reduced in disease-associated dysbiosis (Le Chatelier et al. 2013; Mosca et al. 2016). The stability of the microbiome within a given body region also can have tremendous influence on the potential risk of infections by certain pathogens.

As an example, instability of the vaginal microbiome caused by bacterial vaginosis increases the risk of acquiring human immunodeficiency virus (Atashili et al. 2008). Perturbation of the "healthy" microbiota has also been documented by the overgrowth of pathogenic bacteria within *Enterobacteriaceae* in the gut during enteric infections (Stecher et al. 2013). Alternatively, modification of the disturbed microbiota can be achieved by replacing pathogenic "blooming" bacteria with commensal microorganisms. In fecal microbiota transplantation, the unhealthy gut community of an individual with inflammation caused by *Clostridium difficile* can be improved with the introduction of *Clostridium scindens* (Buffie et al. 2015). Further, there is evidence of recurrent infection caused by antibiotic-resistant *C. difficile* being better controlled when the entire gut community was replaced from a healthy donor (van Nood et al. 2013). Understanding the mechanisms of microbiota competition may allow the development of more targeted interventions against infectious diseases in the future.

1.4.2.2 Nonhuman Vertebrate Hosts

While the human ecosystem has been the most extensively studied, there is increasing empirical evidence of how nonhuman vertebrates respond to microbial communities and their function in response to various disease-causing microorganisms. While this section is not exhaustive, it provides an array of examples that illustrate the dynamic interactions between host and its microbiome and how pathogens exploit mechanisms to increase transmission. The most prevalent vector-borne disease in the Northern Hemisphere is Lyme disease; thus, its disease dynamics and mechanisms have been widely studied. Ticks (*Ixodes* spp.) are primary vectors for transmission to humans, and a tick must feed upon an infected host to acquire the causative microbe *Borrelia burgdorferi*. Classic behavioral studies for this disease have demonstrated that ticks attempting to feed on various host species have a differential response to becoming infected by the bacterium. Specifically, ticks attempting to feed on opossums are less likely to become infected than ticks that feed on mice, because of the grooming habits of the opossums (Keesing et al. 2010). However, there is also the possibility that the opossum microbiome has an impact on the colonization of *Borrelia burgdorferi*, because natural products that inhibit biofilm growth (cyclic lipodepsipeptides) derived from the internal gut microbiota have been detected in the postmortem microbiome of opossums (Motley et al. 2017).

There are also reptile hosts for Lyme disease. There is a significantly reduced diversity in the microbiome of ticks (*Ixodes pacificus*) that feed on western fence lizard hosts compared to blood meals acquired from mammalian hosts (Swei and Kwan 2017). These differences in the gut microbiota between reptilian and mammalian hosts could explain the refractory nature of these lizards against Lyme disease, but additional studies are needed to elucidate cause versus correlation.

Another example of behavior (and potential of microbiome manipulation) affecting disease dynamics is in facial tumors of Tasmanian devils. The pathogen causing these tumors is transmitted through aggressive behaviors, such as biting, that frequently occur during mating and feeding (Deakin and Belov 2012). If these social activities are mediated by the Tasmanian devil microbial communities like that of hyenas, then one would expect that transmission of this disease may be indirectly affected by the host microbiome through behavior modifications.

The occurrence of pathogens and their interactions with the microbiome for nonhuman hosts are not isolated to land-dwelling animals. There is emerging research in hosts that live in aquatic habitats. Bacterial pathogens of teleosts span a wide range of phyla and genera. These pathogens, including *Vibrio*, *Renibacterium*, and *Mycobacterium*, are often found in the normal microbiota of healthy fish but, under certain conditions, such as when experiencing stress, can open the door for opportunistic pathogens to manifest as disease in the host (Llewellyn et al. 2014). The normal microbiota of certain fish has also been demonstrated to act as a natural barrier against fungal pathogens. For example, skin commensals of rainbow trout (*Oncorhynchus mykiss*) have been shown to inhibit *Saprolegnia australis* and *Mucor hiemalis* (Lowrey et al. 2015), which can have important economic impact for aquaculture because of disease-related mortality (Sandoval-Sierra et al. 2014). There are strategies currently being implemented in aquaculture to reduce the prevalence of many types of fish disease, especially in those species that are of global economic and cultural importance, such as *Oncorhynchus*. One strategy is to alter the gut microbiota to increase resistance against pathogens (Tellez et al. 2006), much as is performed with the use of probiotics in mammalian hosts. There is a paucity of data regarding infectious diseases in marine wildlife, even less for how the microbiome is interacting with diseased hosts. Interestingly, recent evidence has demonstrated that Gram-negative anaerobic bacteria of human origin, such as human adapted *Campylobacter*, have been detected in dolphin microbiomes (Godoy-Vitorino et al. 2017). Overall, while there is an ever-expanding amount of research about the microbiome of hosts, vectors, and pathogens, additional studies into the constituents and interactions among micro- and macroorganisms would allow to help better understand why some populations are more competent vectors and how pathogens are transmitted within an ecosystem and ultimately elucidate the mechanisms resulting in disease and effective control strategies to counter the dynamic nature of pathogens. There are many different techniques for studying these aspects of disease ecology, and one of the most recent and potentially transforming methods has been through advances in high-throughput genomic sequencing. This method is allowing scientists to identify thousands of microbial taxa, opening the doors to asking new questions about the role of interkingdom interactions in disease systems.

1.4.3 Current Sequencing Tools

Characterizing a microbiome is becoming more widely available as a tool given the rapid advances in NGS. Depending on the question being investigated, one may use a bevy of approaches to better understand microbiomes and interkingdom interactions in disease ecology; see Table 1.2 for examples. All of the available platforms, however, have the same role of sequencing millions of nucleotide fragments in parallel. Several sequencing platforms can be used for NGS application (e.g., Illumina®, PacBio®, Ion Torrent™), but these platforms are constantly changing in depth of coverage, sequence read length, sensitivity, and cost. One may refer to the Advances in Genome Biology and Technology (AGBT) conference for the most up-to-date information about sequencing technologies.

Currently, Illumina® MiSeq® and HiSeq® are two of the most commonly employed platforms for assessing the structure and function of the microbiome. These are

Table 1.2 Examples of high-throughput genomic sequencing methods and associated potential applications

Sequencing method[a]	Application(s)
Targeted (amplicon based)	Genomics
De novo	Genomics
Nanopore DNA	Genomics
Whole genome	Genomics
Exome	Genomics
Single molecule real-time (SMRT)	Genomics; transcriptomics
Total RNA and mRNA	Transcriptomics
Targeted RNA	Transcriptomics
Small RNA and noncoding RNA	Transcriptomics
Methylation	Epigenomics
Chromatin immunoprecipitation (ChIP)	Epigenomics
Ribosome profiling	Epigenomics

[a]From ten Bosch and Grody (2008), de Magalhães et al. (2010), Liu et al. (2012b), Quail et al. (2012)

versatile platforms that can be used in a wide array of applications based on the gene of interest. To assess the variability in the taxonomic microbiota composition, one may use the 16S rRNA gene, as it is the most frequently used gene to study phylogenic relationships among microbial consortia as it allows for surveys of bacteria and archaea. Because of this frequency, 16S rRNA gene databases are among the most developed. Although other, single-copy genes (e.g., ITS, 18S rRNA) have been used for phylogenetic resolution, the use of these genes for the analysis of environmental amplicons is limited by the fact that the degeneracy of protein sequences makes universal primer design difficult, and the databases are not as extensive when compared to 16S rRNA. Furthermore, after the sequencing has been performed, one must take special consideration to the bioinformatics used to analyze these millions of sequences. Due to the tremendous amount of literature available on the variation in analytical pipelines, filtering considerations, clustering approaches, and statistical and mathematical modeling available for analyzing the sequencing data, these topics will not be discussed. It is recommended that the reader explore the information pertinent to their gene(s) of interest and NGS approach. When assessing a disease system, the goal of research (e.g., Do the microbial communities differ? What is the gene expression in the diseased versus healthy state? Can variant sequences be identified?) is key in determining which sequencing platform will be most useful in providing the desired information about their study system.

1.5 Synthesis for Disease Ecology

The interactions of organisms of at least two taxa, the minimum disease system proposed in this chapter (Fig. 1.1), are inherently a part of disease ecology. However, the interactions of taxa from multiple domains and kingdoms of life go beyond the pathogen, host, reservoirs, and vectors, likely playing functionally important indirect roles that affect disease system dynamics such as pathogen transmission, vector attack rates, and host and reservoir behavior, among others. While much more detailed and mechanistic studies are needed, the literature to date suggests a broad arena for research into how interkingdom interactions outside of the traditional host-pathogen-reservoir-vector organisms may significantly impact the scientific understanding of infectious disease dynamics. Many of the mechanisms of such interactions are understood, but it is clear that many more pathways likely remain undiscovered in both invertebrate and vertebrate hosts, reservoirs, and vectors. Microbial direct and indirect interactions are currently being studied in many animal systems, as these domains of life clearly affect biological outcomes related to host fitness, population dynamics, and their role in communities and ecosystems. While only a few examples of these important interkingdom interactions in disease ecology are discussed in this chapter, the literature suggests that there are likely many more as part of natural ecosystem networks, within which disease systems reside. The future of scientific inquiry into novel interkingdom interactions and their importance in infectious disease ecology is exciting and promising.

Compliance with Ethical Standards

Funding This study was partially funded by two awards from the National Institute of Justice (2014-DN-BX-K008 and 2010-DN-BX-K243).

Conflict of Interest Mark Benbow declares that he has no conflict of interest. Jennifer Pechal declares that she has no conflict of interest. Jeffery K. Tomberlin declares that he has no conflict of interest. Heather Jordan declares that she has no conflict of interest.

Ethical Approval This article does not contain any studies with human participants or animals performed by any of the authors.

References

Ahmed MZ, Breinholt JW, Kawahara AY (2016) Evidence for common horizontal transmission of Wolbachia among butterflies and moths. BMC Evol Biol 16:118

Archibald JM (2015) Endosymbiosis and eukaryotic cell evolution. Curr Biol 25:R911–R921

Arnold JW, Klimpel GR, Niesel DW (1993a) Tumor necrosis factor (TNF alpha) regulates intestinal mucus production during salmonellosis. Cell Immunol 151:336–344

Arnold JW, Niesel DW, Annable CR et al (1993b) Tumor necrosis factor-alpha mediates the early pathology in Salmonella infection of the gastrointestinal tract. Microb Pathog 14:217–227

Atashili J, Poole C, Ndumbe PM et al (2008) Bacterial vaginosis and HIV acquisition: a meta-analysis of published studies. AIDS (London, England) 22:1493

Backhed F, Ding H, Wang T et al (2004) The gut microbiota as an environmental factor that regulates fat storage. Proc Natl Acad Sci USA 101(44):15718–15723

Bader MW, Sanowar S, Daley ME et al (2005) Recognition of antimicrobial peptides by a bacterial sensor kinase. Cell 122:461–472

Bergstrom KS, Kissoon-Singh V, Gibson DL et al (2010) Muc2 protects against lethal infectious colitis by disassociating pathogenic and commensal bacteria from the colonic mucosa. PLoS Pathog 6:e1000902

Bertin Y, Deval C, de la Foye A et al (2014) The gluconeogenesis pathway is involved in maintenance of enterohaemorrhagic *Escherichia coli* O157:H7 in bovine intestinal content. PLoS One 9:e98367

Bordenstein SR, Theis KR (2015) Host biology in light of the microbiome: ten principles of holobionts and hologenomes. PLoS Biol 13:e1002226

Boulangé CL, Neves AL, Chilloux J et al (2016) Impact of the gut microbiota on inflammation, obesity, and metabolic disease. Genome Med 8:42

Buffie CG, Bucci V, Stein RR et al (2015) Precision microbiome reconstitution restores bile acid mediated resistance to *Clostridium difficile*. Nature 517:205–208

Bull MJ, Plummer NT (2014) Part 1: the human gut microbiome in health and disease. Integr Med (Encinitas) 13:17–22

Burdon JJ, Thrall PH (2008) Pathogen evolution across the agro-ecological interface: implications for disease management. Evol Appl 1:57–65

Burkepile DE, Parker JD, Woodson CB et al (2006) Chemically mediated competition between microbes and animals: microbes as consumers in food webs. Ecology 87:2821–2831

Butterworth RF (2015) Pathogenesis of hepatic encephalopathy and brain edema in acute liver failure. J Clin Exp Hepatol 5:S96–S103

Chen C, Brown DR, Xie Y et al (2003) Catecholamines modulate *Escherichia coli* O157:H7 adherence to murine cecal mucosa. Shock 20:183–188

Cho I, Blaser MJ (2012) The human microbiome: at the interface of health and disease. Nat Rev Genet 13:260–270

Chow J, Lee SM, Shen Y et al (2010) Host-bacterial symbiosis in health and disease. Adv Immunol 107:243–274

Chowers MY, Keller N, Tal R et al (1999) Human gastrin: a *Helicobacter pylori*—specific growth factor. Gastroenterology 117:1113–1118

Chowers MY, Keller N, Bar-Meir S et al (2002) A defined human gastrin sequence stimulates the growth of *Helicobacter pylori*. FEMS Microbiol Lett 217:231–236

Christian LM, Galley JD, Hade EM et al (2015) Gut microbiome composition is associated with temperament during early childhood. Brain Behav Immun 45:118–127

Clark A, Mach N (2016) Exercise-induced stress behavior, gut-microbiota-brain axis and diet: a systematic review for athletes. J Int Soc Sports Nutr 13:43

Clay CA, Lehmer EM, Jeor SS et al (2009) Sin nombre virus and rodent species diversity: a test of the dilution and amplification hypotheses. PLoS One 4:e6467

Colston TJ, Jackson CR (2016) Microbiome evolution along divergent branches of the vertebrate tree of life: what is known and unknown. Mol Ecol 25:3776–3800

Cress BF, Englaender JA, He W et al (2014) Masquerading microbial pathogens: capsular polysaccharides mimic host-tissue molecules. FEMS Microbiol Rev 38:660–697

Cui J, Yin Y, Ma Q et al (2015) Comprehensive characterization of the genomic alterations in human gastric cancer. Int J Cancer 137:86–95

Dalebroux ZD, Miller SI (2014) Salmonellae PhoPQ regulation of the outer membrane to resist innate immunity. Curr Opin Microbiol 17:106–113

David LA, Maurice CF, Carmody RN et al (2014) Diet rapidly and reproducibly alters the human gut microbiome. Nature 505:559–563
de la Cruz F, Davies J (2000) Horizontal gene transfer and the origin of species: lessons from bacteria. Trends Microbiol 8:128–133
de Magalhães JP, Finch CE, Janssens G (2010) Next-generation sequencing in aging research: emerging applications, problems, pitfalls and possible solutions. Ageing Res Rev 9:315–323
Deakin JE, Belov K (2012) A comparative genomics approach to understanding transmissible cancer in Tasmanian devils. Annu Rev Genomics Hum Genet 13:207–222
Denis M, Campbell D, Gregg EO (1991) Interleukin-2 and granulocyte-macrophage colony-stimulating factor stimulate growth of a virulent strain of *Escherichia coli*. Infect Immun 59:1853–1856
Dunning Hotopp JC, Clark ME, Oliveira DC et al (2007) Widespread lateral gene transfer from intracellular bacteria to multicellular eukaryotes. Science 317:1753–1756
Eleftherianos I, Atri J, Accetta J et al (2013) Endosymbiotic bacteria in insects: guardians of the immune system? Front Physiol 4:46
Ezenwa VO, Gerardo NM, Inouye DW et al (2012) Animal behavior and the microbiome. Science 338:198–199
Fenn K, Blaxter M (2006) *Wolbachia* genomes: revealing the biology of parasitism and mutualism. Trends Parasitol 22:60–65
Fenn K, Conlon C, Jones M et al (2006) Phylogenetic relationships of the *Wolbachia* of nematodes and arthropods. PLoS Pathog 2:e94
Fischbach MA, Sonnenburg JL (2011) Eating for two: how metabolism establishes interspecies interactions in the gut. Cell Host Microbe 10:336–347
Flores M, Crippen TL, Longnecker M et al (2017) Non-consumptive effects of predatory *Chrysomya rufifacies* larval cues on larval *Cochliomyia macellaria* growth and development. J Med Entomol 54(5):1167–1174
Freestone P (2013) Communication between bacteria and their hosts. Scientifica (Cairo) 2013:361073
Freestone PP, Williams PH, Haigh RD et al (2002) Growth stimulation of intestinal commensal *Escherichia coli* by catecholamines: a possible contributory factor in trauma-induced sepsis. Shock 18:465–470
Freestone PP, Haigh RD, Williams PH et al (2003) Involvement of enterobactin in norepinephrine-mediated iron supply from transferrin to enterohaemorrhagic *Escherichia coli*. FEMS Microbiol Lett 222:39–43
Frohlich KS, Papenfort K (2016) Interplay of regulatory RNAs and mobile genetic elements in enteric pathogens. Mol Microbiol 101:701–713
Gagneux P, Varki A (1999) Evolutionary considerations in relating oligosaccharide diversity to biological function. Glycobiology 9:747–755
Gallio M, Sturgill G, Rather P et al (2002) A conserved mechanism for extracellular signaling in eukaryotes and prokaryotes. Proc Natl Acad Sci USA 99:12208–12213
Garren M, Son K, Raina JB et al (2014) A bacterial pathogen uses dimethylsulfoniopropionate as a cue to target heat-stressed corals. ISME J 8:999–1007
Godoy-Vitorino F, Rodriguez-Hilario A, Alves AL et al (2017) The microbiome of a striped dolphin (*Stenella coeruleoalba*) stranded in Portugal. Res Microbiol 168:85–93
Gray MW (1999) Evolution of organellar genomes. Curr Opin Genet Dev 9:678–687
Gray MW, Burger G, Lang BF (1999) Mitochondrial evolution. Science 283:1476–1481
Greenberg B (1968) Model for destruction of bacteria in the midgut of blow fly maggots. J Med Entomol 5:31–38
Greenberg B (1971a) Flies and disease, vol 1, Ecology, classification and biotic associations. Princeton University Press, Princeton, NJ
Greenberg B (1971b) Flies and disease, vol 2, Biology and disease transmission. Princeton University Press, Princeton, NJ

Greenberg B, Klowden M (1972) Enteric bacterial interactions in insects. Am J Clin Nutr 25:1459–1466

Grice EA, Kong HH, Conlan S et al (2009) Topographical and temporal diversity of the human skin microbiome. Science 324:1190–1192

Groussin M, Mazel F, Sanders JG et al (2017) Unraveling the processes shaping mammalian gut microbiomes over evolutionary time. Nat Commun 8:14319

Han MK, Huang YJ, LiPuma JJ et al (2012) Significance of the microbiome in obstructive lung disease. Thorax 67:456–463

Han BA, Schmidt JP, Bowden SE et al (2015) Rodent reservoirs of future zoonotic diseases. Proc Natl Acad Sci USA 112:7039–7044

Hawver LA, Jung SA, Ng WL (2016) Specificity and complexity in bacterial quorum-sensing systems. FEMS Microbiol Rev 40:738–752

Hendrickson BA, Guo J, Laughlin R et al (1999) Increased type 1 fimbrial expression among commensal *Escherichia coli* isolates in the murine cecum following catabolic stress. Infect Immun 67:745–753

Herman JP, McKlveen JM, Ghosal S et al (2016) Regulation of the hypothalamic-pituitary-adrenocortical stress response. Compr Physiol 6:603–621

Hooper LV (2004) Bacterial contributions to mammalian gut development. Trends Microbiol 12:129–134

Hooper LV, Bry L, Falk PG et al (1998) Host-microbial symbiosis in the mammalian intestine: exploring an internal ecosystem. BioEssays 20:336–343

Hooper LV, Xu J, Falk PG et al (1999) A molecular sensor that allows a gut commensal to control its nutrient foundation in a competitive ecosystem. Proc Natl Acad Sci USA 96:9833–9838

Hooper LV, Wong MH, Thelin A et al (2001) Molecular analysis of commensal host-microbial relationships in the intestine. Science 291:881–884

Hooper LV, Midtvedt T, Gordon JI (2002) How host-microbial interactions shape the nutrient environment of the mammalian intestine. Annu Rev Nutr 22:283–307

Hsiao Elaine Y, McBride Sara W, Hsien S et al (2013) Microbiota modulate behavioral and physiological abnormalities associated with neurodevelopmental disorders. Cell 155:1451–1463

Hughes DT, Sperandio V (2008) Inter-kingdom signalling: communication between bacteria and their hosts. Nat Rev Microbiol 6:111–120

Janzen DH (1977) Why fruits rot, seeds mold, and meat spoils. Am Nat 111:691–713

Jenkins C, Samudrala R, Anderson I et al (2002) Genes for the cytoskeletal protein tubulin in the bacterial genus Prosthecobacter. Proc Natl Acad Sci USA 99:17049–17054

Johnson PTJ, Ostfeld RS, Keesing F (2015) Frontiers in research on biodiversity and disease. Ecol Lett 18:1119–1133

Joseph IM, Kirschner D (2004) A model for the study of *Helicobacter pylori* interaction with human gastric acid secretion. J Theor Biol 228:55–80

Joshi A, Kostiuk B, Rogers A et al (2017) Rules of engagement: the type VI secretion system in *Vibrio cholerae*. Trends Microbiol 25:267–279

Kaneko T, Tabata S (1997) Complete genome structure of the unicellular cyanobacterium *Synechocystis* sp. PCC6803. Plant Cell Physiol 38:1171–1176

Karavolos MH, Khan CM (2014) Multidirectional chemical signalling between mammalian hosts, resident microbiota, and invasive pathogens: neuroendocrine hormone-induced changes in bacterial gene expression. Adv Exp Med Biol 817:241–253

Kaushik M, Knowles SCL, Webster JP (2014) What makes a feline fatal in *Toxoplasma gondii*'s fatal feline attraction? Infected rats choose wild cats. Integr Comp Biol 54:118–128

Keesing F, Belden LK, Daszak P et al (2010) Impacts of biodiversity on the emergence and transmission of infectious diseases. Nature 468:647–652

Kendall MM, Sperandio V (2016) What a dinner party! Mechanisms and functions of interkingdom signaling in host-pathogen associations. MBio 7:e01748

Kim YS, Ho SB (2010) Intestinal goblet cells and mucins in health and disease: recent insights and progress. Curr Gastroenterol Rep 12:319–330
Klasson L, Kambris Z, Cook PE et al (2009) Horizontal gene transfer between *Wolbachia* and the mosquito *Aedes aegypti*. BMC Genomics 10:33
Klein SL (2003) Parasite manipulation of the proximate mechanisms that mediate social behavior in vertebrates. Physiol Behav 79:441–449
Koenig JE, Spor A, Scalfone N et al (2011) Succession of microbial consortia in the developing infant gut microbiome. Proc Natl Acad Sci USA 108:4578–4585
Kondo N, Nikoh N, Ijichi N et al (2002) Genome fragment of *Wolbachia* endosymbiont transferred to X chromosome of host insect. Proc Natl Acad Sci USA 99:14280–14285
Koonin EV, Makarova KS, Aravind L (2001) Horizontal gene transfer in prokaryotes: quantification and classification. Annu Rev Microbiol 55:709–742
Koraimann G, Wagner MA (2014) Social behavior and decision making in bacterial conjugation. Front Cell Infect Microbiol 4:54
Korochkina S, Barreau C, Pradel G et al (2006) A mosquito-specific protein family includes candidate receptors for malaria sporozoite invasion of salivary glands. Cell Microbiol 8:163–175
Kruuk H (1967) Competition for food between vultures in East Africa. Ardea 55:171–193
Lacroix B, Citovsky V (2016) Transfer of DNA from bacteria to eukaryotes. MBio 7(4):e00863-16
Lam K, Babor D, Duthie B et al (2007) Proliferating bacterial symbionts on house fly eggs affect oviposition behaviour of adult flies. Anim Behav 74:81–92
Lam K, Geisreiter C, Gries G (2009a) Ovipositing female house flies provision offspring larvae with bacterial food. Entomol Exp Appl 133:292–295
Lam K, Thu K, Tsang M et al (2009b) Bacteria on housefly eggs, *Musca domestica*, suppress fungal growth in chicken manure through nutrient depletion or antifungal metabolites. Naturwissenschaften 9:1127–1132
Lang BF, Gray MW, Burger G (1999) Mitochondrial genome evolution and the origin of eukaryotes. Annu Rev Genet 33:351–397
Langford-Smith A, Day AJ, Bishop PN, Clark SJ (2015) Complementing the sugar code: role of GAGs and sialic acid in complement regulation. Front Immunol 6:25
Laracuente A, Brown RA, Jobin W (1979) Comparison of four species of snails as potential decoys to intercept schistosome miracidia. Am J Trop Med Hyg 28:99–105
Le Chatelier E, Nielsen T, Qin J et al (2013) Richness of human gut microbiome correlates with metabolic markers. Nature 500:541–546
Lee JH, Del Sorbo L, Khine AA et al (2003) Modulation of bacterial growth by tumor necrosis factor-alpha in vitro and in vivo. Am J Respir Crit Care Med 168:1462–1470
Lee J-H, Wood TK, Lee J (2015) Roles of indole as an interspecies and interkingdom signaling molecule. Trends Microbiol 23:707–718
Lesouhaitier O, Veron W, Chapalain A et al (2009) Gram-negative bacterial sensors for eukaryotic signal molecules. Sensors (Basel) 9:6967–6990
Levy M, Kolodziejczyk AA, Thaiss CA et al (2017) Dysbiosis and the immune system. Nat Rev Immunol 17:219–232
Lin X, Kaul S, Rounsley S et al (1999) Sequence and analysis of chromosome 2 of the plant Arabidopsis thaliana. Nature 402:761–768
Liu CW, Murray JD (2016) The role of flavonoids in nodulation host-range specificity: an update. Plants (Basel) 5(3)
Liu D, Diorio J, Tannenbaum B et al (1997) Maternal care, hippocampal glucocorticoid receptors, and hypothalamic-pituitary-adrenal responses to stress. Science 277:1659–1662
Liu B, Faller LL, Klitgord N et al (2012a) Deep sequencing of the oral microbiome reveals signatures of periodontal disease. PLoS One 7:e37919
Liu L, Li Y, Li S et al (2012b) Comparison of next-generation sequencing systems. J Biomed Biotechnol 2012:11

Liu W, Longnecker M, Tarone AM et al (2016) Response of *Lucilia sericata* (Diptera: Calliphoridae) to compounds from microbial decomposition of larval resources. Anim Behav 115:217–225

Llewellyn MS, Boutin S, Hoseinifar SH et al (2014) Teleost microbiomes: the state of the art in their characterization, manipulation and importance in aquaculture and fisheries. Front Microbiol 5:207

Lloyd-Price J, Abu-Ali G, Huttenhower C (2016) The healthy human microbiome. Genome Med 8:51

Lowery CA, Dickerson TJ, Janda KD (2008) Interspecies and interkingdom communication mediated by bacterial quorum sensing. Chem Soc Rev 37:1337–1346

Lowrey L, Woodhams DC, Tacchi L et al (2015) Topographical mapping of the rainbow trout (*Oncorhynchus mykiss*) microbiome reveals a diverse bacterial community with antifungal properties in the skin. Appl Environ Microbiol 81:6915–6925

Lozupone CA, Stombaugh JI, Gordon JI et al (2012) Diversity, stability and resilience of the human gut microbiota. Nature 489:220–230

Luo G, Niesel DW, Shaban RA et al (1993) Tumor necrosis factor alpha binding to bacteria: evidence for a high-affinity receptor and alteration of bacterial virulence properties. Infect Immun 61:830–835

Lurie-Weinberger MN, Gomez-Valero L, Merault N et al (2010) The origins of eukaryotic-like proteins in *Legionella pneumophila*. Int J Med Microbiol 300:470–481

Lyte M (1992) The role of catecholamines in gram-negative sepsis. Med Hypotheses 37:255–258

Lyte M, Vulchanova L, Brown DR (2011) Stress at the intestinal surface: catecholamines and mucosa-bacteria interactions. Cell Tissue Res 343:23–32

Ma Q, Fonseca A, Liu W et al (2012) *Proteus mirabilis* interkingdom swarming signals attract blow flies. ISME J 6:1356–1366

Maciorkowska E, Panasiuk A, Kondej-Muszynska K et al (2006) Mucosal gastrin cells and serum gastrin levels in children with *Helicobacter pylori* infection. Adv Med Sci 51:137–141

McFall-Ngai M, Hadfield MG, Bosch TCG et al (2013) Animals in a bacterial world, a new imperative for the life sciences. Proc Natl Acad Sci USA 110:3229–3236

Moeller AH, Ochman H (2014) Microbiomes are true to type. Proc Natl Acad Sci USA 111:9372–9373

Mohr RM, Tomberlin JK (2014) Environmental factors affecting early carcass attendance by four species of blow flies (Diptera: Calliphoridae) in Texas. J Med Entomol 51:702–708

Mohr RM, Tomberlin JK (2015) Development and validation of a new technique for estimating a minimum postmortem interval using adult blow fly (Diptera: Calliphoridae) carcass attendance. Int J Legal Med 129:851–859

Molloy PP, Gage Matthew JG (2006) Evolution: vertebrate reproductive strategies get mixed up. Curr Biol 16:R876–R879

Mosca A, Leclerc M, Hugot JP (2016) Gut microbiota diversity and human diseases: should we reintroduce key predators in our ecosystem? Front Microbiol 7:455

Motley JL, Stamps BW, Mitchell CA et al (2017) Opportunistic sampling of roadkill as an entry point to accessing natural products assembled by bacteria associated with non-anthropoidal mammalian microbiomes. J Nat Prod 80(3):598–608

Mumcuoglu KY, Miller J, Mumcuoglu M et al (2001) Destruction of bacteria in the digestive tract of the maggot of *Lucilia sericata* (Diptera: Calliphoridae). J Med Entomol 38:161–166

Munch A, Stingl L, Jung K et al (2008) *Photorhabdus luminescens* genes induced upon insect infection. BMC Genomics 9:229

Murphy PM (1993) Molecular mimicry and the generation of host defense protein diversity. Cell 72:823–826

Myers B, Scheimann JR, Franco-Villanueva A et al (2017) Ascending mechanisms of stress integration: implications for brainstem regulation of neuroendocrine and behavioral stress responses. Neurosci Biobehav Rev 74:366–375

Nayduch D (2017) Special collection: filth fly–microbe interactions. Ann Entomol Soc Am 110:2–5

Nicholson JK, Holmes E, Kinross J et al (2012) Host-gut microbiota metabolic interactions. Science 336:1262–1267

Nikoh N, Tanaka K, Shibata F et al (2008) *Wolbachia* genome integrated in an insect chromosome: evolution and fate of laterally transferred endosymbiont genes. Genome Res 18:272–280

Oh J, Conlan S, Polley EC et al (2012) Shifts in human skin and nares microbiota of healthy children and adults. Genome Med 4:77

Ostfeld RS, Keesing F (2017) Is biodiversity bad for your health? Ecosphere 8:e01676–n/a

Ostfeld RS, Keesing F, Eviner VT (2010) Infectious disease ecology: effects of ecosystems on disease and of disease on ecosystems. Princeton University Press, Princeton

Pacheco AR, Curtis MM, Ritchie JM et al (2012) Fucose sensing regulates bacterial intestinal colonization. Nature 492:113–117

Payie KG, Rather PN, Clarke AJ (1995) Contribution of gentamicin 2′-N-acetyltransferase to the O acetylation of peptidoglycan in *Providencia stuartii*. J Bacteriol 177:4303–4310

Pechal JL, Benbow ME (2016) Microbial ecology of the salmon necrobiome: evidence salmon carrion decomposition influences aquatic and terrestrial insect microbiomes. Environ Microbiol 18:1511–1522

Pechal JL, Crippen TL, Benbow ME et al (2013) The potential use of bacterial community succession in forensics as described by high throughput metagenomic sequencing. Int J Legal Med 128:193–205

Plancak D, Music L, Puhar I (2015) Quorum sensing of periodontal pathogens. Acta Stomatol Croat 49:234–241

Ponnusamy L, Xu N, Nojima S et al (2008) Identification of bacteria and bacteria-associated chemical cues that mediate oviposition site preferences by *Aedes aegypti*. Proc Natl Acad Sci USA 105:9262–9267

Ponting CP, Aravind L, Schultz J et al (1999) Eukaryotic signalling domain homologues in archaea and bacteria. Ancient ancestry and horizontal gene transfer. J Mol Biol 289:729–745

Porat R, Clark BD, Wolff SM et al (1991) Enhancement of growth of virulent strains of *Escherichia coli* by interleukin-1. Science 254:430–432

Price CT, Kwaik YA (2010) Exploitation of host polyubiquitination machinery through molecular mimicry by eukaryotic-like bacterial F-box effectors. Front Microbiol 1:122

Price CT, Al-Quadan T, Santic M et al (2010) Exploitation of conserved eukaryotic host cell farnesylation machinery by an F-box effector of *Legionella pneumophila*. J Exp Med 207:1713–1726

Quail MA, Smith M, Coupland P et al (2012) A tale of three next generation sequencing platforms: comparison of Ion Torrent, Pacific Biosciences and Illumina MiSeq sequencers. BMC Genomics 13:341

Reid G, Brigidi P, Burton JP et al (2015) Microbes central to human reproduction. Am J Reprod Immunol 73:1–11

Ricklefs RE (2010) Host–pathogen coevolution, secondary sympatry and species diversification. Philos Trans R Soc Lond B Biol Sci 365:1139–1147

Robbe C, Capon C, Coddeville B et al (2004) Structural diversity and specific distribution of O-glycans in normal human mucins along the intestinal tract. Biochem J 384:307–316

Rumbaugh KP (2007) Convergence of hormones and autoinducers at the host/pathogen interface. Anal Bioanal Chem 387:425–435

Ruzzini AC, Clardy J (2016) Gene flow and molecular innovation in bacteria. Curr Biol 26:R859–R864

Sampson Timothy R, Mazmanian Sarkis K (2015) Control of brain development, function, and behavior by the microbiome. Cell Host Microbe 17:565–576

Sandoval-Sierra JV, Latif-Eugenin F, Martín MP et al (2014) *Saprolegnia* species affecting the salmonid aquaculture in Chile and their associations with fish developmental stage. Aquaculture 434:462–469

Schreiber KL, Brown DR (2005) Adrenocorticotrophic hormone modulates *Escherichia coli* O157: H7 adherence to porcine colonic mucosa. Stress 8:185–190

Sender R, Fuchs S, Milo R (2016a) Are we really vastly outnumbered? Revisiting the ratio of bacterial to host cells in humans. Cell 164:337–340

Sender R, Fuchs S, Milo R (2016b) Revised estimates for the number of human and bacteria cells in the body. PLoS Biol 14:e1002533

Shivik JA, Clark L (1997) Carrion seeking in brown tree snakes: importance of olfactory and visual cues. J Exp Zool 279:549–553

Skippington E, Ragan MA (2012) Evolutionary dynamics of small RNAs in 27 *Escherichia coli* and *Shigella* genomes. Genome Biol Evol 4:330–345

Sperandio V, Torres AG, Jarvis B et al (2003) Bacteria-host communication: the language of hormones. Proc Natl Acad Sci USA 100:8951–8956

Sprague GF Jr, Winans SC (2006) Eukaryotes learn how to count: quorum sensing by yeast. Genes Dev 20:1045–1049

Stecher B, Maier L, Hardt W-D (2013) 'Blooming' in the gut: how dysbiosis might contribute to pathogen evolution. Nat Rev Microbiol 11:277–284

Swei A, Kwan JY (2017) Tick microbiome and pathogen acquisition altered by host blood meal. ISME J 11:813–816

Tellez G, Higgins S, Donoghue A et al (2006) Digestive physiology and the role of microorganisms. J Appl Poult Res 15:136

ten Bosch JR, Grody WW (2008) Keeping up with the next generation. J Mol Diagn 10:484–492

The Human Microbiome Jumpstart Reference Strains Consortium (2010) A catalog of reference genomes from the human microbiome. Science 328:994–999

Theis KR, Venkataraman A, Dycus JA et al (2013) Symbiotic bacteria appear to mediate hyena social odors. Proc Natl Acad Sci USA 110:19832–19837

Tomberlin JK, Crippen TL, Tarone AM et al (2012) Interkingdom response of flies to bacteria mediated by fly physiology and bacterial quorum sensing. Anim Behav 84:1449–1456

Tomberlin JK, Crippen TL, Tarone AM et al (2017a) A review of bacterial interactions with blow flies (Diptera: Calliphoridae) of medical, veterinary, and forensic importance. Ann Entomol Soc Am 110:19–36

Tomberlin JK, Crippen TL, Wu G et al (2017b) Indole: an evolutionarily conserved influencer of behavior across kingdoms. BioEssays 39:1600203

Toth K, Stacey G (2015) Does plant immunity play a critical role during initiation of the legume-rhizobium symbiosis? Front Plant Sci 6:401

Treseler CB, Maziarz RT, Levitz SM (1992) Biological activity of interleukin-2 bound to *Candida albicans*. Infect Immun 60:183–188

Troyer K (1984) Behavioral acquisition of the hindgut fermentation system by hatchling *Iguana iguana*. Behav Ecol Sociobiol 14:189–193

Turnbaugh PJ, Ley RE, Mahowald MA et al (2006) An obesity-associated gut microbiome with increased capacity for energy harvest. Nature 444(7122):1027–1031

Turnbaugh PJ, Hamady M, Yatsunenko T et al (2009) A core gut microbiome in obese and lean twins. Nature 457:480–484

van Nood E, Vrieze A, Nieuwdorp M et al (2013) Duodenal infusion of donor feces for recurrent *Clostridium difficile*. N Engl J Med 368:407–415

Van Valen L (1973) A new evolutionary law. Evol Theory 3:1

Varki A (2006) Nothing in glycobiology makes sense, except in the light of evolution. Cell 126:841–845

Verbrugghe E, Boyen F, Van Parys A et al (2011) Stress induced *Salmonella typhimurium recrudescence* in pigs coincides with cortisol induced increased intracellular proliferation in macrophages. Vet Res 42:118

Verbrugghe E, Dhaenens M, Leyman B et al (2016) Host stress drives *Salmonella recrudescence*. Sci Rep 6:20849

Verhulst NO, Beijleveld H, Knols BGJ et al (2009) Cultured skin microbiota attracts malaria mosquitoes. Malar J 8:302

Verhulst NO, Andriessen R, Groenhagen U et al (2010a) Differential attraction of malaria mosquitoes to volatile blends produced by human skin bacteria. PLoS One 5:e15829

Verhulst NO, Takken W, Dicke M et al (2010b) Chemical ecology of interactions between human skin microbiota and mosquitoes. FEMS Microbiol Ecol 74:1–9

Verhulst NO, Qiu YT, Beijleveld H et al (2011) Composition of human skin microbiota affects attractiveness to malaria mosquitoes. PLoS One 6:e28991

Wang X, Wang Q, Yang M et al (2011) QseBC controls flagellar motility, fimbrial hemagglutination and intracellular virulence in fish pathogen Edwardsiella tarda. Fish Shellfish Immunol 30:944–953

Weatherbee C, Pechal J, Benbow M (2017) The dynamic maggot mass microbiome. Ann Entomol Soc Am 110(1):45–53

Webster JP (2001) Rats, cats, people and parasites: the impact of latent toxoplasmosis on behaviour. Microbes Infect 3:1037–1045

Webster BL, Diaw OT, Seye MM et al (2013) Introgressive hybridization of *Schistosoma haematobium* group species in Senegal: species barrier break down between ruminant and human schistosomes. PLoS Negl Trop Dis 7:e2110

Webster JP, Borlase A, Rudge JW (2017) Who acquires infection from whom and how? Disentangling multi-host and multi-mode transmission dynamics in the 'elimination' era. Philos Trans R Soc B 372:20160091

Werren JH, Richards S, Desjardins CA et al (2010) Functional and evolutionary insights from the genomes of three parasitoid Nasonia species. Science 327:343–348

Witwer KW, Hirschi KD (2014) Transfer and functional consequences of dietary microRNAs in vertebrates: concepts in search of corroboration: negative results challenge the hypothesis that dietary xenomiRs cross the gut and regulate genes in ingesting vertebrates, but important questions persist. BioEssays 36:394–406

Wong AC-N, Holmes A, Ponton F et al (2015) Behavioral microbiomics: a multi-dimensional approach to microbial influence on behavior. Front Microbiol 6:1359

Wu L, Estrada O, Zaborina O et al (2005) Recognition of host immune activation by *Pseudomonas aeruginosa*. Science 309:774–777

Yamashita K, Kaneko H, Yamamoto S et al (1998) Inhibitory effect of somatostatin on *Helicobacter pylori* proliferation in vitro. Gastroenterology 115:1123–1130

Yanez AJ, Nualart F, Droppelmann C et al (2003) Broad expression of fructose-1,6-bisphosphatase and phosphoenolpyruvate carboxykinase provide evidence for gluconeogenesis in human tissues other than liver and kidney. J Cell Physiol 197:189–197

Zaborina O, Lepine F, Xiao G et al (2007) Dynorphin activates quorum sensing quinolone signaling in *Pseudomonas aeruginosa*. PLoS Pathog 3:e35

Zav'yalov VP, Chernovskaya TV, Navolotskaya EV et al (1995) Specific high affinity binding of human interleukin 1 beta by Caf1A usher protein of *Yersinia pestis*. FEBS Lett 371:65–68

Zhang L, Hou D, Chen X et al (2012) Exogenous plant MIR168a specifically targets mammalian LDLRAP1: evidence of cross-kingdom regulation by microRNA. Cell Res 22:107–126

Zurek L, Denning SS, Schal C et al (2001) Vector competence of *Musca domestica* (Diptera: Muscidae) for *Yersinia pseudotuberculosis*. J Med Entomol 38:333–335

Chapter 2
Biodiversity and Disease Transmission

Serge Morand

Abstract Biodiversity changes associated with the anthropogenic alteration of natural environments have been hypothesized to enhance disease transmission and to facilitate the emergence of infectious diseases. This chapter reviews the various links that may occur between biodiversity and disease transmission on scales ranging from global to local and the likely ecological mechanisms. The consequences of land usage and land cover changes on disease transmission are formulated through the overall effects on biodiversity observed from long-term observatories. Habitat fragmentation should lead to reduced diversity of pathogen species and changes in pathogen prevalence as proposed by the "perturbation hypothesis." However, habitat fragmentation that leads to increased edge, and increasing contacts between different communities of reservoirs and vectors, should increase disease transmission and pathogen prevalence according to the "pathogen pool diversity" hypothesis. Network analyses represent new tools to investigate disease transmission in a changing biodiversity context, i.e., changes in multiple hosts—multiple parasite interactions. Finally, this review advocates for manipulative experiments, theoretical studies, and long-term data collection in ecological observatories that will help in building scenarios of future health.

2.1 Introduction

Abiotic factors (i.e., climate and climate variability) and biotic factors (i.e., host population density, host community structure) have been demonstrated as crucial determinants of disease transmission (Anderson and May 1991; Hudson et al. 2002; Morand et al. 2013). Increasing attention has been given on the role of biodiversity as regulating disease transmission, and studies have questioned how biodiversity changes associated with the anthropogenic alteration of natural environments may

S. Morand (✉)
CNRS-CIRAD Kasetsart University, Bangkok, Thailand
e-mail: serge.morand@univ-montp2.fr; serge.morand@cirad.fr

C. J. Hurst (ed.), *The Connections Between Ecology and Infectious Disease*,
Advances in Environmental Microbiology 5,
https://doi.org/10.1007/978-3-319-92373-4_2

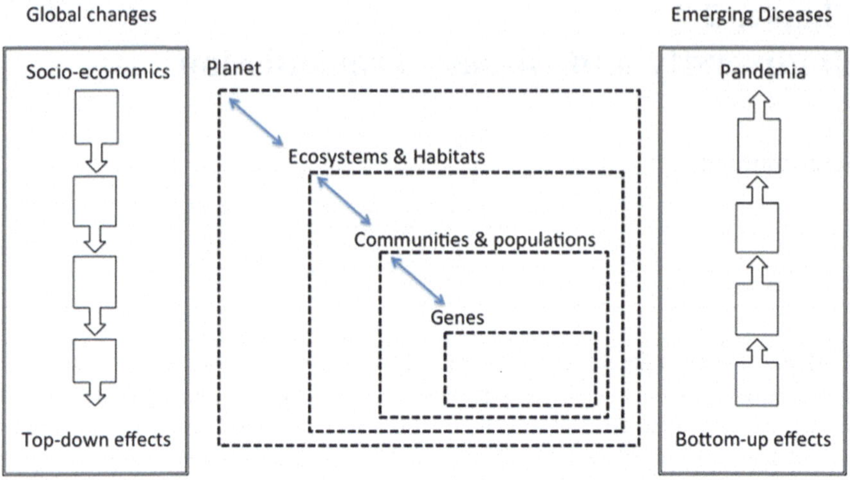

Fig. 2.1 Top-down effects of global changes on biodiversity with bottom-up effects on infectious diseases

enhance disease transmission and potentially facilitate the emergence of new infectious diseases (Keesing et al. 2010). Human modification of habitats leads to changes in the composition of species communities often associated with a major reduction in biodiversity (Dirzo et al. 2014). Several wildlife-borne or vector-borne diseases exhibit an increasing transmission success in these human-altered environments (LoGiudice et al. 2003). This pattern was popularized as the "dilution effect hypotheses" (Schmidt and Ostfeld 2001), although hot debates challenge the potential role of a rich biodiversity as a buffer for disease transmission (Randolph and Dobson 2012; Lafferty and Wood 2013; Keesing et al. 2015).

A first aim of this chapter is to review the various links that may occur between biodiversity and disease transmission on scales ranging from global to local by presenting patterns and likely mechanisms (Fig. 2.1).

A second aim is to present network analyses as a tool for investigating disease transmission in a "biodiversity" context, i.e., multiple hosts—multiple parasites. Finally, this review advocates for long-term data collection in ecological observatories coupled with theoretical epidemiology that will help in building scenarios of future health.

2.2 The Links Between Biological Diversity, Cultural Diversity, and Disease Diversity

Human pathogens are not randomly distributed geographically (Dunn et al. 2010), as their number and diversity increase from higher to lower latitudes (Guernier et al. 2004; Dunn et al. 2010; Bonds et al. 2012; Cashdan 2014; Murray et al. 2015). This pattern follows a similar trend observed for general biodiversity with increasing richness in bird and mammalian species from temperate to tropical zones (Schipper et al. 2008). Among countries, a positive relationship is observed between bird and mammal species richness and the number of human infectious diseases. A country with high biodiversity, in terms of mammals and birds, has a high diversity of human pathogens (Dunn et al. 2010; Morand et al. 2014c).

Cultural diversity appears to mirror biological diversity as emphasized by Maffi (2005). Fincher and Thornhill (2008) showed that countries with high cultural diversity, using the number of spoken languages as a proxy, have greater diversity of infectious diseases. Hence, biodiversity, cultural diversity, and human pathogens' diversity are entangled in their actual geographical distributions due to evolutionary and ecological processes (Morand 2015a) (Fig. 2.2). These large-scale patterns deserve in-depth investigations at regional and local scales. Hence, rather than to investigate at only country level, it would be worth investigating the potential associations among biodiversity, cultural diversity, and health at ecoregion and ecozone levels (Olson and Dinerstein 1998; Bailey 2014), which will allow our scientific progress to overcome nation-state entities.

However, these patterns are affected by the ongoing dramatic global changes affecting biodiversity, the epidemiological environment and the transmission of many if not all infectious diseases (Daily and Ehrlich 1996).

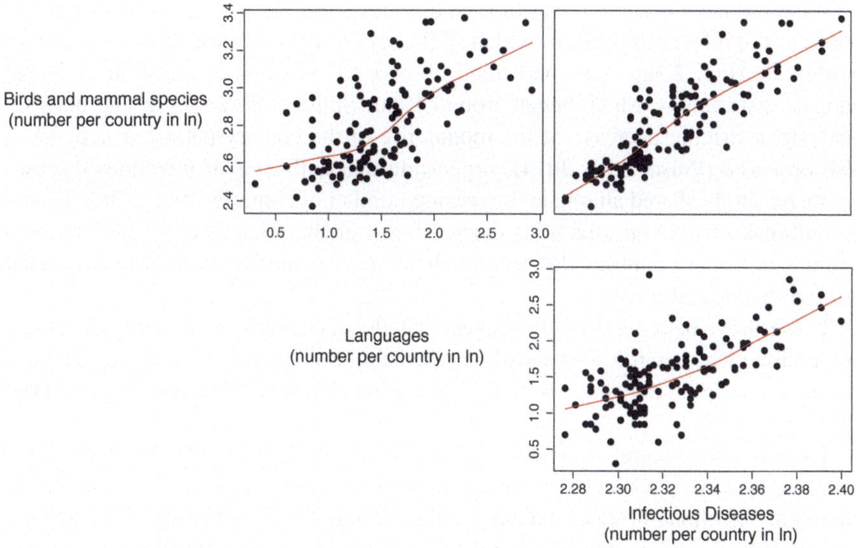

Fig. 2.2 Cross-correlation among number of bird and mammal species, number of languages, and number of infectious diseases per country (redrawn from Morand et al. 2014c; the data are from Bird Life International (http://birdlife.org/datazone/home), the Ethnologue (https://www.ethnologue.com/), IUCN (http://www.iucnredlist.org/), and GIDEON (Global Infectious Diseases and Epidemiology Network, www.gideononline.com)

2.3 Global Changes

Ongoing global changes (climate change, biodiversity loss, land use change, biological invasion) are modifying the epidemiological environment as already mentioned two decades ago by Daily and Ehrlich (1996), with a sharp decrease in the burden of parasitic infection in the developed countries and a still high level in developing countries. Large declines of human parasite burdens and infectious diseases have been observed in developed countries over the last century (Armstrong et al. 1999). However, the last decades have been characterized by a dramatic increase in the number of emerging infectious diseases (Wilcox and Gubler 2005; Jones et al. 2008; Smith et al. 2014). The explanatory factors are generally associated with climate change and climate variability, increased global trade, land use change, overuse of living resources, and biodiversity loss (Chivian and Bernstein 2004; Wilcox and Colwell 2005), which are occurring on an unprecedented scale. Biodiversity loss through altered landscapes due to urbanization and agricultural intensification (Gibbs et al. 2010) appears to be linked to higher disease risks with emergence of novel pathogens resulting from increased contacts between wildlife, domesticated animals, and humans (Lloyd-Smith et al. 2009; Jones et al. 2013; Lindahl and Grace 2015; Hassell et al. 2017). Biodiversity loss also applies to the parasites themselves. Even if more than half of all biodiversity are parasites, when including viruses and pathogenic bacteria, the rate of parasite extinction in the wild is far from being estimated accurately (Dunn et al. 2009, 2010).

What has been observed over the last decades is the increasing number of global outbreaks of infectious diseases with a slight but constant increase of new pathogens worldwide (Fig. 2.3a). A second trend observed is this homogenization of global parasite distribution, which began around 1960 (Smith et al. 2007). Using network analysis, a striking decrease of the modularity of the country-pathogen network is also observed (Poisot et al. 2014), suggesting that outbreaks of infectious diseases are increasingly shared among an increasing number of countries (Fig. 2.3b). Today, an outbreak of a given infectious disease has a greater chance to spread among a larger number of countries due to globalization as countries are highly connected through trades and travels.

These above patterns strongly suggest that global changes are affecting the global epidemiological environment mostly by favoring the spread of infectious diseases among countries and by increasing the risks of pandemics (Tatem and Rogers 2006) (Fig. 2.1).

There is a good correlation between biodiversity and human pathogen diversity. It remains, however, difficult to explain how biodiversity loss can increase newly emergent infectious diseases unless we disentangle the presence of a disease in a country (i.e., its endemicity) and the patterns of epidemics or outbreaks in a country. By analyzing separately the number of diseases and the number of diseases presenting an outbreak, Morand et al. (2014a) were able to show in the Asia-Pacific region that if the number of infectious diseases was related to biodiversity

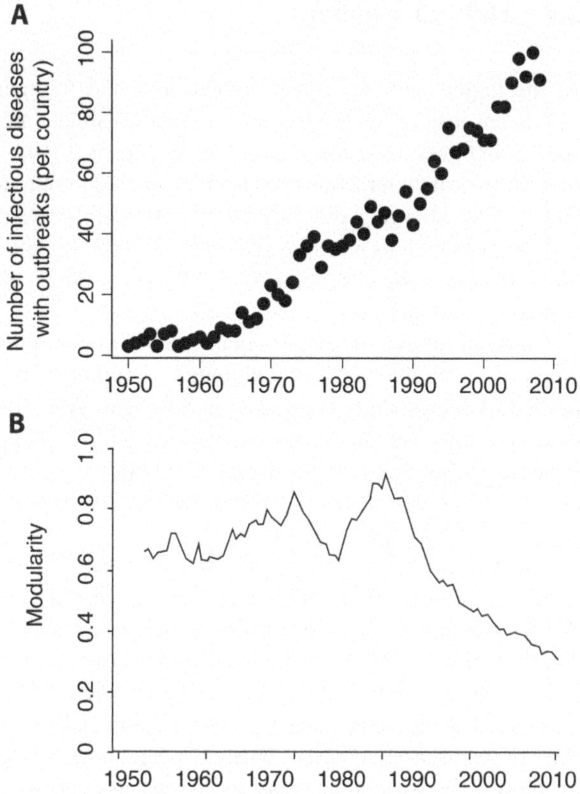

Fig. 2.3 (**a**) Number of infectious diseases presenting outbreaks globally over the last 60 years (redrawn from Morand et al. 2014c, data from GIDEON (Global Infectious Diseases and Epidemiology Network, www.gideononline.com)). (**b**) Change in values of modularity of country-infectious diseases over time. Modularity is a measure of division of a network into modules (or groups or communities) with a high value of modularity indicating the presence of a high number of modules. The decrease of modularity values since the beginning of the 1960s depicts a global homogenization of infectious diseases (redrawn from Poisot et al. 2014)

among countries, the number of diseases presenting an outbreak was related to biodiversity loss. There was a positive association between the number of zoonotic diseases presenting an outbreak and the number of mammal and bird species at threat (taking into account potential confounding effects such as gross national product, population size, and disease surveillance and surveys). Interestingly, the number of languages at threat was also found positively associated with the number of bird and mammal species at threat (Morand and Lajaunie 2017). Hence, correlative and comparative approaches suggest (1) that biodiversity is a source of pathogen diversity but (2) that biodiversity at threat is a source of epidemics including new emerging zoonotic diseases. This may also apply to disease prevalence. For example, Derne et al. (2011) showed a negative correlation between the number of mammal species and the annual incidence of human leptospirosis among nations and territory islands worldwide. This represents a pattern that should be investigated for more zoonotic diseases. The link between transmission ecology of diseases and biodiversity should be investigated at local level, i.e., remnant habitats, than regional or global levels, i.e., among nations (Johnson et al. 2015).

2.4 Habitat Changes

At the larger landscape level, habitat loss and fragmentation generally lead to a significant loss of species because remnant habitats are too small and isolated for species to either persist or to recolonize (Morris 2010). The effect of fragmentation on host-parasite interactions has been examined in relation to conservation medicine and also to emerging diseases by investigating prevalence or species richness of pathogens as well as parasites (Allan et al. 2003; Lo Giudice et al. 2003; Wells et al. 2007; Gillespie and Chapman 2008; Mbora and Mc Peek 2009; Hussain et al. 2013; Fenoglio et al. 2012).

Considering parasite species richness or number of diseases, no clear results have emerged from the literature, although a comparative or meta-analysis study is needed. A recent study conducted in Southeast Asia (Bordes et al. 2015), using an extensive data set on 16 rodent species and 29 helminth species from several localities of Southeast Asia, showed that both the estimated degree of fragmentation and the rate of deforestation through the last 20–30 years have no influence on both parasite and host species richness.

On the other hand, parasite prevalence appears to be higher in fragmented places. A trend confirmed by Brearley et al. (2013), who analyzed 19 studies for changes in disease prevalence to assess the directional influence of human-modified landscapes. Nearly half of the studies examined by Brearley et al. (2013), (53%), indicated an increase in disease prevalence due to human-induced landscape change, while 21% identified a decrease in disease prevalence and 26% found varied disease prevalence (i.e., some disease increased, whereas others not). Studies in urban landscapes tend to show a consistent increase in disease prevalence (see Hassell et al. 2017). Gottdenker et al. (2014) analyzing a larger number of studies showed a similar trend with 56.9% of the studies documented increased pathogen transmission in response to anthropogenic change, and 10.4% of studies observed a decreasing pathogen transmission, although 30.4% had variable and complex pathogen responses (2.4% showed no detectable changes). The pathogen responses to anthropogenic changes should be investigated in relation to the types of land use/land cover changes, which have been rarely done.

Murray and Daszak (2013) proposed two non-exclusive hypotheses for infectious disease emergence due to land use change, which are based on different mechanisms. The first one, the "perturbation hypothesis," assumes that land use change perturbs disease dynamics in multi-host disease systems by disrupting the cross-species transmission rate. The second one, the "pathogen pool hypothesis," assumes that land use change allows exposure of novel hosts to a rich pool of pathogen diversity, which ultimately influences cross-species transmission rate.

The framework developed by Haddad et al. (2015) allows linkage between the consequences of land usage and land cover changes, and specifically habitat fragmentation, on disease transmission through the overall effects on biodiversity (such as reservoirs and vectors) (Fig. 2.4).

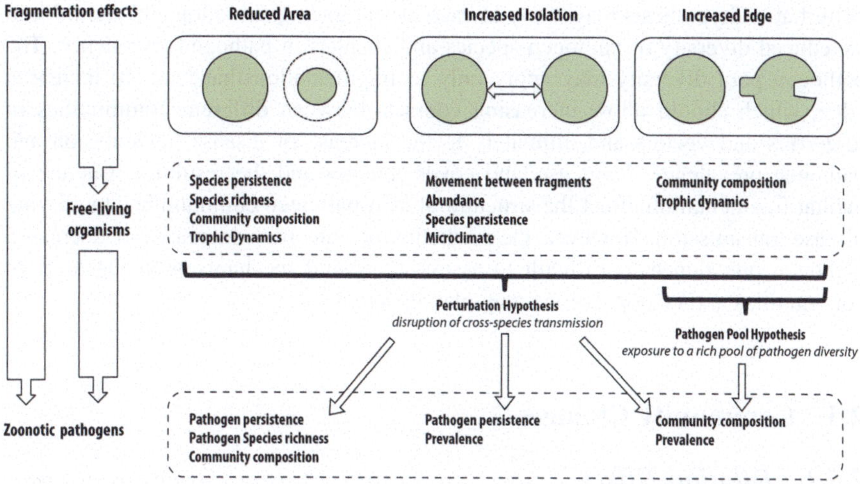

Fig. 2.4 The framework of Haddad et al. (2015), linking the different aspects of habitat fragmentation on biodiversity, is related to infectious disease emergence through the two hypotheses developed by Murray and Daszak (2013): the "perturbation" hypothesis and the "pathogen pool diversity" hypothesis. The broad arrows on the left of the figure show the consequences of habitat fragmentation on zoonotic pathogens, directly on free-living infectious stages or through effects on free-living organisms (i.e., vectors and reservoirs). Across the top is an indication of how areal reduction, increased isolation, and increased edge can affect the diversity and dynamics of free-living species according to long-term observations (Haddad et al. 2015). Reduced area affects species persistence, species richness, community composition, and trophic dynamics, or increased isolation affects movement between fragments, species abundance, and species persistence as well as microclimate conditions. Increased edge affects mostly community composition and trophic dynamics. Across the bottom, the consequences of habitat changes are related to the "perturbation" hypothesis and the "pathogen pool diversity" hypothesis. The "perturbation" hypothesis could apply to all kinds of areal changes (reduced area, increased isolation, increased edge), whereas the "pathogen pool diversity" hypothesis should apply only to habitat fragmentation leading to increased edge. Reduced area and increased isolation should affect pathogen persistence and prevalence. Increased edge might affect the community composition of pathogens, with an increasing pathogen species richness, and pathogen prevalence

In their review on experimental nature, Haddad et al. (2015) showed that fragmentation consistently reduced species richness and community composition. They also showed that reduced territorial fragment size and increased proportion of edge habitat led to important changes in tree communities' composition, with subsequent impacts on the community composition of animals. Then, whatever the aspects of fragmentation, whether they might be either reduced fragment area, increased isolation, or increased edge, those aspects will have strong impacts on species richness, community composition, species abundance, and movement (Fig. 2.4). The consequences of habitat fragmentation, through its different aspects, can be linked to disease transmission using Murray and Daszak's hypotheses (Fig. 2.4). The perturbation hypothesis may apply to each aspect of fragmentation and should lead to reduced diversity of pathogen species and changes in pathogen prevalence. The pathogen pool diversity may apply only to fragmentation that leads to increased edge, which should allow increasing contacts between different communities of reservoirs and vectors and ultimately to the increase of disease transmission and pathogen prevalence. Land use/land cover changes and the resulting structure of habitat fragmentation affect the structure of host-pathogen communities, modifying disease transmission. However, the prediction of outcomes, increased or decreased pathogen prevalence, is difficult to assess as several mechanisms at the level of communities and/or populations may play different roles.

2.5 Community Changes

2.5.1 Dilution Effect

Elton (1958) pioneered the field of disease transmission and biodiversity with his "disease diversity hypothesis." As quoted by Johnson et al. (2015) Elton emphasized that "outbreaks most often happen on cultivated or planted land. . .that is, in habitats and communities very much simplified by man." According to Elton, biological diversity homogenization should favor outbreaks of diseases. The "dilution effect hypothesis" has been proposed for zoonotic or vector-borne diseases (Schmidt and Ostfeld 2001). This hypothesis suggests that host species richness and diversity act as a buffer to pathogen spread, very similar to the "biotic resistance hypothesis" formulated in biological invasions. Host communities composed of a large ratio of noncompetent host species to competent hosts should contribute to reduced parasite transmission through wasted transmissions, in comparison to host communities with high proportion of competent hosts. The transmission of a pathogen to a noncompetent host creates a dilution with the loss of infective stages, which lowers the prevalence of infection in the competent hosts. If these competent hosts act as reservoirs for human infectious diseases, the dilution effect decreases their transmission risks to humans. This phenomenon has been reported in several vector-borne diseases such as Lyme disease (LoGiudice et al. 2003) or West Nile disease (Swaddle and Calos 2008) but also for rodent-borne diseases such as hantavirus (Carver et al. 2011).

A first meta-analysis has tested the importance of dilution hypothesis in disease transmission (Salkeld et al. 2013). The meta-analysis performed on 13 studies, following their criteria, showed a weak relationship between biodiversity and disease transmission suggesting that the dilution hypothesis for disease transmission was merely idiosyncratic. The review of Johnson et al. (2015) cited 90 studies published between 2012 and mid-2014, and most of them supported the existence of a dilution effect for various diseases affecting humans, wildlife, livestock, or plants. Civitello et al. (2015) conducted another meta-analysis using 61 parasite species, which provided evidence that host diversity inhibits parasite abundance. This abundance inhibition appeared independent of host density, study design, and type and specialization of parasites. Civitello et al. (2015) argued that a dilution effect was robust, but its magnitude was more related to frequency of focal host species than to their population density.

As a pattern of dilution can be observed in many kinds of transmission, including both direct and indirect such as those using vectors, it is likely that different epidemiological mechanisms can be operating. First, host population densities are a key parameter in parasite and disease transmission with higher host densities linked to both higher parasite and disease prevalence (Anderson and May 1979; Begon et al. 2002) and also to higher parasite species richness (Kamiya et al. 2014; Morand 2015b). Several empirical studies have emphasized the importance of this host parameter in parasite and pathogen transmission. For example, in a critical review, Jonsson et al. (2010) showed that the density of rodent populations appears to be the key factor that correlates with human infections by hantavirus. Werden et al. (2014) demonstrated the importance of rodent host densities, as compared to their diversity, on the prevalence of *Ixodes scapularis* tick nymphs infected with *Borrelia burgdorferi*. By controlling for various confounding factors, Werden et al. (2014) showed that when the relative abundance of mice was high in a site, higher small mammal species richness had little effect on nymph infection. This study suggests that high host diversity does not always reduce the prevalence of parasite infection, but host density of competent host does matter.

2.5.2 Invasion

Species invasion offers the opportunity to analyze the epidemiological consequences of changes in host-pathogen communities. The consequences of disease introductions in invaded environments may have dramatic effects on local species and communities, which may affect ecosystem functioning as observed for both myxomatosis and rinderpest (Dobson 1995). Diseases that are introduced either directly or through invasive species are of major concern in conservation biology with ongoing global trade and escalating new threats. As an example, 10 of 64 invasive bird species cause health impacts in Europe (Shirley and Kark 2009).

Invasion can decrease transmission of a native parasite. For instance, Telfer et al. (2005) observed a decline in the prevalence of two intracellular bacteria *Bartonella birtlesii* and *Bartonella taylorii* in native wood mouse populations concurrent with increase in the density of an invading bank vole. This prevalence decline was explained by the reduced infection having been vectored by fleas on wood mice.

2.5.3 Trophic Web

Predation on host reservoir species belonging to lower trophic levels may affect disease transmission. Orrock et al. (2011) in a comparative study showed that the prevalence of Sin Nombre virus (SNV) in deer mice was higher on islands harboring fewer rodent predators. The empirical and theoretical study of Levi et al. (2012) suggested that the increase in Lyme disease in recent years in the Northeastern and Midwestern United States correlates with the decline of a key small mammal predator, the red fox. The wide decline of the red fox was explained by the competitive expansion of coyote populations, which was also found to predict Lyme disease distribution in New York State.

2.5.4 *Multi-infection*

Parasite interspecific interactions also play a critical role in the dynamics of parasite transmission, whereas most studies have investigated only a single pathogen or parasite (Bordes and Morand 2011). Several studies conducted in rodents (Telfer et al. 2010; Bordes et al. 2015) showed that the very diversity of multi-pathogen infections affect the dynamics of each pathogen leading potentially to the generation of super-shedders (Lass et al. 2013). However, such multi-pathogen studies are still rare, although they would provide a better understanding of the epidemiological dynamics altered by biodiversity loss and habitat alterations.

2.5.5 *Synanthropic Species*

Several studies on rodent communities in tropical disturbed landscapes have shown a shift from species-rich communities comprised mainly of native species to reduced communities dominated by generalist species, which are often invasive and may be synanthropic (Charles and Ang 2010; McFarlane et al. 2012). Increase in the density of one or more synanthropic species does appear to be a common pattern (McFarlane et al. 2012; Morand et al. 2015). As a result, the proportion of recognized zoonotic diseases is often higher in populations of those generalist rodent species inhabiting areas that are considered to have been either disturbed or urbanized (Suzán et al. 2009; Palma et al. 2012). Blasdell et al. (2011) showed that hantavirus-infected rodents in Southeast Asia primarily were rodent species associated either with settlements or agricultural environments, i.e., synanthropic rodents, and suggested a likely relationship between human-dominated habitats and the increase in rodent-borne disease occurrence. Another study showed that hantavirus prevalence for the same generalist species varied between habitats differing in their level of disturbance (Goodin et al. 2006). Moreover, by carrying and disseminating parasites across multiple habitats, synanthropic species could enhance both host-switching and spillover to other reservoirs and to humans (Bordes et al. 2017).

2.6 Investigating Mechanisms Using Network Analyses

Finally, all these studies, and in particular comparative and meta-analyses, plead to "move beyond debates over the generality of the dilution effect and toward a mechanistic, predictive framework for biodiversity–disease interactions" as emphasized by Civitello et al. (2015).

In contrast to ecological networks (i.e., trophic web, plant-pollinator network), host-parasite networks have been recently investigated (Hagen et al. 2012) for several host-parasite systems such as fish-parasite, mammal-flea, primate-pathogen, domestic animal-zoonotic disease, and rodent-helminth networks (Poulin 2010; Vazquez et al. 2005; Morand et al. 2014b; Pilosof et al. 2015) (Fig. 2.5).

Using a large food web data set, Chen et al. (2008) demonstrated that the number of parasite species harbored by a given host species is related to the position of this given host species in the host-parasite network. A host species with high parasite diversity, often characterized by a wide diet range, occupies a network position close to many prey species, which helps at accumulating parasites from species at lower trophic levels.

Network-based approaches have been widely used to study transmission heterogeneity (Bansal et al. 2007; Bogich et al. 2013). In particular, network topology has been investigated to determine pathogen transmission across human and wildlife populations (Gómez et al. 2013). Network topology analysis helps to visualize the overall interactions between hosts and pathogens and to estimate potential host sources of parasites/pathogens. Network centrality is a measure of the connection among each host species in the network. A central host, characterized by high value of centrality in a host-pathogen network, is the one that is infected by many pathogens that infect many other hosts in the network. The centrality of a given host species is a good estimate of its potential to be a source of pathogens to other species (Gómez et al. 2013).

Morand et al. (2014b) investigated the structure of the interactions between human and their domestic animals and the shared diseases. They used centrality, a measure of the connection among each host species (humans and domestic animals) in the network, through the sharing of parasites/pathogens, and they showed that the most central hosts in the network (i.e., high value of centrality) are the ones that are infected by many pathogens that infect many other hosts in the network. Moreover, they showed that these central domesticated hosts in the network, which favor pathogen transmission not only to humans but also to all other domesticated animals, were associated a long time ago with humans.

The application of ecological network theory to the study of host-parasite interactions has identified several patterns in disease transmission such as the existence of a negative correlation between network connectance (or link numbers) and parasite species richness and the likely nested and modular host-parasite networks.

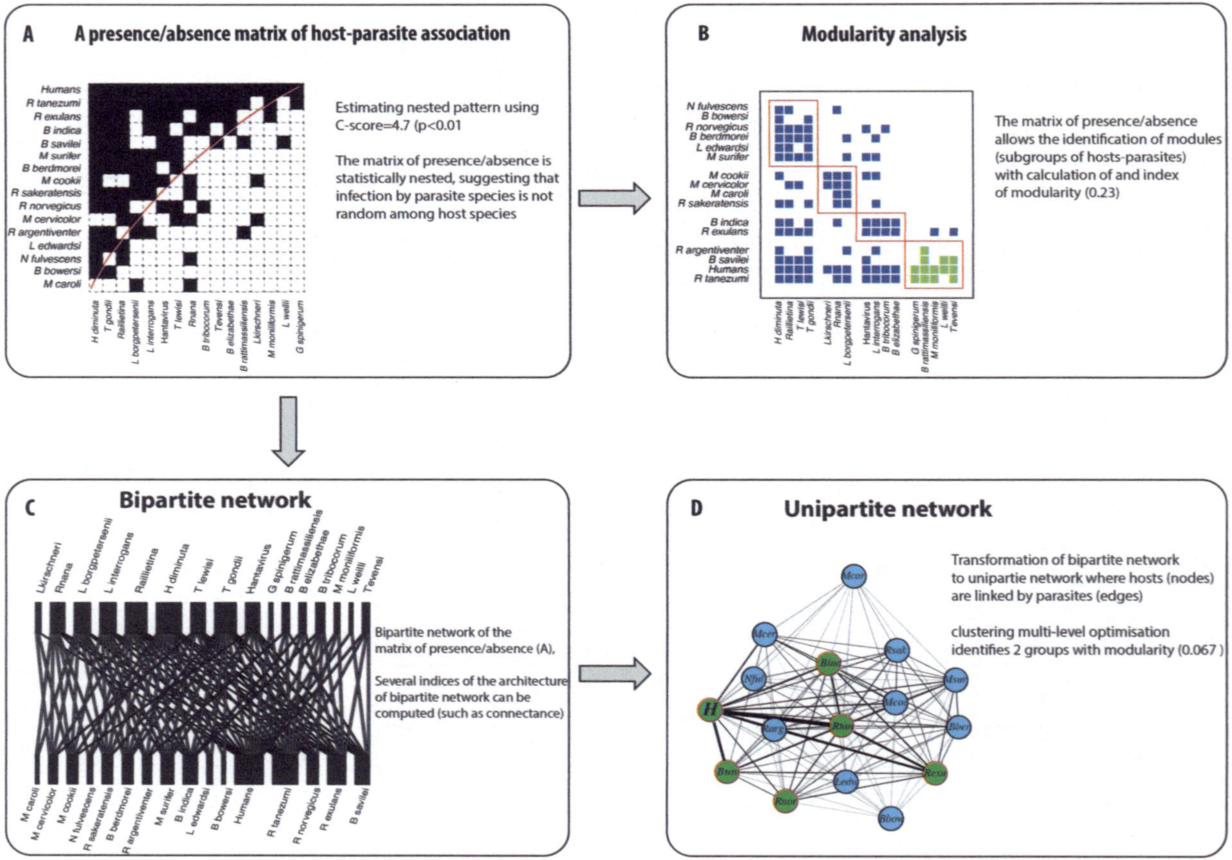

Fig. 2.5 Example of the use of network analysis. (**a**) Matrix of presence/absence of parasite species among host species and nestedness with the filled spaces representing presence. A nested pattern is depicted as a nesting of a given subset into a larger one. Several measurements of nested pattern are available such as

Bordes et al. (2015), studying an extensive data set on 16 rodent species and 29 helminth species from several localities of Southeast Asia and using network analysis, observed that rapid habitat fragmentation impacted host-parasite interactions as the rodent-helminth network becomes less connected and thus more modular, suggesting that parasite sharing among host species may become more difficult to maintain with the increase of habitat disturbance. That study also suggested it takes time to observe changes in both reservoir and pathogen species richness in responses to habitat changes.

Few studies have tried to link how habitat fragmentation may alter host-pathogen network properties. However, an improved understanding of the properties of network structures could provide insights into the response of interacting host-pathogen communities to anthropogenic disturbances. Using field data acquired from rodent species in various habitats in Southeast Asia, Bordes et al. (2017) investigated networks based on shared parasites, shared pathogens, and shared habitats among rodent species and humans. They investigated the architectures of bipartite and unipartite networks using modularity, subgroup partitioning, and node centrality in order to assess the relative epidemiological importance of particular rodent species (Fig. 2.5). They found that three synanthropic species were found to be members of subgroups including humans in both unipartite and bipartite networks on zoonotic agents and shared habitats. Moreover, these three synanthropic rodent species showed high values of centrality in shared zoonotic agents and high centrality values of shared habitats for two of these rodent species. These three synanthropic rodent species showed high degree of connectivity with humans, and they represent a high risk for direct zoonotic spillover. Moreover, when a rodent species is central in habitat-sharing, it may also play a key role as a bridge host for pathogen spread among habitats. Finally, these rodent species identified using the network approach are some of the potential reservoir hosts for new emerging arenaviruses in similar disturbed environment of Southeast Asia (Blasdell et al. 2016).

Fig. 2.5 (continued) C-score, or checkerboard score, which gives the randomness' statistics of the distribution of parasite presence among host species in the matrix of presence/absence (a low C-score indicates a high randomness). (**b**) Matrix of presence/absence of parasite species among host species and modularity. Modularity anlysis can help to depict modules, i.e., occurrences where there are more connected hosts and parasites in a subgroup relative to other subgroups. These modules can also be depicted in a bipartite network. (**c**) Bipartite network showing topology of (**a**) the matrix of presence/absence of parasite species among host species. Calculation of several indices, such as connectance and modularity, allow the investigation of the bipartite network architecture. (**d**) Transformation of topology from a bipartite network to a unipartite network, where the links (shared parasites) among nodes (host species) of unipartite networks depict shared parasites among hosts. Modularity analysis is used to identify subgroups of hosts that shared more parasites among them than among other subgroups. Other indices can be used such as centrality

2.7 Conclusion: Disease Ecology in the Anthropocene Defaunation

Anthropogenic factors have been identified as drivers of emerging diseases and have been found to favor the persistence and transmission of certain parasites (Rohr et al. 2011; Morand et al. 2014a, b, c).

The importance of human-dominated habitats in disease spread is of major concern as the risk of disease spread should be higher in human-dominated habitats. The ongoing "Anthropocene defaunation" is leading to empty tropical forests at least of large-bodied species (Dirzo et al. 2014). The sharp decline in abundance observed in many populations should have implications for zoonotic diseases by both decreasing and increasing risks. Pathogen diversity will decrease with their host diversity, but the prevalence of remaining pathogens specially those using synanthropic species as reservoirs or bridge species will increase. Indeed, rodents and small mammals in general will be the dominant mammal clades found not only in tropical forests but also in almost all human-modified environments. This may result in an increase in pathogen spillover into humans as already observed in Southeast Asian environments with critical ongoing defaunation.

The next step in research should better integrate field manipulative experiments (Suzán et al. 2009) and theoretical studies (Mihaljevic et al. 2014; White et al. 2015). Indeed, Gottdenker et al. (2014) emphasized that few studies were experimental in nature (relative to reviews). A second step is to build scenarios of future health in the face of biodiversity loss, and toward this goal we need long-term data collection such as the NEON (National Ecological Observatory Network) initiative (Springer et al. 2016), which will allow us to better explore disease transmission in relation to global and biodiversity changes.

Acknowledgments This work was part of the BiodivHealthSEA project (http://www.biodivhealthsea.org) funded by the French ANR programme CP&ES (grant number ANR 11CPEL 002) and supported by the RTPI-CNRS INEE "Biodiversity, Health and Societies in Southeast Asia."

Compliance with Ethical Standards

Funding This study was funded by French ANR project FutureHealthSEA (grant number ANR-17-CE35-0003-01).

Conflict of Interest Serge Morand declares that he has no conflict of interest.

Ethical Approval This article does not contain any studies with human participants or animals performed by the author.

References

Allan BF, Keesing F, Ostfeld RS (2003) Effect of forest fragmentation on Lyme disease risk. Conserv Biol 17:267–272

Anderson RM, May RM (1979) Population biology of infectious diseases: part I. Nature 280:361–367

Anderson RM, May RM (1991) Infectious diseases of humans: dynamics and control. Oxford University Press, Oxford

Armstrong GL, Conn LA, Pinner RW (1999) Trends in infectious disease mortality in the United States during the 20th century. J Am Med Assoc 281:61–66

Bailey RG (2014) Ecoregions: the ecosystem geography of the oceans and continents. Springer, New York

Bansal S, Grenfell BT, Meyers LA (2007) When individual behaviour matters: homogeneous and network models in epidemiology. J R Soc Interface 4:879–891

Begon M, Bennett M, Bowers RG, French NP, Hazel SM, Turner J (2002) A clarification of transmission terms in host-microparasite models: numbers, densities and areas. Epidemiol Infect 129:147–153

Blasdell K, Cosson JF, Chaval Y, Herbreteau V, Douangboupha B, Jittapalapong S, Lundqvist A, Hugot JP, Morand S, Buchy P (2011) Rodent-borne hantaviruses in Cambodia, Lao PDR, and Thailand. EcoHealth 8:432–443

Blasdell K, Duong V, Eloit M, Chretien F, Ly S, Hul V, Deubel V, Morand S, Buchy P (2016) Evidence of human infection by new arenaviruses endemic to SEA. elife 5:e13135

Bogich TL, Funk S, Malcolm TR, Chhun N, Epstein JH, Chmura AA, Kilpatrick AM, Brownstein JS, Hutchison OC, Doyle-Capitman C, Deaville R, Morse SS, Cunningham AA, Daszak P (2013) Using network theory to identify the causes of disease outbreaks of unknown origin. J R Soc Interface 10:20120904

Bonds MH, Dobson AP, Keenan DC (2012) Disease ecology, biodiversity, and the latitudinal gradient in income. PLoS Biol 10:e1001456

Bordes F, Morand S (2011) The impact of multiple infections on wild animal hosts: a review. Infect Ecol Epidemiol 1:1–10. https://doi.org/10.3402/iee.v1i0.7346

Bordes F, Blasdell K, Morand S (2015) Transmission ecology of rodent-borne diseases: new frontiers. Integr Zool 10:424–435

Bordes F, Caron A, Blasdell K, de Garine Wichatitsky M, Morand S (2017) Forecasting potential emergence of zoonotic diseases in South-East Asia: network analysis identifies key rodent hosts. J Appl Ecol 54(3):691–700. https://doi.org/10.1111/1365-2664.12804

Brearley G, Rhodes J, Bradley A, Baxter G, Seabrook L, Lunney D, Liu Y, McAlpine C (2013) Wildlife disease prevalence in human-modified landscapes. Biol Rev Camb Philos Soc 88:427–442

Carver S, Kuenzi A, Bagamian KH, Mills JN, Rollin PE, Zanto SN, Douglass R (2011) A temporal dilution effect: hantavirus infection in deer mice and the intermittent presence of voles in Montana. Oecologia 166:713–721

Cashdan E (2014) Biogeography of human infectious diseases: a global historical analysis. PLoS One 9:e106752

Charles JK, Ang BB (2010) Non volant small mammal community responses to fragmentation of keang forests in Brunei Darussalam. Biodivers Conserv 19:543–561

Chen HW, Liu WC, Davis JA, Jordan F, Hwang MJ, Shao KT (2008) Network position of hosts in food webs and their parasite diversity. Oikos 117:1847–1855

Chivian E, Bernstein AS (2004) Embedded in nature: human health and biodiversity. Environ Health Perspect 112:12–13

Civitello DJ, Cohen J, Fatima H, Halstead NT, Liriano J, McMahon TA, Ortega CN, Sauer EL, Sehgal T, Young S, Rohr JR (2015) Biodiversity inhibits parasites: broad evidence for the dilution effect. Proc Natl Acad Sci USA 112:8667–8671

Daily GC, Ehrlich PR (1996) Global change and human susceptibility to disease. Annu Rev Energy Environ 21:125–144

Derne BT, Fearnley EJ, Lau CL, Paynter S, Weinstein P (2011) Biodiversity and leptospirosis risk: a case of pathogen regulation? Med Hypotheses 77:339–344
Dirzo R, Young HS, Galetti M, Ceballos G, Isaac NJB, Collen B (2014) Defaunation in the Anthropocene. Science 345:401–406
Dobson AP (1995) The ecology and epidemiology of rinderpest virus in Serengeti and Ngorongoro crater conservation area. In: Sinclair ARE, Arcese P (eds) Serengetti II. Research management and conservation of an ecosystem. University of Chicago Press, Chicago, pp 485–505
Dunn RR, Harris NC, Colwell RK, Koh LP, Sodhi NS (2009) The sixth mass coextinction: are most endangered species parasites and mutualists? Proc R Soc Lond B 276:3037–3045
Dunn RR, Davies TJ, Harris NC, Galvin MC (2010) Global drivers of human pathogen richness and prevalence. Proc R Soc Lond B 277:2587–2595
Elton CS (1958) The ecology of invasions by animals and plants. Methuen, London
Fenoglio MS, Srivastava D, Valladares G, Cagnolo L, Salvo A (2012) Forest fragmentation reduces parasitism via species loss at multiple trophic levels. Ecology 93:2407–2420
Fincher C, Thornhill R (2008) A parasite-driven wedge: infectious diseases may explain language and other biodiversity. Oikos 9:1289–1297
Gibbs HK, Ruesch AS, Achard F, Clayton MK, Holmgren P, Ramankutty N, Foley JA (2010) Tropical forests were the primary sources of new agricultural land in the 1980s and 1990. Proc Natl Acad Sci USA 107:16732–16737
Gillespie TR, Chapman CA (2008) Forest fragmentation, the decline of an endangered primate and changes in host-parasite interactions relative to an unfragmented forest. Am J Primatol 70:222–230
Gómez JM, Nunn CL, Verdú M (2013) Centrality in primate-parasite networks reveals the potential for the transmission of emerging infectious diseases to humans. Proc Natl Acad Sci USA 110:7738–7741
Goodin DG, Koch DE, Owen RD, Chu Y-K, Hutchinson JMS, Jonsson CB (2006) Land cover associated with hantavirus presence in Paraguay. Glob Ecol Biogeogr 15:519–527
Gottdenker NL, Streicker DG, Faust CL, Carroll CR (2014) Anthropogenic land use change and infectious diseases: a review of the evidence. EcoHealth 11:619–632
Guernier V, Hochberg ME, Guégan JF (2004) Ecology drives the worldwide distribution of human diseases. PLoS Biol 2:740–746
Haddad NM, Brudvig LA, Clobert J, Davies KF, Gonzalez A, Holt RD, Lovejoy TE et al (2015) Habitat fragmentation and its lasting impact on earth's ecosystems. Sci Adv 1:e1500052
Hagen M, Kissling WD, Rasmussen C, De Aguiar MAM, Brown LE et al (2012) Biodiversity, species interactions and ecological networks in a fragmented world. Adv Ecol Res 46:89–210
Hassell JM, Begon M, Ward MJ, Fèvre EM (2017) Urbanization and disease emergence: dynamics at the wildlife–livestock–human interface. Trends Ecol Evol 32(1):55–67. https://doi.org/10.1016/j.tree.2016.09.012
Hudson PJ, Rizzoli A, Grenfell BT et al (2002) The ecology of wildlife diseases. Oxford University Press, Oxford
Hussain S, Ram MS, Kumar A, Shivaji S, Umapathy G (2013) Human presence increases parasitic load in endangered Lion-Tailed Macaques (*Macaca silenus*) in its fragmented rainforest habitats in Southern India. PLoS One 8:e63685
Johnson PTJ, Ostfeld RS, Keesing F (2015) Frontiers in research on biodiversity and disease. Trends Ecol Evol 18:1119–1133
Jones KE, Patel NG, Levy MA et al (2008) Global trends in emerging infectious diseases. Nature 451:990–994
Jones BA, Grace D, Kock R, Alonso S, Rushton J, Said MY et al (2013) Zoonosis emergence linked to agricultural intensification and environmental change. Proc Natl Acad Sci USA 110:8399–8404
Jonsson CB, Figueiredo LTM, Vapalahti O (2010) A global perspective on hantavirus ecology, epidemiology, and disease. Clin Microbiol Rev 23:412–441
Kamiya T, O'Dwyer K, Nakagawa S, Poulin R (2014) What determines species richness of parasitic organisms? A meta-analysis across animal, plant and fungal hosts. Biol Rev 89:123–134

Keesing F, Belden LK, Daszak P, Dobson A, Harvell CD, Holt RD, Hudson P, Jolles A, Jones KE, Mitchell CE, Myers SS, Tiffany B, Ostfeld RS (2010) Impacts of biodiversity on the emergence and transmission of infectious diseases. Nature 468:647–652

Keesing F, Ostfeld RS, Saale R (2015) Is biodiversity good for your health? Science 349:235–236

Lafferty KD, Wood CL (2013) It's a myth that protection against disease is a strong and general service of biodiversity conservation: response to Ostfeld and Keesing. Trends Ecol Evol 28:503–504

Lass S, Hudson PJ, Thakar J, Saric J, Harvill E, Albert R, Sarah E, Perkins SE (2013) Generating super-shedders: co-infection increases bacterial load and egg production of a gastrointestinal helminth. J R Soc Interface 10:20120588

Levi T, Kilpatrick MA, Mangel M, Wilmers CC (2012) Deer, predators, and the emergence of Lyme disease. Proc Natl Acad Sci USA 109:10942–10947

Lindahl J, Grace D (2015) The consequences of human actions on risks for infectious diseases: a review. Infect Ecol Epidemiol 5:11

Lloyd-Smith JO, George D, Pepin KM, Pitzer VE, Pulliam JRC, Dobson AP et al (2009) Epidemic dynamics at the human-animal interface. Science 326:1362–1367

LoGiudice K, Ostfeld RS, Schmidt KA, Keesing F (2003) The ecology of infectious disease: effects of host diversity and community composition on Lyme disease risk. Proc Natl Acad Sci USA 100:567–571

Maffi L (2005) Linguistic, cultural and biological diversity. Annu Rev Anthropol 29:599–617

Mbora DNM, Mc Peek MA (2009) Host density and human activities mediate increased parasite prevalence and richness in primates threatened by habitat loss and fragmentation. J Anim Ecol 78:210–218

McFarlane R, Sleigh A, McMichael T (2012) Synanthropy of wild mammals as a determinant of emerging infectious diseases in the Asian-Australasian region. EcoHealth 9:24–35

Mihaljevic JR, Joseph MB, Orlofske SA, Paull SH (2014) The scaling of host density with richness affects the direction, shape, and detectability of diversity-disease relationships. PLoS One 9: e97812

Morand S (2015a) Diversity and origins of human infectious diseases. In: Muehlenbein MP (ed) Basics in human evolution. Elsevier, New York, pp 405–414

Morand S (2015b) (macro-)Evolutionary ecology of parasite diversity: from determinants of parasite species richness to host diversification. Int J Parasitol Parasites Wildl 4:80–87

Morand S, Lajaunie C (2017) Biodiversity conservation in Southeast Asia: challenges in a changing environment. Routledge EarthScan, Oxon

Morand S, Owers K, Waret-Szkuta A, McIntyre KM, Baylis M (2013) Climate variability and outbreaks of infectious diseases in Europe. Nat Sci Rep 3:1774

Morand S, Jittapalapong S, Supputamongkol Y, Abdullah MT, Huan TB (2014a) Infectious diseases and their outbreaks in Asia-Pacific: biodiversity and its regulation loss matter. PLoS One 9:e90032

Morand S, McIntyre KM, Baylis M (2014b) Domesticated animals and human infectious diseases of zoonotic origins: domestication time matters. Infect Genet Evol 24:76–87

Morand S, Owers K, Bordes F (2014c) Biodiversity and emerging zoonoses. In: Akio Yamada A, Kahn LH, Kaplan B, Monath TP, Woodall J, Conti L (eds) Confronting emerging zoonoses: the one health paradigm. Springer, Tokyo, pp 27–41

Morand S, Bordes F, Blasdell K, Pilosof S, Cornu J-F, Chaisiri K, Chaval Y, Cosson J-F, Claude J, Feyfant T, Herbreteau V, Dupuy S, Tran A (2015) Assessing the distribution of disease-bearing rodents in human-modified tropical landscapes. J Appl Ecol 52:784–794

Morris RJ (2010) Anthropogenic impacts on tropical forest biodiversity: a network structure and ecosystem functioning perspective. Philos Trans R Soc Lond B 365:3709–3718

Murray KA, Daszak P (2013) Human ecology in pathogenic landscapes: two hypotheses on how land use change drives viral emergence. Curr Opin Virol 3:79–83

Murray KA, Preston N, Allen T, Zambrana-Torrelio C, Hosseini PR, Daszak P (2015) Global biogeography of human infectious diseases. Proc Natl Acad Sci USA 112:12746–12751

Olson DM, Dinerstein E (1998) The Global 200: a representation approach to conserving the Earth's most biologically valuable ecoregions. Conserv Biol 12:502–515

Orrock JL, Allan BF, Drost CA (2011) Biogeographic and ecological regulation of disease: prevalence of Sin Nombre virus in island mice is related to island area, precipitation, and predator richness. Am Nat 177:691–697

Palma RE, Polop JJ, Owen RD, Mills JM (2012) Ecology of rodent-associated Hantaviruses in the Southern cone of South America: Argentina, Chile, Paraguay and Uruguay. J Wildl Dis 48:267–281

Pilosof S, Morand S, Krasnov BR, Nunn CL (2015) Potential parasite transmission in multi-host networks based on parasite sharing. PLoS One 10:e0117909

Poisot T, Nunn C, Morand S (2014) Ongoing worldwide homogenization of human pathogens. BioRxiv. https://doi.org/10.1101/009977

Poulin R (2010) Network analysis shining light on parasite ecology and diversity. Trends Parasitol 26:492–498

Randolph SE, Dobson ADM (2012) Pangloss revisited: a critique of the dilution effect and the biodiversity-buffers-disease paradigm. Parasitology 139:847–863

Rohr JR, Dobson AP, Johnson PT, Kilpatrick AM, Paull SH, Raffel TR, Ruiz-Moreno D, Thomas MB (2011) Frontiers in climate change-disease. Trends Ecol Evol 26:270–277

Salkeld DJ, Padgett K, Jones JH (2013) A meta-analysis suggesting that the relationship between biodiversity and risk of zoonotic pathogen transmission is idiosyncratic. Ecol Lett 16:679–686

Schipper J, Chanson JS, Chiozza F, Cox NA, Hoffmann M, Katariya V et al (2008) The status of the world's land and marine mammals: diversity, threat, and knowledge. Science 322:225–230

Schmidt KA, Ostfeld RS (2001) Biodiversity and the dilution effect in disease ecology. Ecology 82:609–619

Shirley SM, Kark S (2009) The role of species traits and taxonomic patterns in alien bird impacts. Glob Ecol Biogeogr 18:450–459

Smith KF, Sax DF, Gaines SD, Guernier V, Guégan JF (2007) Globalization of human infectious disease. Ecology 88:1903–1910

Smith KF, Goldberg M, Rosenthal S, Carlson L, Chen J, Chen C, Ramachandran S (2014) Global rise in human infectious disease outbreaks. J R Soc Interface 11:20140950

Springer YP, Hoekman D, Johnson PTJ et al (2016) Continental scale surveillance of infectious agents: tick-, mosquito-, and rodent-borne parasite sampling designs for NEON. Ecosphere e01271:7

Suzán G, Marcé E, Giermakowski JT, Mills JN, Ceballos G, Ostfeld RS, Armien B, Pascale JM, Yates TL (2009) Experimental evidence for reduced rodent diversity causing increased hantavirus prevalence. PLoS One 4:e5461

Swaddle JP, Calos SE (2008) Increased avian diversity is associated with lower incidence of human West Nile infection: observation of the dilution effect. PLoS One 3:e2488

Tatem AJ, Rogers DJ (2006) Global transport networks and infectious disease spread. Adv Parasitol 62:293–343

Telfer S, Bown KJ, Sekules R, Begon M, Hayden T, Birtles R (2005) Disruption of a host-parasite system following the introduction of an exotic host species. Parasitology 130:661–668

Telfer S, Lambin X, Birtles R, Beldomenico P, Burthe S, Paterson S, Begon M (2010) Species interactions in a parasite community drive infection risk in a wildlife population. Science 330:243–246

Vazquez DP, Poulin R, Krasnov BR, Shenbrot GI (2005) Species abundance and the distribution of specialization in host–parasite interaction networks. J Anim Ecol 74:946–955

Wells K, Smales LR, Kallo EKV, Pfeiffer M (2007) Impacts of rain-forest logging on helminth assemblages in small mammals (Muridae, Tupaiidae) from Borneo. J Trop Ecol 23:35–43

Werden L, Barker IK, Bowman J, Gonzales EK, Leighton PA, Lindsay LR, Jardine CM (2014) Geography, deer, and host biodiversity shape the pattern of lyme disease emergence in the Thousand Islands archipelago of Ontario, Canada. PLoS One 9:e85640

White LA, Forester JD, Craft ME (2015) Using contact networks to explore mechanisms of parasite transmission in wildlife. Biol Rev 92:389–409

Wilcox BA, Colwell RR (2005) Emerging and reemerging infectious diseases: biocomplexity as an interdisciplinary paradigm. EcoHealth 2:244–257

Wilcox BA, Gubler DJ (2005) Disease ecology and the global emergence of zoonotic pathogens. Environ Health Prev Med 10:263–272

Part II
The Ecology of Infectious Diseases Affecting Humans

Chapter 3
Understanding and Estimating the Risk of Waterborne Infectious Disease Associated with Drinking Water

Christon J. Hurst

Abstract This chapter presents an understanding of how drinking water treatment reduces the risk of communitywide infectious disease. The presentation begins with data that demonstrate how a reduction of the typhoid risk in American cities paralleled the progressive implementation of filtration technology at the start of the twentieth century. Typhoid, which is caused by the bacteria *Salmonella enterica* serovar Typhi, is one of numerous infectious diseases associated with the presence of microbial contaminants in drinking water. Drinking water may serve as a primary infection route for these pathogens and the pathogens often can be transmitted within human populations by other, secondary routes. A compartment disease transmission model is presented to illustrate how the use of treatment technologies both to prevent primary acquisition of infections from drinking water and to prevent secondary spread of those infections within the community can effectively reduce the extent of a waterborne infectious disease outbreak. Additionally, a risk estimation technique is presented which has been demonstrated to accurately calculate the probability of community gastrointestinal illness in human populations based upon the numbers of pathogenic bacteria, protozoa and virus in drinking water.

3.1 Introduction

Permalum bibit cave (Beware of bad water!). A reduction of the typhoid risk in United States cities at the start of the twentieth century paralleled the progressive implementation of community drinking water filtration technology. From 1900 through 1913, as the percentage of city population served by water filtration utilities increased from 8.7% to 48%, the annual typhoid death rates in those same cities decreased from 36 to 16 per 100,000 population (Johnson 1916). Typhoid death rates

C. J. Hurst (✉)
Universidad del Valle, Cali, Colombia

Cincinnati, OH, USA
e-mail: christonjhurst@fuse.net

C. J. Hurst (ed.), *The Connections Between Ecology and Infectious Disease*, Advances in Environmental Microbiology 5,
https://doi.org/10.1007/978-3-319-92373-4_3

in the United States continued that downward trend, falling to 5.3 per 100,000 population in 1921 (Editorial 1923). By 1960, the typhoid death rate in the United States had been reduced to less than 1 per 100,000 (Centers for Disease Control and Prevention 2018a). Eventually, by 2006 the incidence of typhoid in the United States had fallen to only a few hundred cases per year, and those often were associated with travelers to South Asian (Lynch et al. 2009). Drinking water is only one of many sources from which typhoid, caused by the bacteria *Salmonella enterica* serovar Typhi, and numerous other infectious diseases typically associated with the presence of microbial contaminants in drinking water can be aquired. Of the four diseases which may cause the greatest risk of suffering and death in underdeveloped parts of the world: cholera, hepatitis, malaria and typhoid, malaria is the only one that is not transmitted by ingestion of water. Infection with the water-transmissible *Orthohepevirus orthohepevirus A*, formerly called Hepatitis E virus, is particularly onerous because it causes a high incidence of death in pregnant women (Khuroo 1991; Ramalingaswami and Purcell 1988).

Not all community-distribution water supplies are treated. Often, this represents either a matter of cost or else an avoidance of disinfected water because of taste preference. At other times, the lack of community drinking water treatment may result from a mistaken assumption that water from streams, wells, or springs is by nature intrinsically pure and so water from these sources can be consumed without treatment. Even when treated drinking water is available through community distribution systems, people still may prefer to drink water obtained from untreated sources such as springs or wells. Community drinking water treatment and plumbing distribution networks are not foolproof systems and the supplied drinking water often does contain pathogenic microorganisms for which the associated health risks will be examined in this chapter. Unfortunately, the popular alternative of consuming bottled water instead of community-supplied water based upon the idea of safety may not always be a good choice from a health standpoint since commercially bottled water can contain pathogens. The current and very dangerous natural health food fad of intentionally consuming "Raw water" is a mistake which results from the philosophy that any presented untreated natural water somehow is more pure and healthful than is treated water, and that treatment removes the natural healthfulness of the water just as treatment may remove vitamins from processed food. The presumption that all things natural are more healthy than are anythings processed often is not valid. Water from all sources can be contaminated with pathogenic microorganisms and the consumption of contaminated water will produce waterborne disease.

Most of the microorganisms present in aquatic environments seem to have no effect on human health. However, some clearly do represent a public health risk and for this reason are considered to be disease causing and are referred to as pathogens. The true definition of the term "pathogen" is "something that causes pathological changes" in the appearance of cells and tissues. Often the "pathogen" is the entire organism viewed as an infectious agent. In other cases, however, the pathological change is due to a toxic substance produced by a noninfectious organism, and some toxins, such as those produced by algae and cyanobacteria, can bioaccumulate within the food chain to extremely high concentrations. This chapter addresses organisms which are infectious for humans, and while some of these represent aquatic contaminants that come from either human or animal reservoirs, others represent organisms which are naturally present in the environment.

Drinking water treatment can be performed with varying degrees of effectiveness on a household basis. Household treatments include heating water to a temperature that is below the boiling point but high enough to destroy the microbial contaminants, heating water still higher to achieve the boiling point, and perhaps even using distillation. Unintentional water treatment is a side benefit from heating and boiling water for the purpose of preparing foods and beverages. Household cartridge filtration systems can effectively remove microbial pathogens from water. Laboratory evaluation has indicated that the practice of using clay as a coagulant for treating water on a household basis may have some success (Lund and Nissen 1986). Chemical coagulants such as alum also can have some degree of effectiveness in the treatment of drinking water (Ahmad et al. 1984). Powdered plant materials also have been evaluated as coagulants that could be used on a household basis both alone and in conjunction with other chemicals. Chemical disinfectants which can be used effectively to treat water on a small scale include chlorine and iodine. Chemical coagulants can be used in conjunction with chlorine compounds, and the later type of combined mixture is commercially available as tablets.

Two additional concepts need to be mentioned before delving further into this chapter. The first concept is the idea of primary versus secondary transmission. Primary transmission represents the initial introduction of an infectious agent into a group of susceptible individuals. Secondary transmission represents the subsequent transfer of the infectious agent from the initially infected individuals to other members of the group. Primary and secondary transmission of a disease can occur by different routes. An example used in this chapter is the disease typhoid, which often has water as the vehicle of primary transmission into a population, and for which food contaminated by handlers may then serve as a vehicle of secondary transmission within the population. The second concept is prevention of disease, which is the hallmark of public health. Prevention of infectious disease can be divided into three separate phases, termed primary, secondary, and tertiary. Primary prevention represents preclusion of disease either by reducing the exposure of new susceptible "host" individuals to the disease agent or by altering susceptibility of the host. Secondary prevention consists of early detection of the disease which hopefully leads to successful early treatment. Tertiary treatment is aimed at minimizing and hopefully eliminating long-term impairments and disabilities, and sometimes extending an individuals survival. This chapter will consider primary and secondary prevention.

3.2 Disease Transmission Routes

For the purpose of this chapter, the term 'disease transmission' refers to infectious disease. There are many mechanisms, both environmental and non-environmental, by which pathogenic microorganism are transmitted and these are termed disease transmission routes. This chapter focuses on water as a source of microbial pathogens that cause infectious disease, but many infectious diseases are aquired by direct physical contact with either a human or animal harboring an infectious agent, and other transmission routes exist as well.

Disease transmission routes can be divided into two broad categories, vertical and horizontal. Vertical disease transmission, which is passage of an infection from a mother to her embryos and offspring either prior to or during birth, is considered to represent a form of transmission by direct contact. Horizontal transmission of disease can also occur via direct contact mechanisms and encompasses those organisms transferred in saliva, transplanted body tissues, blood and blood products. The touching of contaminated skin and wounds can also result in transmission by direct contact. As such, the category of transmission by direct contact includes infections acquired from sexual partners among which are those caused by pathogens carried in semen. Included in the latter category are bacterial diseases such as syphilis, viral diseases such as hepatitis B and herpes simplex, and the chlamydial disease caused by *Chlamydia trachomatis*. For the purpose of this chapter, diseases transmitted by bites from humans, vertebrate animals, and arthropods, are also considered to represent transmission by direct physical contact.

Examples of illnesses transmitted from vertebrate animals to humans by direct physical contact with contamination on the animals body tissues including skin are the bacterial disease salmonellosis and the viral disease caused by simian immunodeficiency virus. Rabies is an example of a disease whose transmission results from animals biting humans. Diseases that can be transmitted from animals to humans are called zoonoses and animals are the natural host for those causative microorganisms. Humans usually serve only as incidental hosts in zoonoses and the humans are non-essential for the normal cycle of transmission.

The large number of diseases transmitted via arthropod vectors often involve chains of transmission that include physical contact with either humans or other vertebrates as the reservoir, invertebrates serving as the vector, and humans being the susceptible host. A variety of insects are known to serve as arthropod vectors and some of the causative microorganisms are capable of replicating in their vectors. Chiggers, which are perhaps the best known of the mite family Trombiculidae, are capable of transmitting the rickettsial disease scrub typhus. Fleas are perhaps most notorious because they transmit the bacterial disease plague. Biting flies can transmit bacterial diseases such as tularemia. Mosquitos are responsible for the transmission of many viral diseases such as dengue and encephalitis. Ticks are associated with the transmission of bacterial illnesses such as Lyme disease and tularemia, as well as the viral disease caused by tick-borne encephalitis virus. Flies which do not bite also may be associated with the transmission of gastrointestinal diseases as will be discussed later in this chapter. In these latter instances, the flies serve only as passive carriers or vehicles when they transfer microorganisms from feces to food.

As alluded to above, many viral diseases which some people consider to be directly transmitted are in fact normally transmitted through the environment. These include rotaviral gastroenteritis, smallpox, and those so-called "childhood diseases" known as chickenpox, measles, and mumps.

3.2.1 Disease Transmission by Environmental Water Routes

Water serves as the vehicle for a large proportion of environmentally transmitted disease. Bacterial diseases spread by this route include the notorious typhoid (Johnson 1916; Kehr and Butterfield 1943) and cholera. In particular, outbreaks of cholera often take the form of massive, wide-ranging epidemics, termed pandemics, which have caused fear in human populations since at least 1817. A large cholera pandemic which occurred in the Americas resulted when the causative bacteria perhaps arrived there from another part of the world in contaminated ship ballast water (McCarthy and Khambaty 1994). That pandemic resulted in approximately 350,000 cases of illness and nearly 4,000 deaths (Swerdlow and Ries 1992, 1993). An even larger cholera outbreak occurred in Haiti with the pathogen having been introduced there by international aid workers from Nepal (Transnational Development Clinic 2013). These two bacterial illnesses, cholera and typhoid, were the driving force behind the development of formal drinking water treatment utilities during the middle and end of the nineteenth century. Yet, even as we now are well into the twenty-first century, we have not escaped the historical fact that disease microorganisms consumed in conventionally treated drinking water may account for 35% of the gastrointestinal illnesses among consumers even in developed countries (Payment et al. 1991).

Public health emphasis is often placed on diseases associated with the ingestion of water, since this route of transmission has resulted in massive outbreaks. However, it is important to remember that ingesting water is not the only way in which we acquire diseases that are transmitted via contaminated water. We also acquire diseases from occupational and recreational activities performed in water, as well as from the consumption of water-contaminated shellfish and food crops, and from inhalation of water aerosols. Figure 3.1 represents water-related environmental routes by which infectious agents are transmitted to susceptible individuals. The animal reservoirs would include cows as a source of cryptosporidiosis, and rodents as a source of both giardiasis and campylosis. Table 3.1 Lists some examples of infectious diseases associated with waterborne pathogens. Drinking water often may serve as a primary infection route for these pathogens and the pathogens often can be transmitted within human populations by other, secondary routes. This chapter addresses waterborne infections affecting humans. But, it is important to remember that waterborne infections also affect other animal species including those upon which we rely as livestock, and the pathogens causing those infections would flow through the same pathways as shown in Fig. 3.1.

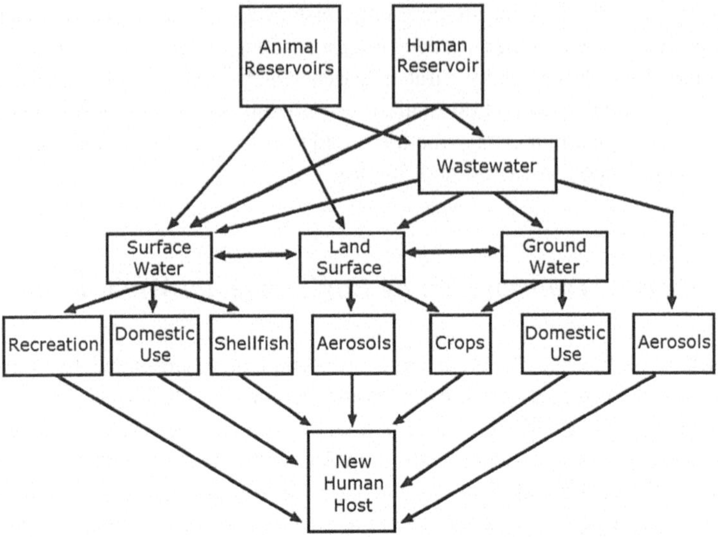

Fig. 3.1 Water-related environmental routes by which infectious agents are transmitted to susceptible individuals. The animal reservoirs symbolically would include cows as a source of cryptosporidiosis, and rodents as a source of both giardiasis and campylosis

3.2.1.1 Animal Reservoirs of Waterborne Microbial Contaminants

Continuing our examination of Fig. 3.1, we notice that animals can be an important source of microbial contaminants in surface water via two different pathways. Generally we consider vertebrate animals in this regard. The first of these pathways is fecal material and urine that animals deposit directly into the water. The second pathway represents organisms in fecal material and urine which are deposited on the land and then subsequently washed into surface waters by overland flow. Surface flow also carries its load of contaminants into groundwater via percolation through the soil and subsurface matrixes, and also water flows from the land surface into subterranean environments through geologic structures such as caves and sinkholes.

Those waterborne bacterial diseases of humans which may originate from animal reservoirs include many that are caused by microbes whose reservoirs are mammals, including campylosis contributed by not only beavers but also migratory birds, muskrats, and other rodents, and leptospirosis contributed by livestock. In humans, examples of water-associated protozoa are those microorganisms responsible for causing giardiasis and cryptosporidiosis. These protozoans are capable of cross infecting a variety of animals (Plutzer et al. 2018) including beavers and livestock, with a result that those animals may serve as reservoirs and transmit the diseases to humans via surface waters. Reoviruses and rotaviruses, which cause waterborne gastroenteritis, can cross infect humans and a variety of animals including cattle. Thus, not only wildlife but also infected livestock living in watershed areas could be reservoirs of waterborne viral gastroenteritis. Additional microbial contaminants can

Table 3.1 Examples of infectious diseases associated with waterborne pathogens

Exposure route	Pathogen group	Type of disease	Causative microorganism(s)[a]
Ingestion (includes contaminated drinking water and other beverages plus ice and water-associated contamination of foods)			
	Bacterial		
		Enteric fever	*Salmonella* (especially *Salmonella enterica* subsp. enterica serovar Typhi, which causes typhoid fever)
		Enteritis	*Campylobacter*, *Shigella* (causes bacterial dysentery), *Vibrio* (especially *V. cholerae*, which causes cholera)
		Febrile syndrome	*Francisella tularensis*
		Septicemia	*Vibrio vulnificus*
	Protozoan		
		Enteritis	*Cryptosporidium parvum*, *Entamoeba histolytica* (causes amebic dysentery), *Giardia intestinalis*(includes the formerly named *Giardia lamblia*)
	Viral		
		Encephalitits	*Enterovirus*
		Gastroenteritis	*Alphacoronavirus*, *Mamastrovirus*, *Norovirus*, *Rotavirus*, *Vesivirus*
		Hepatitis	*Vesivirus*, *Hepatovirus*
		Meningitis	*Enterovirus*
Body Surface Contact (usually associated with either recreational or aquatic occupational activities)			
	Bacterial		
		Enteritis	*Vibrio cholerae*
		Nephritis	*Leptospira interrogans*
		Wound infections	*Vibrio parahemolyticus*, *Vibrio vulnificus*
	Metazoan		
		Worm infestation	*Schistosoma*
	Protozoan		
		Encephalitis	*Naegleria*
		Enteritis	*Entamoeba histolytica*

(continued)

Table 3.1 (continued)

Exposure route	Pathogen group	Type of disease	Causative microorganism(s)[a]
	Viral		
		Encephalitis	*Enterovirus*
		Gastroenteritis	*Alphacoronavirus*, *Mamastrovirus*, *Norovirus*, *Rotavirus*, *Vesivirus*
		Meningitis	*Enterovirus*
		Pharyngoconjunctival fever	*Mastadenovirus*
Inhalation			
	Bacterial		
		Pneumonic fever	*Legionella pneumophila*

[a]If an organism is indicated by both its genus and species names, then the disease association is with that particular species. If only a genus name is given, then the disease association is with more than one species belonging to that genus

come from other warm blooded animals such as birds, as well as from cold blooded animals like reptiles and fish.

3.2.1.2 Human Reservoirs of Waterborne Microbial Contaminants

We humans carry microorganisms both on and within our bodies, and we leave a trail of them behind us. Those microorganisms that cause intestinal illnesses are of greatest concern with regard to infections caused by ingestion of water. Humans can become infected by these same organisms from sources other than water, such as food. Because these organisms are naturally present in feces they tend to be transmitted by the fecal oral route, meaning that infection is acquired either through oral contact with or ingestion of fecally contaminated materials. The subsequent excretion of these organisms in human fecal wastes represents a reservoir in the cycle of waterborne disease transmission, as also represented in Fig. 3.1. The existance of these organisms of human origin as contaminants in wastewater is a factor that holds true for all categories of microorganisms, including bacteria (Amos et al. 2014) and virus (Melnick 1947; Melnick et al. 1954; Paul et al. 1940). Discharge of wastewater into the environment then results in microbial contamination of environmental waters that may subsequently be used as sources of potable water. Drinking water from these sources thereby completes the cycle of infection, from humans to water, and back to humans. While it is the microorganisms that are shed in human fecal material which make wastewater a major source of microbial contaminants, additional animals such as rats which dwell in sewerage collection systems will directly contribute contaminants to the wastewater. Humans also contribute microbial contaminants directly to water during the course of recreational activities.

Inhalation of water aerosols represents another route of exposure to microbial contaminants contained in wastewater. Liquid (droplet) aerosols notably are generated during wastewater aeration and also during the spray application of wastewater including sludge suspensions onto land. It is possible that aerosols generated during wastewater treatment might serve as a source of disease in wastewater workers.

3.2.1.3 The Fate of Microbial Contaminants in Water

The ability of microbial contaminants to survive after they enter water is a major issue with regard to waterborne infectious disease. Microbial contaminants that are released into water will die-off if they are unable to replicate in that environment. The results of mathematical analyses have revealed that water temperature is a major factor affecting the survival of microorganisms in water, with survival increasing at lower temperatures (Hurst 1991; Hurst et al. 1992; Kehr and Butterfield 1943). Other important factors, at least from the standpoint of virus, are: the amount of nutrients available in the water which could support growth of bacterial organisms, generally exhibiting a detrimental effect upon viral survival; the level of turbidity in the water, with higher turbidity generally being beneficial to viral survival; and water hardness, likewise generally having a beneficial effect upon viral survival (Hurst 1991).

3.2.1.4 Aquatic Microbes as a Reservoir of Waterborne Microbial Contaminants

There are microbes which naturally reside in water and are pathogenic for mammals. Perhaps most notably among these are *Legionella pneumophila*, which exists as an infection within free living amoeba and typically is acquired by inhalation, and *Vibrio cholerae* which resides on the chitinous shells of microcrustaceans and macrocrustaceans and typically is acquired by ingestion. The natural aquatic existance of *Vibrio cholerae* may be one of the reasons why the disease cholera can exist as an endemic focus even in areas which do not experience outbreaks of the disease (Eichold et al. 1993; Weissman et al. 1974).

But, in addition to *Legionella* and *Vibrio*, there are a large number of other bacterial genera which contain member species that naturally are found in water. Some of these species are either pathogenic or opportunistically pathogenic for mammals including humans, and the list of these genera includes: *Achromobacter*, *Acinetobacter*, *Actinobaculum*, *Aeromonas*, *Afipia*, *Alcaligenes*, *Burkholderia*, *Chromobacterium*, *Chryseobacterium*, *Citrobacter*, *Clostridioides*, *Clostridium*, *Corynebacterium*, *Edwardsiella*, *Enterobacter*, *Enterococcus*, *Eubacterium*, *Ewingella*, *Hafnia*, *Helicobacter*, *Histophilus*, *Klebsiella*, *Kytococcus*, *Leclercia*, *Lelliottia*, *Leptospira*, *Mannheimia*, *Morganella*, *Mycobacterium*, *Mycoplasma*, *Neorickettsia*, *Plesiomonas*, *Proteus*, *Providencia*, *Pseudomonas*, *Rahnella*, *Rhodococcus*, *Serratia*, *Shewanella*, *Shigella*, *Sphingobacterium*, *Sphingomonas*, *Stenotrophomonas*, *Streptococcus*, *Tsukamurella*, and *Yersinia*. Information on the ecology of these bacteria can be found in the chapter "Opportunistic Bacteria Associated with Mammalian Livestock Disease" (Hurst 2018).

Aside from the bacteria, there also are a large number other microbial genera which contain species that naturally are found in water and similarly are either pathogenic or opportunistically pathogenic for mammals including humans. These include the fungi genera *Hortaea*, and *Verruconis*, the oomycete genus *Pythium*, and the Ichthyosporea genus *Rhinosporidium*. Information on the ecology of these genera can be found in Hurst (2016b). Additionally, there also are many groups of aquatic protozoa that infect mammals including humans.

3.2.1.5 The Interconnected Environmental Flow of Water and Its Microbial Contaminants

Microbes may not stay where we place them! Microbial contaminants are not necessarily stationary within the environment and the interconnected flow of water along with its microbial contaminants, as diagrammed in Fig. 3.1, causes microbial contaminants to end up in places that we may not have intended. Wastewater and the pathogens which it contains often are disposed of by direct discharge into surface water, and once there, these contaminants in surface water are joined by microbial contaminants that initially may have been placed onto the land surface. Land disposed microbial contaminants include those contained in wastewater used for irrigation. And, we also must remember that microbial contaminants present in fecal solids can end up on the land surface either by having been directly excreted onto the land by animals including humans or by fecal solids having been applied onto the land surface intentionally as fertilizer. The collected solids from wastewater often just simply are discarded onto land with a hope that the solids then will be forgotten and either naturally or magically disappear, at which point a new doctoral student may get assigned the smelly task of studying the viruses contained in those solids (Hurst et al. 1978)! But, nothing applied to the land surface is assured of remaining there. If by no other means, then rainwater, melting snow and ice, can wash the land surface applied material into surface waters. Importantly, contaminated surface applied water can move its contaminants into the subsurface where they join with existing groundwater. In addition to the natural process of surface water percolating into and recharging groundwater aquifers, wastewater also can carry its contaminants into groundwater more directly via either engineered infiltration or direct injection with both of those processes often done for the purpose of recharging aquifers. Groundwater with its contaminants flows toward and often discharges into surface water. Groundwater also can return its contaminants to the land surface through seeps and springs. Surface water can return its contaminants to the land surface very dramatically and, at the particular moment while I am writing this chapter, the Ohio River and its tributary the Little Miami River here in southwestern Ohio, are flooding and doing exactly that! Humans may subsequently encounter these repositioned pathogens either by contact with, or consumption of, contaminated surface waters and ground waters. When surface waters return to the land surface during floods there additionally is attendant concern about mosquito transmitted illnesses (Anders et al. 1994; Atchison et al. 1993; Cotton 1993), because flooding increases water surface area, which in turn supports the life cycle of mosquitos.

Another major issue with surface water is that the natural transport of microbial contaminants during the course of surface water flow may convey microorganisms over long distances, taking the microbes far from visible sources of contamination (Dahling and Safferman 1979). Interestingly, long distance aquatic transport of microbes also may be aided by human activities, with one possible example having been the transport of causative microbial agents of waterborne disease between different parts of the world as contaminants of ship ballast water (McCarthy and Khambaty 1994).

3.2.1.5.1 Diseases Acquired from Microbially Contaminated Surface Water

The domestic use of contaminated water clearly results in disease transmission (Craun 2012; Hurst 2003) and it is possible to find microbial contaminants associated with water taps even when the water lines feeding the taps apparently are free of the suspect organism. This latter type of contamination has been described by Grundmann et al. (1993). The most commonly suspected mode of transmission associated with domestic water use is through ingestion of the contaminated water. Bacterial diseases acquired in this way include campylosis, cholera (NSF International 2011), typhoid (Johnson 1916), and tularemia (Stewart 1991). Viral diseases acquired from drinking water include diarrhea, hepatitis and poliomyelitis (Mosley 1966). Recently, the most notable protozoan diseases acquired by the ingestion of water seem to be cryptosporidiosis and giardiasis (Baldursson and Karanis 2011; Efstratiou et al. 2017; Mahmoudi et al. 2017; Rosado-García et al. 2017). Another interesting route by which diseases can be acquired from tap water is through the use of contaminated water when washing wounds (Lowry et al. 1991). Prevention of these diseases is the reason that we practice drinking water treatment at community levels.

Participation in recreational surface water activities is frequently associated with the possibility for acquisition of infectious disease caused by bacteria (Leonard et al. 2015, 2018a, b, c). Protozoan contaminants contributed in this way include the causative agents of cryptosporidiosis and giardiasis. Such disease can be contracted even from water which has no known sewage input (Sorvillo et al. 1992), suggesting that humans themselves contaminate water through the course of recreational activities. Viral contaminants contributed to water during recreational activities include causative agents of gastroenteritis and pharyngoconjunctival fever. These health problems can result from microbial pollution of both swimming pools and beaches.

Ingestion of shellfish harvested from waters that have been contaminated by discharged wastewater represents another means through which people acquire microbial illnesses from polluted water. Crustaceans can accumulate pathogenic microorganisms from the environment and then potentially pass these organisms onward to humans who consume the crustaceans. Molluscs are divided into two categories. The first is bivalves, such as oysters and hardshell clams. The second group is gastropods, such as conch. Human pathogenic viruses, such as the species *hepatovirus A* (formerly called hepatitis A virus) of the genus *Hepatovirus*, can be detected in bivalve molluscs, and the ingestion of contaminated bivalves has caused outbreaks not only of viral disease but also of bacterial diseases such as vibriosis (Centers for Disease Control and Prevention 2018c). It is presumed that bivalve molluscs accumulate these microbial contaminants during the natural process called filter feeding. Gastropod molluscs do not filter feed, and thus would not be expected to accumulate pathogens by this same mechanism.

3.2.1.5.2 Diseases Acquired from Microbially Contaminated Land Surfaces

Wastewater is sometimes used for the irrigation of vegetable crops. As mentioned previously, this practice presents a health hazard, because it may contaminate those crops with potentially pathogenic microorganisms that have been shed in feces. Human consumption of crops that were irrigated with wastewater has been implicated in the spread of cholera and it may prove possible to ameliorate this problem by disinfecting sewage effluents that are intended for agricultural usage. Particulate aerosols may be generated from wastewater sludges that have been applied onto land surfaces either for the purpose of composting that sludge or for final sludge disposal. These aerosols present a pathogenic hazard because they can contain both bacterial and fungal contaminants and may be transported by the wind.

3.2.1.5.3 Diseases Acquired from Microbially Contaminated Ground Water

It is often thought that ground water is intrinsically pristine and pure, but groundwater can be contaminated with microorganisms that are pathogenic for humans. Domestic use of contaminated ground water has resulted in at least one outbreak of the bacterial illness cholera (Blake et al. 1977). Ingestion of contaminated ground water also has resulted in outbreaks of the protozoan illnesses cryptosporidiosis and giardiasis and the viral illnesses gastroenteritis and hepatitis. A heavy reliance is often placed on the use of groundwater for crop irrigation and this reliance might result in the microbial contamination of crops.

The above diseases, along with a number of others transmitted through environmental routes associated with contaminated water, are summarized in Table 3.1 along with the names of the causative microorganisms.

3.2.2 Microbial Contaminants Get Transferred Around the House, Health Care Settings, and Even in Space Travel Environments

Microbial contaminants can be transferred from one type of contaminated material to another creating a cycle of contamination. An interesting example of this is provided by an outbreak of giardiasis for which it appears that raw food initially became contaminated by washing it in tap water, after which the contaminated food was chopped on a cutting board. Other foods appear to then have become contaminated when they were later prepared on that same cutting board, which had never adequately been washed (Grabowski et al. 1989). Using microbially safe water to wash the initial raw food would have prevented the contamination process. Adequate washing of the cutting board to physically remove contaminating microorganisms using microbially safe water, or even disinfection of the cutting board, would have broken the cycle of contamination. Households are in fact broadly contaminated with fecal microbes (NSF International 2011). An edited volume in this series (Hurst 2017) recently addressed modeling the transmission and prevention of infectious disease in a variety of environmental settings, including those which are terrestrial and others which are orbital, along with an understanding of the microbial reservoirs.

3.3 The Concept of Modeling Transmission of Disease Through Host Populations: Epidemic Versus Endemic

Figure 3.2 presents a very basic example of the models developed for analyzing transmission of infectious disease through populations of animals and humans. Such models are commonly referred to as "Disease transmission models", and represent a type of compartmental model. In Fig. 3.2, the individual compartments are shown as rectangular boxes and they representing segments of the host population being studied. The different compartments are connected by solid arrows that represent the rates at which individual members of the population move from one compartment to another. These rates of movement are sometimes described as the "force" of flow through the model. The model presented in Fig. 3.2 has four compartments: susceptible, meaning those individuals who are susceptible to infection by the pathogenic microorganism whose affect is being studied; infectious, meaning those individuals who have become infected and are in a state where they can transmit the infectious agent to other individuals; immune, meaning those individuals who have successfully completed convalescence from the infection and who are, for at least a time, resistant to reinfection; and removed, representing individuals who at least temporarily are excluded from the population under study. In this example, it is presumed that the removed individuals are those who have died from the infection. Figure 3.2 also shows the point in the model at which transferral of an infectious agent actually occurs, represented as an arrow with a dashed line.

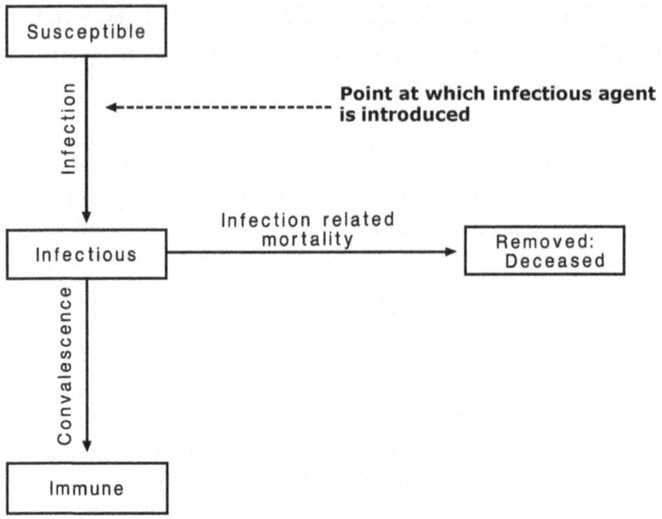

Fig. 3.2 Basic compartment model appropriate for describing the epidemic propagation of disease through a population and the response of individuals to that disease. Models of this type are commonly termed "Disease Transmission Models". This model also shows the point at which an infectious agent (causative microorganism) is transmitted to susceptible individuals

The model presented in Fig. 3.2 was developed for epidemic disease transmission, and accordingly contains relatively few compartments. Figure 3.3 presents a model developed for endemic disease transmission and takes us a large step forward to a model that is more complex and more complete. In Fig. 3.3, live individuals can move into and out from three removed categories; susceptible, infectious, and immune. The epidemic disease model presented in Fig. 3.2 allows for disease-related mortality. The endemic disease model shown in Fig. 3.3 likewise includes disease-related mortality, and additionally allows for both natural mortality and vaccine related mortality. The model shown in Fig. 3.3 also enables the addition of susceptible individuals to the population through new births, here termed "host progeny", and returns immune individuals to the susceptible compartment through the eventual waning of immunity. The models used in this chapter pertain to the entire human population of a community.

A more detailed description of using compartment models to describe disease transmission including the transmission of disease in livestock animals has been presented by Hurst and Murphy (1996). Disease transmission models found in the literature can vary with respect to both the compartments they contain and the indicated movement of individuals. Such variation reflects the intended application of a given model, i.e., whether it is for endemic versus epidemic disease, and whether it represents a disease exposure which is pertinent to only a subset of the population, such as infants, or to the entire population (Hurst and Murphy 1996). There are considered to be four major stages of a disease. The first of those is the incubation period which occurs prior to the demonstration of symptoms. The second is a

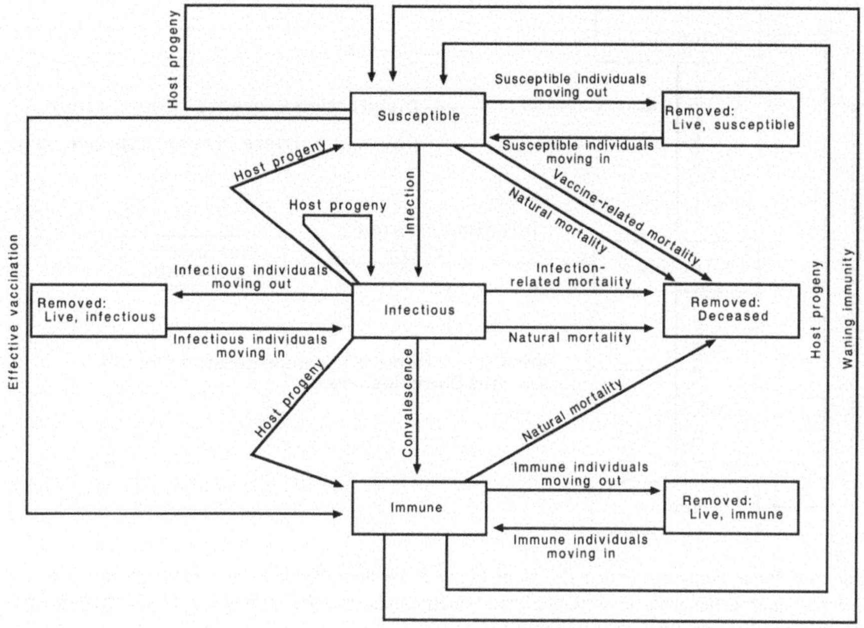

Fig. 3.3 Compartment model appropriate for describing endemic disease propagation in a dynamic population

prodromal period when the symptoms are not severe and may not be specific, during which time the hosts usual functions still can be performed although perhaps with some amount of either distres or discomfort. The prodromal phase is followed by an illness period during which customary symptomatology may be displayed. The fourth stage is a hoped period of convalescense. These four stages can be represented by different compartments.

Figure 3.1 also could be used as a compartment model but its variables would be different and their parameter values would be harder to estimate.

3.4 Prevention Is the Best Solution for Infectious Disease

To quote Benjamin Franklin (1735), although out of proper context, "In the first Place, as an Ounce of Prevention is worth a Pound of Cure ...". Franklin actually was addressing fire safety, but we often use that quote when referring to health. Microbial illnesses are an important cause of morbidity and mortality in human populations. There are several means by which we try either to prevent disease transmission or to lessen the severity with which disease impacts upon the affected individuals. A compartment disease transmission model as shown in Fig. 3.4 can be used to illustrate the implementation of treatment technologies both to prevent initial

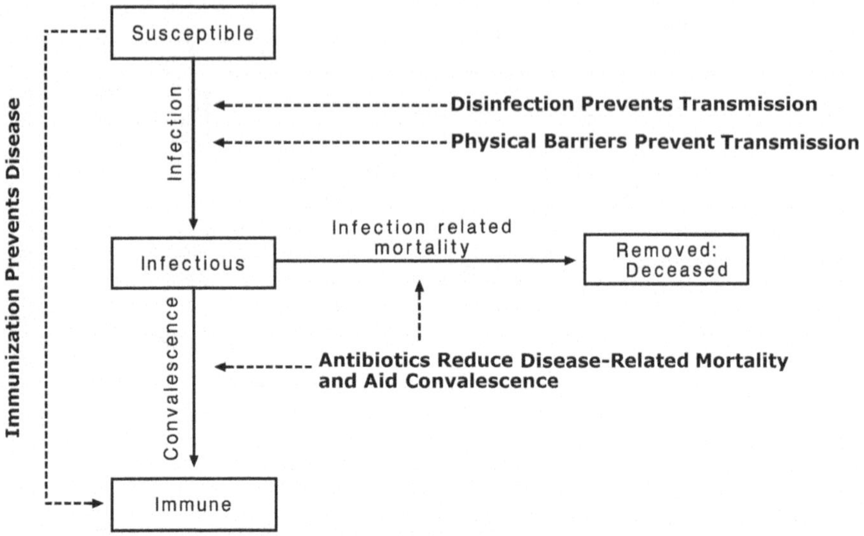

Fig. 3.4 Basic compartment model from Fig. 3.2, modified to demonstrate the points at which different approaches may be used singly and in combination either to prevent disease transmission or to lessen the consequences of disease

acquisition of infections from drinking water and then to prevent secondary spread of the disease within the community.

Table 3.2 lists some of the categories of barriers that can be used to prevent transmission of microbial pathogens (Hurst 2016a). Figure 3.4 illustrates the points at which immunization, disinfection, physical barriers, and antibiotics are employed to reduce the transmission of pathogenic microorganisms.

3.4.1 Using Immunization to Prevent Infectious Disease

Immunization is a very potent technique by which we try to both prevent the transmission of disease and to reduce deleterious post-transmission outcome for the diseased recipient. Immunization typically is used either prior to infection or, if there is a long incubation period such as is the case for rabies, then immunization can effectively be used during the incubation period. Immunization of a susceptible individual will not actually prevent transmission of the pathogen to that individual, but immunization is used in the hope of subsequently increasing the rapidity of the individuals immune response, thereby reducing replication of the pathogen within the newly infected individual. The goal is that quickly and effectively reducing reproduction of the causative microbe in this individual will prevent the acquired infection from progression to the point that the individual demonstrates overt disease

Table 3.2 Categories of barriers used to prevent transmission of microbial pathogens

Physical barriers		
	Thermal	
	Acoustic (usually ultrasonic)	
	Pressure	
		Barometric
		Hydrostatic
		Osmotic
	Radiation	
		Electronic
		Neutronic
		Photonic
		Protonic
	Impaction (includes gravitational)	
	Adhesion (adsorption)	
	Electrostatic	
	van der Waals	
	Filtration (size exclusion)	
	Geographic features	
	Atmospheric factors (includes such meterological aspects as humidity, precipitation, and prevailing winds)	
Chemical barriers		
	Ionic (includes pH and salinity)	
	Surfactant	
	Oxidant	
	Alkylant	
	Desiccant	
	Denaturant	
Biological barriers		
	Immunological (includes specific as well as nonspecific)	
		Naturally induced (intrinsic response to microbial exposure)
		Naturally transferred (lacteal, transovarian, transplacental, etc.)
		Artificially induced (includes cytokine injection and vaccination)
		Artificially transferred (includes injection with antiserum and tissue transfers such as transfusion and grafting)
	Biomolecular resistance (not immune-related)	

(continued)

Table 3.2 (continued)

Physical barriers		
		Lack of receptor molecules
		Molecular attack mechanisms (includes nucleotide-based restrictions)
		Antibiotic compounds (metabolic inhibitors, either intrinsic or artificially supplied)
	Competitive (other species in ecological competition with either the microbe, its vectors, or its hosts)	

and thereby reduce the likelihood that the immunized individual will transmit that infection to other susceptible individuals.

Ability to mount a protective immune response naturally varies in relationship to normal factors such as increasing age or pregnancy (Gove et al. 1987). Other, unnatural conditions which can reduce the bodys ability to generate an adequate immune response thus resulting in a more severe disease outcome are underlying malnutrition (Crawford and Vermund 1988) and genetic immunosuppressive disorders (Saulsbury et al. 1980), as well as immunosuppressive viral diseases including acquired immune deficiency syndrome (Clifford et al. 1990) and measles (Crawford and Vermund 1988).

3.4.1.1 Passive Versus Active Immunization

There are two basic approaches employed for immunizing individuals. The first of these is passive immunization, which typically results from the administration of immune globulin preparations by injection. Preformed antibodies present in the injected material provide immediate but short lived protection. Passive immunization sometimes is performed before an individual is exposed to the disease-causing microorganism, in which case it represents primary prevention. Passive immunization also may be performed following exposure, in which case it represents secondary prevention. Pre-exposure treatment is most likely to be done in those instances where the anticipated disease has a short incubation period. Post-exposure treatment is used in those instances where the disease either has a long incubation period, or the exposed individual is considered to be at risk of a severe disease outcome.

The second approach is active immunization, often termed "vaccination". This involves administering antigenic materials which cause the recipient to develop an immune response. Active immunization normally is used before exposure to the causative microorganism as a means of primary prevention. The protection resulting from active immunization usually lasts for a much longer time period than does the protection afforded by passive immunization. The United States Government has recently published an informative advisory document on immunization practices for humans (Centers for Disease Control and Prevention 2018b).

There are, however, drawbacks to reliance upon vaccination for disease control. The process of developing, commercially producing, distributing and successfully administering vaccines is both complex and costly. The complexity of this process limits the number of diseases against which we can attempt to protect people. The cost factor sadly also limits the extent to which vaccination programs are utilized. Problems can result from vaccination, with instances of vaccine-related illness and even deaths known to occur among vaccine recipients (Centers for Disease Control 1991). Also, live attenuated organisms that may be administered for vaccination purposes can occasionally revert to virulent forms (Abraham et al. 1993) and cause illness instead of preventing illness. Immunity acquired through vaccination is not necessarily permanent, and can wane with time such that immunized individuals again become susceptible to illness caused by the targeted pathogen. There often are religious objections against vaccination (Ahmed et al. 2018) and sometimes the people who administer vaccines are attacked and killed because of those religious objections (McGirk 2015).

3.4.1.2 Separating the Concepts of Immunity Against Infection, Which Is Short Term, and Immunity Against Illness Which May Be Longer Term

In order to understand the risk values developed from the modeling efforts presented in this chapter, it is important to realize that infection with a given species of microorganism does not offer assurance against subsequent reinfection by that same species at some later time. This impermanence of protection allows both previously ill and also vaccinated individuals to return to the pool of susceptibles. Immunity against either infection or reinfection may last only a few months or less. Reinfection can occur even if the individual has developed a measurable immune response to that organism as determined by antibody titers (Cash et al. 1974). Studies have shown that it is possible for individuals to be reinfected with the same subgroup (Simhon et al. 1981) and even the same strain of enteric virus (Koprowski 1956), with the same strain of enteric protozoa (Nash et al. 1987), or the same phage type of enteric bacteria (Hornick et al. 1970b). It is possible for reinfection to occur as soon as 12 weeks after the initial infection (Nash et al. 1987). No information was found which indicated an upper limit to the number of times an individual could be reinfected by the same species, serotype, or strain of microorganism.

The basis for the phenomenon of naturally acquired immunity assumes that after a sufficient number of antigenic exposures to, or infections by, the same serotype of microorganism, an individual can develop a level of immune response which will be sufficiently strong and rapid so as to protect not against the possibility of subsequent infections and reinfections by that microbe, which can happen repeatedly, but rather whether the residual capacity for response can protect against the possibility that subsequent infection and reinfection events by that microbe would cause illness.

Thusly, when people mention the idea of protective long term immunity they actually should understand themselves not to be considering an "all or none" level of protection provided by their residual capability to produce an immune response. Not even naturally acquired immunity can be presumed protective against infection. Immunity is protective only against the likelihood and severity of resulting illness. While the duration of post-disease protection against future illness caused by either the same or a similar pathogen may last for perhaps 20 years, that duration period is variable and could be considerably shorter.

Natural acquired immunity has not been taken into account by the modeling exercises presented in this chapter for three reasons. First, different strains of even the same species of microorganism will vary in their immunogenicity, which is defined as the ability to illicit an immune response. Second, the duration in terms of years for the protective effect attributed to naturally acquired immunity remains uncertain and probably is variable. Third, it has been proven that reinfection of an individual by even the same strain of microorganism can result in sequential episodes of illness (Hornick et al. 1970b; Simhon et al. 1981).

This understanding about immunity and immunization is the reason why the wording in Fig. 3.4 includes "Immunization prevents disease" and not "Immunization prevents infection".

3.4.2 *Infection Control Practices*

There are other approaches, aside from vaccination, that can be used to prevent infection. Among these are two general categories of infection control practices. The first represents establishment of physical barriers for preventing the transfer of infectious agents. The second consists of disinfection. Disinfection can include some physical barrier techniques such as irradiation, in addition to chemical barriers, and both categories of infection control practices represent primary prevention.

3.4.2.1 Use of Physical Barriers to Prevent Disease Transmission

The United States Department of Health and Human Services divides physical barriers for laboratory situations into two groups, primary barriers and secondary barriers. According to this classification scheme, primary barriers protect individuals and their immediate environment. As such, primary barriers would consist of protective clothing like gloves and masks, as well as personal safety equipment like filtration equipped respirators. Secondary barriers include physical containment practices and design factors intended to limit dispersion of the pathogens, among which are the use of separate rooms for handling clean versus contaminated materials, washing and other decontamination techniques that are used for work surfaces and implements, and disinfection practices including the sterilization of contaminated materials.

These barriers are applicable in different ways, some against the microbes themselves, some against vehicles, and some are used against vectors and infected hosts. Many of these same barrier concepts are applicable at different size levels. The concept of filtration, for example, applies to the use of size excluding filter materials that are designed with pores sufficiently small so as to physically retain the individual microbes themselves. Size exclusion filtration also is applied as a barrier in the form of such things as sand filters used for water treatment, and while such filters may not have a pore diameter sufficiently small as to remove individual microbes, these filters nevertheless can effectively retain particle-associated microbes and small vectors. Latex gloves have pores that similarly may not retain the microbes themselves but those pores are small enough that they do act as size exclusion barriers by retaining droplets of liquid containing the microbes, and condoms similarly prevent the transmission of sexually transmitted diseases. Protective clothing, window screens and bed nets are effective as size exclusion barriers against many insect vectors and other small arthropods, while simultaneously reducing our exposure to any disease-causing pathogens which those animals may be carrying. Fences, doors, gates, and walls are used as size exclusion barriers against numerous other vectors and also against infected hosts. The concept of physical barriers also includes quarantine practices, whereby infected individuals are isolated from susceptible individuals. Quarantine isolation has been used for humans, most notably for diseases such as tuberculosis, leprosy, influenza, and ebola. Quarantine can be effective, but may be expensive and is considered stigmatizing. These and other physical barrier concepts can be applied for public health purposes to prevent disease transmission in our daily living situations.

Table 3.3 shows the usefullness of filtration barriers for treating drinking water by illustrating the relationship between increase in percentage of United States city population supplied with community filtered drinking water and decrease in overall city population typhoid death rates. The cities used for compiling that table of data were selected for having reliable records as to cause of death (Johnson 1916). A presumption made when presenting this information is that some of the cases of typhoid illness may have been contracted directly from pathogens contained in the drinking water, and that secondary transmission of the illness subsequently may have resulted when infections initially acquired from water then were spread from the ill individuals to other, susceptible people. Additional sources of the pathogen would have including contaminated food.

Preventing disease transmission by using physical barrier techniques also can be done at the household level for treating drinking water. As an example, ceramic water filters can be used on the household level, acting by size exclusion to protect against bacteria and protozoa in contaminated water plus those filters can prevent the passage of virus that are associated with larger suspended particles. Home sandfilters were used to treat drinking water before ceramic filtration became popular, and sand filters remain in use at household levels. The efficiency of commercially designed drinking water filtration devices often is aided by particle agglomeration and sedimentation using either natural or synthetic coagulants.

Table 3.3 Relation between increase in percentage of United States city population supplied with filtered drinking water and decrease in overall city population typhoid death rates

Year	Percentage of city population served by water filtration utilities	Typhoid fever[a] death rates for same cities
1900	8.7	36
1901	10.8	34
1902	11.9	37
1903	13.3	38
1904	16.0	35
1905	17.4	30
1906	20.5	33
1907	23.2	32
1908	23.3	25
1909	30.1	21
1910	34.6	24
1911	37.2	20
1912	42.4	16
1913	48.0	16

Data is presented for those United States cities having accurate registration as to cause of human death
Source: Johnson (1916)
[a]Death rates given per 100,000 population per year

3.4.2.2 Use of Disinfection to Prevent Disease Transmission

Disinfection, the destruction of pathogenic microorganisms, is something which, when given sufficient time, can occur under natural environmental conditions that do not allow growth of the pathogenic microorganisms (Hurst 1991; Hurst et al. 1992; Kehr and Butterfield 1943) and this can be termed natural disinfection. Sunlight can be an effective disinfectant as are manufactured sources that generate either microwave, ultraviolet, or gamma radiations. Disinfection usually represents primary prevention. Under special medical situations, where an individual already is infected, disinfection could be considered a means of secondary prevention by either reducing the extent of infection or precluding dispersal of the pathogens.

Using disinfectants to destroy pathogens in drinking water is a practice that can be performed either at the municipal level with formal treatment facilities, or at individual and household levels. Disinfection is also performed on wastewaters to protect the quality of receiving waters.

Disinfection of drinking water on an individual or household basis can be accomplished by either boiling of the water, or distilling the water, or adding oxidizing chemicals such as chlorine and iodine.

Disinfection practices performed on an individual basis also include the washing of hands, dishes and eating implements. Many washing agents and detergents can be effective in aiding either the physical removal or destruction of microorganisms. Washing with an artificial detergent, like ordinary soap, can provide some chemical disinfection activity (Baker et al. 1941). Although we often use soap as an aid when washing, using just water alone can have some beneficial effect (Mbithi et al. 1993). Interestingly, Schurmann and Eggers (1985) have shown that at least in the case of decontaminating skin, the use of moistened sand may be even more effective than is washing with ordinary soap. The simple practice of using chlorine bleach when washing clothing also has a disinfection effect.

3.4.3 Use of Antibiotics to Preclude or Lessen the Severity of Disease

The successful development of antibiotic drugs represents one of the miracles of modern science and medicine (Taylor 1942). These drugs normally are used for secondary prevention, to reduce the severity of illness in diseased patients. However, in some instances antibiotic drugs are used prophylactically as primary prevention to preclude the initiation of an infection. Despite the miracle that these antibiotic compounds represent, microorganisms are capable of evolving drug resistance (Coronado et al. 1993). Thus, time has proven Taylor (1942) to have been overly optimistic when he stated that with discovery of the organic arsenic compound Salvarsan, and the sulphonamides, humankind had successfully conquered bacteria.

3.4.4 Demonstrating the Effectiveness of Techniques for Reducing the Transmission of Infectious Agents

I will present two approaches, compartment disease transmission modeling and risk estimation modeling.

3.4.4.1 Compartment Disease Transmission Modeling for Waterborne Infectious Disease

Drinking water treatment will be used as an example of our effectiveness at reducing the incidence of infectious disease in human populations. The bacterial disease typhoid is well known to be transmitted by the ingestion of contaminated water, which often represents its primary route of transmission into a population of susceptible individuals. This organism may be harbored by human carriers of the disease, and those are individuals who experience prolonged infection without necessarily displaying the symptoms normally associated with this illness. Human carriers may represent a secondary route of transmission for typhoid, the means by which the infectious agent is transferred within a human population following primary waterborne transmission. These carriers can pass the pathogen on to other members of a population through vehicles such as contaminated food. Typhoid caused extremely high incidences of waterborne disease and death worldwide in the nineteenth century and early part of the twentieth century. While this disease still occurs both endemically and epidemically throughout the world, in developed countries its occurrence has almost completely been staunched by the introduction of adequately operated drinking water treatment facilities. Table 3.3 presented the change in urban typhoid death rates in the United States during the beginning of the twentieth century, serving as an example of the effectiveness of community drinking water treatment for reducing waterborne disease. In this case, the treatment used was water filtration, which is a form of physical barrier. In some of the cities, the water may have been chlorinated following filtration which is a combined-treatment practice whose usage began during the latter part of the time period represented in Table 3.3 (Holmquist 1924). Interestingly, the initial use for chlorination of drinking water was to control the development of water odor within distribution systems and only later was it discovered that chlorination also reduces the level of pathogenic microbes in the water. The data listed in Table 3.3 are from a publication by Johnson (1916) for cities with accurate reporting as to cause of human death. These data show that in the year 1900, for the represented cities, the percentage of urban population served by water filtration utilities was 8.7% and the concurrent annual typhoid death rate was 36 per 100,000 population. During the 13 years which followed, the percentage of population served by water filtration in those cities gradually increased to 48% and simultaneously the annual typhoid death rate gradually decreased to 16 per 100,000 population. These data dramatically suggest how drinking water treatment can be effective for reducing waterborne disease even when secondary transmission of the disease can occur within the community.

The first approach that I will present for demonstrating the potential effectiveness of drinking water treatment is the disease transmission modeling represented in Tables 3.4 and 3.5 (Hurst and Murphy 1996). The basic disease transmission model presented in Fig. 3.1 was used for this purpose, with exclusion of mortality, to estimate the social impact of a waterborne outbreak in a community of 10,000 people. It was assumed for the purpose of this modeling that: (1) all of the individuals in the community were susceptible to infection at the time of the community's initial exposure to the causative microorganism; (2) an individual became infectious one day after becoming infected; (3) the probability of an infectious individual recovering from the infection and subsequently becoming immune was 20% on any given day during their period of infectiousness; and (4) on each day that an individual was infectious, they had a 2% chance of transmitting the infection to another individual. As the percentage of susceptible individuals in the population decreased during the epidemic, transmission of the disease also became reduced. Running of the model was discontinued when the movement of individuals from the infectious category to the immune category had become less than 0.5 per day, at which point movement would have rounded down to zero individuals per day. For the models presented in Tables 3.4 and 3.5 it was assumed that exposure to the primary route of transmission, presumably microbial contaminants in the drinking water, occurred for a single day and that the initial infection rate was 10%.

Table 3.4 Model of community wide waterborne disease epidemic caused by single day ingestion exposure to contaminated water with subsequent person to person secondary transmission

	Number of individuals[a]		
Day number	Susceptible	Infectious	Immune (recovered)
1	10,000	0	0
2	9,000	1,000	0
3	8,982	818	200
4	8,967	669	364
5	8,955	547	498
6	8,945	448	607
7	8,937	366	697
8	8,930	300	770
9	8,925	245	830
10	8,921	200	879
11	8,917	164	919
12	8,914	134	952
13	8,912	109	979
14	8,910	89	1,001
15	8,908	73	1,019
16	8,907	59	1,034
17	8,906	48	1,046
18	8,905	39	1,056
19	8,904	32	1,064
20	8,903	27	1,070
21	8,903	22	1,075
22	8,903	18	1,079
23	8,903	14	1,083
24	8,903	11	1,086
25	8,903	9	1,088
26	8,903	7	1,090
27	8,903	6	1,091
28	8,903	5	1,092
29	8,903	4	1,093
30	8,903	3	1,094
31	8,903	2	1,095
	Total cases of infection = 1,097		

[a]Assumptions: (1) No initial immunity within community. (2) Infected persons become infectious on next day. (3) Each day 20% of previous days infectious individuals have recovered, calculated as follows: (number recovered on day n) = (number recovered on day $n-1$) + (0.2 × number infectious on day $n-1$). (4) Each day 2% of the infectious individuals transmit the infection to a single contact, assuming that contact is susceptible, calculated as follows: (number infectious on day n) = (0.8 × number infectious on day $n-1$) + [(0.02 × number infectious on day $n-1$) × (portion of initial population still susceptible on day $n-1$ expressed as a decimal value)]. This equation is valid only if there are individuals in the infectious category for day $n-1$. (5) On any given day, everyone who is neither infectious nor immune is considered susceptible, calculated as follows: (number susceptible on day n) = [10,000 – (number infectious on day n + number immune on day n)]. (6) Reinfection does not occur during the course of the epidemic

Table 3.5 Model of the same waterborne disease outbreak as in Table 3.4 with assumption that both initial transmission and secondary transmission were reduced by 80% achieved through use of physical barrier and disinfection practices

	Number of individuals[a]		
Day number	Susceptible	Infectious	Immune (recovered)
1	10,000	0	0
2	9,800	200	0
3	9,799	161	40
4	9,798	130	72
5	9,797	105	98
6	9,797	84	119
7	9,797	67	136
8	9,797	54	149
9	9,797	43	160
10	9,797	34	169
11	9,797	27	176
12	9,797	22	181
13	9,797	18	185
14	9,797	14	189
15	9,797	11	192
16	9,797	9	194
17	9,797	7	196
18	9,797	6	197
19	9,797	5	198
20	9,797	4	199
21	9,797	3	200
22	9,797	2	201
	Total cases of infection = 203		

[a]Assumptions:
1–3, 5, and 6 same as stated for Table 3.3
4. Each day 0.4% of the infectious individuals transmit the infection to a single contact, assuming that contact is susceptible calculated as follows: (number infectious on day n) = ($0.8 \times$ number infectious on day $n-1$) + [($0.02 \times 0.2 \times$ number infectious on day $n-1$) $\times$ (portion of initial population still susceptible on day $n-1$ expressed as a decimal value)]. This equation is valid only if there are individuals in the infectious category for day $n-1$

The model shown in Table 3.4 assumes that physical barrier techniques and disinfection practices were not employed to reduce either primary (waterborne) or secondary transmission of the disease. In that case, 1,000 individuals initially would become infected. The epidemic ends with a total of 1,097 individuals having been infected, 2 of which remain infected at day 31. Ninety seven of these infections resulted from secondary transmission of the infectious agent within the community. Table 3.5 shows what would happen to this same population assuming that drinking water treatment was 80% effective at reducing the incidence of initial infection within the population, and that the use of physical barriers and disinfection practices was similarly 80% effective at reducing secondary transmission. Using these assumptions for the efficiency of physical barriers and disinfection, Table 3.5 indicates that a total of only 203 individuals would become infected during this outbreak, with 2 of those individuals remaining infected at day 22, and only 3 of the individuals having been infected as a result of secondary transmission.

3.4.4.2 Risk Estimation Modeling for Waterborne Infectious Disease

Several items of information are helpful in order for us to estimate the risk of infectious disease caused by ingesting microbially contaminated water. One of these is the levels of pathogenic microorganisms present in water that is likely to be consumed by individuals. A second item is a means of calculating the probability of acquiring infection from drinking that water, along with the accompanying probabilities that infection will lead to illness, and illness ultimately progress to death. Having this information available for calculating the risk of aquiring infections from ingesting water, we can then determine the extent to which these risks could be reduced by using treatment processes to remove microorganisms before the water is consumed.

Several previous efforts have been made at estimating the disease risk to humans from consuming drinking water that contained microbial contaminants. These efforts have a long history, dating back at least to 1943 (Kehr and Butterfield 1943). The risk estimation technique being represented here has a demonstrated ability to accurately calculate the probability of community gastrointestinal illness based upon the known levels of pathogenic virus and protozoa in a communitys drinking water. The levels of pathogenic bacteria in that water had not been determined by the researchers, and so those levels of bacteria are being inferred based upon a presumption that the infectious risk not accounted for by either virus or protozoa contained in the water would have been attributable to pathogenic bacteria in the water.

3.4.4.2.1 Defining Probability Values for the Health Risks Associated with Infectious Disease

The first task is to define the probabilities of infection, illness, and death associated with each microbial group, genus, species or serovar. A problem to avoid is the potentially misleading idea of curve fitting, which can alter value estimates for the minimum infectious dose of a microorganism as exampled by the information listed in Table 3.6.

Instead, I will present the suggestion (Hurst et al. 1996) that the risks be estimated as defined in Eqs. (3.1)–(3.3) and use information which came from published literature representing numbers for microorganisms and individuals. These equations determine the probability of infection per ingested pathogenic microorganism, the probability that the acquired infection will progress to illness, and that the illness will progress to death.

Table 3.6 How the use of curve fitting can alter value estimates for the minimum infectious dose of a microorganism

Microorganism examined for infectivity	Experimentally determined minimum infectious dose[a] (reference)	Minimum infectious dose as estimated by curve fittings[b]	Underestimation of risk which would result from using curve-fitting values[c]
Enterovirus enterovirus B (was Human echovirus 12)	17 (Schiff et al. 1984)	500.0	29.4
Enterovirus enterovirus C (was Human poliovirus 1)	20 (Minor et al. 1981)	110.2	5.5
Enterovirus enterovirus C (was Human poliovirus 3)	0.5[d] (Katz and Plotkin 1967)	1.93	3.9
Rotavirus	0.9 (Ward et al. 1986)	1.62	1.8
Giardia	10.0 (Rendtorff 1954)	49.27	4.9

[a]The number of viral organisms was determined by infectivity in cultured cells. The number of *Giardia* was determined by direct microscopic enumeration

[b]Data from Regli et al. (1991). This column contains estimated minimum infectious dose values derived by performing curve-fitting operations on the actual experimental data. The curve-fitting values listed in Regli et al. (1991) were presented as the estimated concentration of microorganisms per liter of drinking water which would give an annual risk of one infection per 10,000 individuals and assumed that each individual ingested 2 liters of water per day. Therefore, backtracking to obtain the estimated minimum infectious dose values which had been derived by curve fitting was done by multiplying the published values by 7.3×10^6, which was calculated as follows: (2 liters of water/day) × (365.25 days/year) × (10,000 individuals)

[c]If the calculation of risk were done using values derived by curve fitting in place of the actual experimentally determined values, an underestimation of risk would result. The underestimation of risk (as a factor of error) which would be associated with using those curve-fitting values was calculated as follows: (the estimated minimum infectious dose as determined by Regli et al. 1991 through curve fitting) ÷ (the experimentally determined minimum infectious dose for that organism)

[d]Actual dose was 1 $TCID_{50}$, divided by 2 to yield an estimated $TCID_{100}$ as reported in this table, using the technique described by Hurst et al. (1988)

$$\text{Probability of Infection} = \frac{1}{\text{Minimum number of microorganisms required to produce infection}} \tag{3.1}$$

$$\text{Probability of Illness} = \frac{\text{Number of individuals who showed symptoms of illness}}{\text{Total number of individuals infected}} \tag{3.2}$$

$$\text{Probability of Death} = \frac{\text{Number of individuals who died from that illness}}{\text{Total number of individuals who showed symptoms of that illness}} \quad (3.3)$$

The probability of infection estimates made by Hurst et al. (1996) were for humans experimentally ingesting typically water-associated pathogens and they are presented in Table 3.7. The probability of illness estimates made by Hurst et al. (1996) for humans experimentally ingesting typically water-associated pathogens are presented in Table 3.8. The probability of death estimates made by Hurst et al. (1996) for humans in outbreak situations caused by typically water-associated pathogens are presented in Table 3.9. The use of median values for bacteria, protozoa and virus has been chosen because the conditions, either experimental or natural, differed between those studies for which the data has been summarized. The fact that differences existed between the studies precludes averaging those data to generate mean values.

From examining the calculated median values listed in Table 3.9, it can be estimated that the overall probability of infection leading to death for illnesses caused by either enteric bacteria or virus is approximately 0.01 to 0.02, or from 1 to 2%. There are notable exceptions! As one example, human illness caused by *Orthohepevirus orthohepevirus A*, previously designated Hepatitis E virus, normally has a 0.028 probability of death (which also can be stated as either 2.8%, or 1 in 36 meaning 1 death per 36 individuals, or 1/36). However, the probability of death for women who are in either their second or third trimester of pregnancy at the time when they become ill from this virus typically ranges from approximately 0.17 (17%, or 1 in 6, Gove et al. 1987) to 0.22 (22%, Khuroo 1991). The death rate for hepatitis E illness in pregnant women can be as high as 39% (Khuroo 1991) and as many as 46% of the individuals who die during a hepatitis E outbreak situation may be pregnant women (Gove et al. 1987). As another example, Hoxie et al. (1997) observed that the overall risk of death from waterborne *Cryptosporidium* associated with a large general population of humans who experienced an outbreak of infectious illness caused by that genus in Milwaukee, Wisconsin, may have been 0.0001 (0.01%, or 1 in 10,000). Unfortunately, numbers from a publication by Crawford and Vermund (1988) indicate that the probability of cryptosporidial illness leading to death can be 0.14 (14%, or 1 in 7, or 2/14) in the case of underlying malnutrition. The risk of death from *Cryptosporidium* can be 0.20 in the case of underlying measles (20%, or 1 in 5) (Crawford and Vermund 1988) representing a 2,000-fold increase over the rate of death from *Cryptosporidium* infections in the general population. By itself, illness caused by the measles virus *Morbillivirus measles morbillivirus* is almost never fatal. However, this member of the *Morbillivirus* genus seems possibly to be the most powerfully immunosuppressive virus known to infect humans (Hurst and Adcock 2000). The single value found that represented a general population and seemed sufficiently reliable for being used in estimating the probability that illness from an enteric protozoan infection would lead to death was from a waterborne outbreak of cryptosporidiosis for which the estimated probability of death is 0.0002, or 0.02% (Marchione 1994; Mac Kenzie et al. 1994). I have used

Table 3.7 Minimum infectious dose of enteric microorganisms for humans as determined by ingestion

Microorganism examined	Lowest number of microorganisms required to cause infection	Probability of infection per individual microorganism ingested	Reference
Bacteria[a]			
Salmonella enterica subsp. enterica serovar Typhi (was *Salmonella typhi*, Quailes strain)	100,000	1.0×10^{-5}	Hornick et al. (1970a)
Vibrio cholerae (type 01)	1,000	1.0×10^{-3}	Levine et al. (1984)
Vibrio cholerae (type non-01) strain A	1,000,000	1.0×10^{-6}	Morris et al. (1990)
Vibrio cholerae (type non-01) strain B	10,000,000	1.0×10^{-7}	Morris et al. (1990)
Vibrio cholerae (type non-01) strain C	100,000	1.0×10^{-5}	Morris et al. (1990)
		Median for bacteria = 0.00001 [1.0×10^{-5}]	
Protozoans[b]			
Cryptosporidium parvum	30[c]	0.033	DuPont et al. (1995)
Giardia intestinalis (was Giardia lamblia)	10	0.1	Rendtorff (1954)
		Median for protozoans = 0.067 [6.7×10^{-2})	
Virus[d]			
Enterovirus enterovirus B (was Human echovirus 12)	17	0.059	Schiff et al. (1984)
Enterovirus enterovirus C (was Human poliovirus 1, SM strain)	2	0.5	Koprowski (1955)
Enterovirus enterovirus C (was Human poliovirus 1, SM strain)	2	0.5	Katz and Plotkin (1967)
Enterovirus enterovirus C (was Human poliovirus 1, (strain not identified)	20	0.05	Minor et al. (1981)

(continued)

Table 3.7 (continued)

Microorganism examined	Lowest number of microorganisms required to cause infection	Probability of infection per individual microorganism ingested	Reference
Enterovirus enterovirus C (was Human poliovirus 3, Fox strain)	0.5[e]	2.0	Katz and Plotkin (1967)
Rotavirus	0.9	1.1	Ward et al. (1986)
		Median for virus = 0.5 [5.0×10^{-1}]	

Footnotes for Table 3.7:
Infection was determined serologically
[a]Number of organisms was determined by culturing
[b]Number of organisms was determined by microscopy
[c]Lower dosages were not tested
[d]Number of organisms was determined by infectivity in cultured cells
[e]Actual dose was 1 $TCID_{50}$, divided by 2 to yield estimated ID_{100} of 0.5 as reported in this table, using the technique described by Hurst et al. (1988)

this value to represent probability of death from protozoan illness for the purpose of these modeling exercises. However, I understand that use of this value for protozoan illness may represent a severe understatement when it comes to estimating the risk for specific subpopulations of individuals.

There have been reports of concurrent enteric infections in individuals caused by multiple genera of either protozoa (Cristino et al. 1988) or bacteria (Mata et al. 1983), and of simultaneous infections with protozoa and virus (Grohmann et al. 1993; Crawford and Vermund 1988; Taylor et al. 1985), as well as with bacteria and virus (Mata et al. 1983). It is even possible to become concurrently infected with two different enteric virus of the same genus, as has been noted for infections either by rotaviruses (Rodriguez et al. 1983) or enteroviruses (Koprowski 1955; Tambini et al. 1993). In fact, the capability of simultaneously being infected with multiple species of the genus *Enterovirus* is what allows effective administration of oral poliovirus vaccine in a trivalent form, containing all three human poliovirus in an infectious form (Abraham et al. 1993).

I considered it important to use only data developed in humans for this modeling exercise. The reason for this decision is that the minimum number of microorganisms required to cause an infection can differ between humans and other animals by as much as several hundred fold (Koprowski 1955).

Table 3.8 Likelihood of infection resulting in illness for enteric microbial diseases

Causative microorganism	Ratio of illnesses to total number of infections	Probability of illness	Reference
Bacteria			
Salmonella enterica subsp. enterica serovar Typhi (was *Salmonella typhi*)	34/69	0.4928	Hornick et al. (1970a)
Vibrio cholerae (type 01)	15/55	0.2727	Levine et al. (1984)
Vibrio cholerae (type non-01)	8/15	0.5333	Morris et al. (1990)
		Median for bacteria = 0.4928 [4.9×10^{-1}]	
Protozoans			
Cryptosporidium	36/65	0.5538	Bongard et al. (1994)
Cyclospora	11/50	0.2200	Ortega et al. (1993)
Giardia intestinalis (was *Giardia lamblia*)	11/34	0.3235	Lopez et al. (1980)
Giardia intestinalis (was *Giardia lamblia*)	3/5	0.6000	Nash et al. (1987)
Giardia[a]	26/133	0.1955	Islam (1990)
Giardia	977/1,211	0.8068	Birkhead and Vogt (1989)
		Median for Giardia = [0.4617 = 4.6×10^{-1}] Median for protozoans = 0.4386 [4.4×10^{-1}]	
Viruses			
Enterovirus enterovirus C (was Human coxsackievirus A21)	24/26	0.9231	Couch et al. (1965)
Enterovirus enterovirus B (was Human echovirus 12)	3/59	0.0508	Schiff et al. (1984)
Rotavirus	4/5	0.8000	Kapikian et al. (1983)

(continued)

Table 3.8 (continued)

Causative microorganism	Ratio of illnesses to total number of infections	Probability of illness	Reference
Rotavirus	17/30	0.5667	Ward et al. (1986)
Rotavirus	15/26	0.5769	Ward et al. (1989)
		Median for virus = 0.5769 [5.8×10^{-1}]	

Infection was defined serologically
[a]Data cited here was that listed for India

3.4.4.2.1.1 Using "The Blindfolded Bowler Analogy" to Understand the Calculation for Probability of Infection

It is important to note that infection and illness are not synonymous terms. While an individual cannot become ill without first becoming infected, one can become infected without that infection progressing to illness. Studies have been done in which microorganisms were fed to otherwise healthy individuals, and those individuals were then examined for evidence of infection. Two criteria can be used for determining the existance of an intestinal infection. The first, and most important, criterion is to look for the development of a response in the individuals antibody titer against the ingested organism (Bernstein et al. 1986; Cash et al. 1974; Clark et al. 1986). Development of either a detectable antibody titer when none previously existed, or an increase in an existing antibody titer, indicates that the individual was infected. The second criterion is to determine whether the number of indicated organisms shed in the individuals feces is greater than could be accounted for by the dose that the individual was fed. Finding appreciable numbers of the same type of organism in the feces suggests that the individual was infected, if those organisms are obligate intracellular parasites such as virus and some protozoans. This technique cannot accurately be used to assess infection by organisms which live in the lumen of the intestine, such as other protozoans and bacteria.

For the purpose of the modeling exercise presented in this chapter, it is assumed that any single microorganism is capable of causing infection, and that the differences in minimum infectious dose between the various microorganisms simply represent the random statistical probabilities that the necessary combination of events leading to an infection by those microorganisms would occur. It can be seen from the information presented in Table 3.7 that the lowest number of microorganisms required to cause infection is generally greater for bacterial organisms than for either protozoans or virus. Also, it can be seen that the minimum infectious dose varies at both the genus and species level for all three categories of microorganisms examined, bacteria, protozoa, and virus. The viral data indicate that there are additional strain dependent differences regarding the minimum infectious dose for microorganisms

Table 3.9 Likelihood of illness resulting in death for enteric microbial diseases

Causative microorganism	Ratio of deaths to total cases of illness	Probability of death	Reference
Bacteria			
Campylobacter	2/6,441	0.0003	Tauxe et al. (1988)
Escherichia coli (type 0111)	1/9	0.1111	Caprioli et al. (1994)
Escherichia coli (type 0157-H7)	1/10	0.1000	Turney et al. (1994)
Escherichia coli (type 0157-H7)	4/243	0.0165	Geldreich et al. (1992)
Salmonella enterica subsp. enterica serovar Typhi (was *Salmonella typhi*)	20,000/ 300,000[a]	0.0667	Johnson (1916)
Salmonella enterica subsp. enterica serovar Typhi (was *Salmonella typhi*, Zermatt, phage type E1)	3/437	0.0069	Bernard (1965)
Shigella flexneri	1/610	0.0016	Centers for Disease Control and Prevention (1994)
Vibrio cholerae (Burundi, Zimbabwe type 01)	292/1,177	0.2481	Ndayimirije et al. (1993)
Vibrio cholerae (West Africa type 01)	20,000/ 150,000	0.1333	Swerdlow and Ries (1993)
Vibrio cholerae (Bangladesh type non-01)	1,473/ 107,297	0.0137	Cholera Working Group (1993)
Vibrio cholerae (Portugal type 01)	48/2,467	0.0195	Blake et al. (1977)
Vibrio cholerae (Americas type 01)	6,323/ 731,312	0.0086	Centers for Disease Control and Prevention (1993)
		Median for bacteria = 0.0180 [1.8×10^{-2}]	
Protozoans			
Cryptosporidium parvum	100/403,000	0.0002	Marchione (1994), Mac Kenzie et al. (1994)
		Median statistic not meaningful [2.0×10^{-4}]	
Viruses			
Enterovirus enterovirus A (was Human coxsackievirus A2)	2/403	0.0050	Assaad and Borecka (1977)
Enterovirus enterovirus A (was Human coxsackievirus A3)	2/27	0.0741	Assaad and Borecka (1977)
Enterovirus enterovirus A (was Human coxsackievirus A4)	3/580	0.0052	Assaad and Borecka (1977)

(continued)

Table 3.9 (continued)

Causative microorganism	Ratio of deaths to total cases of illness	Probability of death	Reference
Enterovirus enterovirus A (was Human coxsackievirus A6)	2/204	0.0098	Assaad and Borecka (1977)
Enterovirus enterovirus A (was Human coxsackievirus A10)	6/351	0.0171	Assaad and Borecka (1977)
Enterovirus enterovirus A (was Human coxsackievirus A16)	2/1,654	0.0012	Assaad and Borecka (1977)
Enterovirus enterovirus B (was Human coxsackievirus A9)	8/3,039	0.0026	Assaad and Borecka (1977)
Enterovirus enterovirus B (was Human echovirus 6)	14/4,774	0.0029	Assaad and Borecka (1977)
Enterovirus enterovirus B (was Human echovirus 9)	14/5,237	0.0027	Assaad and Borecka (1977)
Enterovirus enterovirus C (was Human coxsackievirus A19)	3/18	0.1667	Assaad and Borecka (1977)
Enterovirus enterovirus C (was Human poliovirus 1)	72/8,074	0.0089	Assaad and Borecka (1977)
Enterovirus enterovirus C (was Human poliovirus 1)	95/1,031	0.0921	Kim-Farley et al. (1984)
Enterovirus enterovirus C (was Human poliovirus 2)	43/2,360	0.0182	Assaad and Borecka (1977)
Enterovirus enterovirus C (was Human poliovirus 3)	27/2,427	0.0111	Assaad and Borecka (1977)
Enterovirus enterovirus C (was Human poliovirus 3)	1/54	0.0185	van Wijngaarden et al. (1992)
Enterovirus enterovirus C (were Human polioviruses 1, 2, & 3, data had been combined)	11/204	0.0539	Langmuir et al. (1956)
Hepatovirus hepatovirus A (was Hepatitis A virus)	–[b]	0.0060[a]	Centers for Disease Control (1990)
Orthohepevirus orthohepevirus A (was Hepatitis E virus, Kashmir)	–[b]	0.0280[c]	Ramalingaswami and Purcell (1988)
Orthohepevirus orthohepevirus A (was Hepatitis E virus, Somalia)	87/2,000	0.0435	Gove et al. (1987)
Orthohepevirus orthohepevirus A (was Hepatitis E virus, Delhi)	90/97,600	0.0009	Viswanathan (1957)
Rotavirus	2/222	0.0090	Fang et al. (1989)
Rotavirus	7/3,439	0.0020	Foster et al. (1980)
		Median for virus = 0.0094 [9.4×10^{-3}]	

[a]Relationship between number of deaths from illness to total cases of illnesses was estimated by the cited author

[b]Values were not given in the reference

[c]Value reported for adult males

belonging to the same species. It was gratifying to notice that when values were available from two different research studies for the same strain within a single species of microorganism, the SM strain of Human poliovirus 1, both studies reported the same minimum infections dose level. The median value for probability of infection by viruses would not be changed by deleting either one of these two values for the SM strain.

When looking at the lowest number of viruses required to cause infection, it can be seen that in two instances, Human poliovirus 3 results published by Katz and Plotkin (1967), and *Rotavirus* results published by Ward et al. (1986), the infectious dose is less than one infectious unit as measured by cultured cells. This may reflect the fact that these virus evolved to replicate within cells of the intestinal wall as part of a living organ, preceded by exposure to the chemical environment within the intestines. As such, these virus are not as well suited to replicating under the conditions extant in the laboratory, where cultured cells were used to grow and quantitate the virus.

For many of the microorganisms listed in Table 3.7, the minimum infectious dose is much greater than one. This is particularly true in the case of bacteria, where the minimum infectious doses reported in Table 3.7 range from 1,000 to 10,000,000 organisms. Higher minimum infectious dose values can be found in the literature for many of the bacteria and protozoa listed in Table 3.7. However, in those other studies, it generally is the case that lower doses of microorganisms were not examined. If a unit of 100 microorganisms is sufficient to cause an infection, then so would 1,000 or even 1,000,000. For this reason, I have tried to cite studies which determined not only a minimum infectious dose, but which also tested lower dosage levels of the same organism and found the lower dosages insufficient to cause an infection.

Hurst et al. (1996) created an analogy which they named "The Blindfolded Bowler Analogy" to explain the basis for Eq. (3.1). For this analogy, it is assumed that a bowler is placed into a bowling alley and blindfolded. He is then pointed in the direction of a bowling lane and successively given objects that he is asked to roll down the lane. If the bowler is given 10 balls all of the same size, and not told whether any of them resulted in a strike until he has rolled all ten, he may find that on the whole, 10 balls is not sufficient to result in a strike. Thus, for that set of conditions, the minimum "unit" number of balls required to generate a strike is greater than 10. If allowed to use a larger number of balls, perhaps either 20, 1,000 or even 100,000 balls, the bowler will eventually find the minimum "unit" number of balls required to generate a strike. If, for example, under a defined set of conditions, there was a reliable chance that a unit of 100 balls could consistently yield a strike, and an assumption can be made that each individual ball was physically capable of producing the strike, then the probability of any individual ball producing a strike would be 0.01, the numerical inverse of the unit number of balls. It may be expected that using larger diameter balls would decrease the minimum unit number of balls required to cause a strike, and that using smaller diameter balls might increase the minimum unit number required to cause a strike. If the bowler instead were given cylinders to roll down the lane, and the length of those cylinders was greater than the diameter of the bowling balls, it might be the case that a unit of only 20 cylinders would be required to produce a strike. In that case, the probability of a strike would be 0.05 per object rolled down the lane. Rolling cylinders whose length was equal to the width of the bowling lane might reduce the minimum unit number to 1, giving a probability of 1.0 per cylinder.

This analogy is not used to suggest that size, or overall shape, determines the minimum infectious dose for microorganisms. Instead, factors which likely influence the probability of infection per ingested organism include the number of susceptible cells in the intestinal wall, the relative number and accessibility of receptor sites on those cells, plus random genetic and phenotypic variation both on the part of the microorganism and the host. We can, of course, remove the bowlers blindfold, which enables him to understand the situation. We cannot, however, remove our own blindfold with respect to the variation in minimum unit number of virus, protozoans, or bacteria required to cause infection if ingested by a human. For the purposes of this risk modeling effort, I have again applied the Blindfolded Bowler Analogy and assumed that the probability of infection per individual microorganism ingested is the numerical inverse of the minimum infectious dose.

If we were to place a barrier across the lane used by our blindfolded bowler, we would severely decrease the probability of a rolled object producing a strike. The acidity of stomach secretions likewise represents a barrier to microorganisms which infect cells lining the intestines. Some organisms, such as enteroviruses, have evolved a natural resistance to low pH exposures, which enables them to get past the barrier posed by the stomach. Protozoan cysts and oocysts also seem to have evolved some resistance to degradation by acids. Ingesting food along with pathogenic microorganisms seems to help the microorganisms survive their trip through the stomach (Blake et al. 1980). This finding is important because humans often ingest food along with water and other beverages. Individuals who consume antacids eliminate this barrier from their stomach, a particular problem if they consume microbially contaminated water along with those antacids. Individuals who have the disease hypochlorhydria, and therefore produce an abnormally low amount of stomach acid, are correspondingly at greater risk of enteric infectious diseases such as cholera (Nalin et al. 1978).

The organism *Vibrio cholerae* has an interesting means of natural protection from stomach acids. In addition to causing massive outbreaks of disease associated with the consumption of contaminated water, this microorganism is known to cause illness associated with the consumption of macrocrustaceans, such as crabs. The natural habitat of this bacterial organism is the chitinous shells of crustaceans (Grimes 1991). Copepods and other taxa of microcrustaceans can reside as natural microfauna within community drinking water distribution systems (van Lieverloo et al. 2012). The association of *Vibrio* organisms with the chitinous shells of micro- and macro-crustaceans may facilitate survival of the microbes during passage through the stomach, and could be an important factor in waterborne disease if the *Vibrio cholerae* in source waters were attached to microcrustacean copepods and these microcrustaceans inadvertently ingested along with the water (Nalin 1976). This latter knowledge suggests that microcrustaceans may represent a potentially supportive environment for *Vibrio* contaminants that gain entrance to water distribution systems.

3.4.4.2.1.2 Understanding the Calculations for Probabilities of Illness and of Death

There are five main variables that need to be included when estimating the risk of acquiring infectious disease by ingestion of water. These are the concentration of pathogenic organisms in water which are infectious by the route of ingestion, the amount of water ingested by an individual per unit time, and the probabilities of infection, illness, and death associated with ingesting those pathogenic organisms.

The approach used in this chapter presumes that the risks of illness and death from ingesting pathogenic microbes in water are standard inverse calculations based upon knowing the numbers of infected individuals who become ill and the numbers of ill individuals who die. Illnesses that do not lead to death cover a broad range in terms of the severity of symptoms. The risk modeling exercise presented in this chapter does not address severity of illness, except in distinguishing the likelihood of infection leading to illness (Table 3.8) from the likelihood that illness subsequently results in death (Table 3.9).

There is a great deal of confidence in the numerical values used for estimating the probability of infection per ingested organism (Table 3.7) and the probability of an infection event resulting in illness (Table 3.8), because the data used for those calculations had been generated in carefully controlled prospective studies. There is somewhat less confidence in the numerical values available for calculating the probability of illness progressing to death (Table 3.9), because the latter values represent statistics that generally were compiled retrospectively from epidemiological surveys. There must also be some uncertainty as to whether the cited epidemiological studies considered the total number of illness cases attributed to a particular pathogen, or only those cases for which severe symptomatology was demonstrated. Given a fixed number of deaths for a particular disease, varying the criteria used in establishing the number of cases of illness will, in turn, change the calculated probability of illness progressing to death. The 1955–1956 waterborne disease outbreak in Delhi, India, caused by Hepatitis E virus represents an example of this variation in reporting the number of cases of illness. Viswanathan (1957) reported that 90 deaths were attributable to hepatitis contracted during this outbreak. I have chosen to cite his estimate of 97,600 cases of illness for this outbreak, since he was an epidemiologist who studied that epidemic at the time when it occurred (Viswanathan 1957). Some more recent publications (Khuroo 1991; Ramalingaswami and Purcell 1988) cited 29,300 cases of illness for this same outbreak, a number which represents only those individuals whose illness was sufficiently severe that it resulted in jaundice. If I instead had used the lower number of cases of illness in this risk modeling exercise, then the probability of illness progressing to death for that outbreak would have risen from 0.0009 to 0.0031. The choice of using overall median values to represent the probability of illness progressing to death for the categories of bacterial, protozoan, and viral diseases, rather than using corresponding mean values, helps to lessen the impact upon this modeling exercise of any cited study that may represent a particularly high or low probability value.

3.4.4.2.1.3 Six Important Considerations

Risk of infection, illness and death can be influenced by other factors in addition to these six whose implications must be understood when making risk assessment estimations for infectious diseases. The first three of these are issues that specifically apply to estimations for drinking water. The last three apply not only to drinking water but also to any infectious disease risks regardless of the source of the pathogenic organisms. I have not tried to account for these considerations when making the modeling efforts presented in this chapter.

1. Estimations of the levels of microorganisms in samples of water inherently are underestimations. This is due to the deficiencies of detection methodology.
2. To some extent, estimations of disease risk based on the levels of microbial contaminants in the municipal water distribution mains will be overestimations. This is because some of the water which people ingest is treated by heating techniques prior to its consumption, including cooking and the preparation of beverages that are consumed hot. This heating generally will reduce the level of viable microbial contaminants in the water. However, heating water and then maintaining it warm for a long time, as occurs with water-heating tanks in homes and commercial buildings including nursing facilities, can change the risk in that some pathogenic organisms such as bacteria of the genus *Legionella* can grow in water that is kept warm. Disease caused by *Legionella* usually results from inhalation of aerosolized water, rather than from ingestion.
3. Directly estimating the level of disease risk associated with pathogenic bacteria present in water is best done by enumerating those bacteria which actually are pathogenic. Most "indicator bacteria" are not considered pathogenic, and information regarding the level of non-pathogenic indicator organisms in water cannot directly be used for making risk assessment estimates.
4. If death does not occur as the outcome of an initial infection by a particular strain of a specific microorganism, then each subsequent reinfection of the same person by the same strain may carry a decreased risk that the infection would progress to illness as mentioned above.
5. Immunosuppression can increase the probability that an infection will have a severe outcome. There are some natural conditions such as pregnancy, and abnormal conditions such as malnutrition, that can result in suppression of immune system function. The result of this immunosuppression is that when some individuals become infected they may have a greater probability of death from that infection (see Sect. 3.4.4.2.1) than will other members of the general population.
6. Concurrent infections, which are incidences when a person is infected simultaneously by more than one pathogenic organism, can drastically increase the probability of death associated with any one of the involved pathogens (see Sect. 3.4.4.2.1) and many people are likely to be simultaneously infected by more than a single pathogen.

3.4.4.2.2 Validation of This Risk Estimation Technique

Restating from Sect. 3.4.4.2.1.2, five main variables must be considered when estimating the risk of acquiring infectious disease by ingestion of water. These are the concentration of pathogenic organisms in water which are infectious by the route of ingestion, the amount of water ingested by an individual per unit time, and the probabilities of infection, illness, and death associated with ingesting those pathogenic organisms.

If we know the concentrations of pathogenic bacteria, protozoans and virus in drinking water, then it is possible to estimate the risks of infection, illness, and death associated with drinking the water that contains those pathogenic microbes. Two validation exercises are being presented for this risk estimation technique. Equations (3.4)–(3.7) present the format which is being used for this chapter to estimate the individual human health risks of infection, illness, and death associated with ingested microorganisms. These are based upon using Eq. (3.1) for the probability of infection per ingested organism, Eq. (3.2) for the probability of illness per incidence of infection, and Eq. (3.3) for the probability of death per incidence of illness.

$$\begin{array}{l}\text{Number of}\\\text{organisms}\\\text{ingested}\\\text{annually}\end{array} = \begin{array}{l}\text{Number of}\\\text{organisms}\\\text{per liter}\\\text{of water}\end{array} \times \begin{array}{l}\text{Number of}\\\text{liters}\\\text{ingested}\\\text{per day}\end{array} \times \begin{array}{l}\text{365.25 days}\\\text{of water}\\\text{ingestion}\\\text{per year}\end{array} \tag{3.4}$$

$$\begin{array}{l}\text{Annual}\\\text{risk of}\\\text{infection}\end{array} = \begin{array}{l}\text{Number of}\\\text{organisms}\\\text{ingested}\\\text{annually}\end{array} \times \begin{array}{l}\text{Probability}\\\text{of infection}\\\text{per ingested}\\\text{organism}\end{array} \tag{3.5}$$

$$\begin{array}{l}\text{Annual}\\\text{risk of}\\\text{illness}\end{array} = \begin{array}{l}\text{Number of}\\\text{organisms}\\\text{ingested}\\\text{annually}\end{array} \times \begin{array}{l}\text{Probability}\\\text{of infection}\\\text{per ingested}\\\text{organism}\end{array} \times \begin{array}{l}\text{Probability}\\\text{of illness}\\\text{per incidence}\\\text{of infection}\end{array} \tag{3.6}$$

$$\begin{array}{l}\text{Annual}\\\text{risk of}\\\text{death}\end{array} = \begin{array}{l}\text{Number of}\\\text{organisms}\\\text{ingested}\\\text{annually}\end{array} \times \begin{array}{l}\text{Probability}\\\text{of infection}\\\text{per ingested}\\\text{organism}\end{array} \times \begin{array}{l}\text{Probability}\\\text{of illness}\\\text{per incidence}\\\text{of infection}\end{array} \times \begin{array}{l}\text{Probability}\\\text{of death}\\\text{per incidence}\\\text{of illness}\end{array} \tag{3.7}$$

This validation effort requires either knowing or estimating the levels of pathogenic bacteria, protozoa, and virus in community supplied drinking water. These risks will be calculated separately for bacterial, protozoan, and viral contaminants. The next item of information needed for this exercise is the amount of water ingested daily per individual. The value used in Eq. (3.4) as an estimate for the volume of water ingested on a daily basis per individual consumer is 2 liters per day. This volume is the estimate recommended by the United States Environmental Protection Agency for use in calculating the risks from ingestion of water (United States Environmental Protection Agency 1997). More recent risk factor publications by the EPA are available but their contents are highly nuanced and do not give such a definitive number. Daily water consumption may vary greatly due to factors such as environmental temperature, differences in individual body mass, and the amount of physical exercise performed. Other estimates for water ingestion are available, including one by Payment et al. (1991) that consumers drink the equivalent of 46.5–47.3 glasses of water per week. Making an assumption that the average glass of water represents 250 ml, the value published by Payment et al. (1991) would represent approximately 1.7 liters of water ingested per day. It is possible, and perhaps likely, that not all of this ingested water will be consumed plain. Instead, some water will be used in soups, sauces, and as the basis for other beverages (Birkhead et al. 1993).

For these modeling exercises, the probability of each individual occurrence of infection, termed here to be an "infection event", is assumed to be independent of any other infection event caused by either the same or another microorganism. The probability of an infection event leading to illness is conditionally dependent upon an infection having occurred, there can be no microbially induced illness without infection although there can be infections that do not result in illness. The probability of illness progressing to death is in turn dependent upon an infection event having led to illness. It is possible that the probability of individual infection events leading to illness, and the accompanying severity of illness including progression to death, may decrease with subsequent reinfections by the same species, serotype, or strain of microorganism (Nash et al. 1987). This was discussed above in Sect. 3.4.1.2, which attempted to separate the concepts of immunity against infection and immunity against illness.

3.4.4.2.2.1 First Risk Validation Exercise

An attempt to validate the risk modeling approach described in this chapter can be done using data which Payment and colleagues have published for a section of Quebec province, Canada. Results of an epidemiological study (Payment et al. 1991) indicated that the annual risk of consumers acquiring gastroenteritis from ingesting microorganisms contained in the community-distributed tap water was 0.26. In an earlier study, it was determined that the average level of enteric viruses in the raw water entering that area's drinking water treatment plants was 3.3 cell culture infectious units per liter (Payment et al. 1985). In that same 1985 publication, it was indicated that the average cumulative viral reduction achieved by those community drinking water treatment plants was 99.97%. Table 3.10 shows that by substituting into Eqs. (3.4)–(3.6) this reported value for the level of viruses in the raw water, and taking into account the reduction in viral level achieved by community drinking water treatment, the result is a predicted individual annual risk of viral illness equal to 0.2086. The calculation used for this prediction is as follows: [(3.3 viral organisms per

Table 3.10 For validation exercise 1: calculating the estimation of risk due to infectious, pathogenic virus in the municipality's water

Concentration of organisms (virus that are infectious for mammalian cells, enumerated as infectious units) per liter of raw water	3.3
Liters of water ingested per person per day	× 2.0
Days of water ingestion per year	× 365.25
Probability of infection (per ingested virus capable of infecting mammalian cells)	× 0.5
Probability of illness (per incidence of viral infection)	× 0.5769
Estimate of residual disease risk following municipal treatment of the water (this value is included because the data provided for viral levels were determined for the water prior to processing at the treatment plant), calculated as 1.0000 - 0.9997	× 0.0003
Estimated annual risk of virus-induced illness per individual consumer	= 0.208606

liter of raw water) × (2.0 liters of water ingested per day) × (365.25 days of water ingestion per year) × (5.0×10^{-1} probability of infection per ingested virus) × (5.8×10^{-1} probability of illness per viral infection) × (1.0000–0.9997 as the estimate of residual viral disease risk following community drinking water treatment)].

An estimate for the amount of disease risk represented by pathogenic protozoan cysts and oocysts in the community-treated drinking water for that same geographical area can be made using results published later by Payment and Franco (1993). In that 1993 publication, the authors presented information regarding the levels of protozoans belonging to two genera, *Giardia* and *Cryptosporidium*, in the finished (treated) water produced by three water treatment plants. The levels of these protozoa in finished water from two of the treatment plants was below the researchers' limit of detection. However, the level of protozoans in finished water from the third treatment plant was 0.02 *Cryptosporidium* oocysts and 0.02 *Giardia* cysts per 100 l. In Tables 3.11 and 3.12 we see that substituting these levels of protozoa into Eqs. (3.4) through (3.6) yields an estimated annual protozoan disease risk from *Cryptosporidium* of 0.00267, plus an estimated annual risk of protozoan disease risk from *Giardia* of 0.006745.

Looking at Table 3.13 we can see that by adding together the estimated 0.2086 annual risk of viral illness per individual consumer, and the estimated annual risks from protozoan illness per consumer, gives an annual risk estimate of 0.2180 for illnesses attributed to these two categories of microorganisms. This indicates that by using the risk modeling approach presented in this chapter, we are able to account for

Table 3.11 For validation exercise 1: calculating the estimation of risk due to *Cryptosporidium* in the municipality's water[a]

Concentration of organisms (*Cryptosporidium* oocysts, enumerated by visual inspection following immunofluorescent staining) per liter of water following municipal treatment of the water	0.0002
Liters of water ingested per person per day	× 2.0
Days of water ingestion per year	× 365.25
Probability of infection (per ingested *Cryptosporidium* oocyst)	× 0.033
Probability of illness (per incidence of cryptosporidial infection)	× 0.5538
Estimated annual risk of *Cryptosporidium*-induced illness per individual consumer	= 0.00267

[a]This calculation does not include a value for estimation of residual disease risk following municipal treatment of the water, because the data provided for *Cryptosporidium* oocyst levels were determined for the water following processing at the treatment plant

Table 3.12 For validation exercise 1: calculating the estimation of risk due to *Giardia* in the municipality's water[a]

Concentration of organisms (*Giardia* cysts, enumerated by visual inspection following immunofluorescent staining) per liter of water following municipal treatment of the water	0.0002
Liters of water ingested per person per day	× 2.0
Days of water ingestion per year	× 365.25
Probability of infection (per ingested *Giardia* cyst)	× 0.1
Probability of illness (per incidence of giardial infection)	× 0.4617
Estimated annual risk of *Giardia*-induced illness per individual consumer	= 0.006745

[a]This calculation does not include a value for estimation of residual disease risk following municipal treatment of the water, because the data provided for *Giardia* cyst levels were determined for the water following processing at the treatment plant

Table 3.13 For validation exercise 1: overall summary of risk estimations for the known levels of infectious, pathogenic virus and protozoans with the remaining risk assumed to be due to infectious, pathogenic bacteria

Total epidemiologically determined risk of illness per person per year associated with microbial contaminants in the municipally supplied tap water		0.26 (100.0%)
Estimated proportion of that risk attributable to virus in the municipally supplied tap water	0.20861 (80.2%)	
Estimated proportion of that risk attributable to *Cryptosporidium* in the municipally supplied tap water	0.00267 (1.0%)	
Estimated proportion of that risk attributable to *Giardia* in the municipally supplied tap water	0.00674 (2.6)%	
Total proportion of the epidemiologically determined risk which could be accounted for by the levels of virus, *Cryptosporidium*, and *Giardia* present in the municipally supplied tap water	0.21802 (83.9%)	0.21802 (83.9%)
Difference between the actual epidemiologically determined risk of illness and the level of risk predicted by this estimation technique using data for levels of virus and protozoa in the water		0.04198 (16.1%)

84% of the actual epidemiologically-determined risk. If we are able to make the assumption that the remaining 0.042 level of annual illness risk is due to bacterial contaminants in the distribution system water, then we can perform a backwards calculation to estimate the associated level of pathogenic bacteria in the distribution system water as shown in Table 3.14. According to this modeling approach, the residual individual annual illness risk of 0.042 would represent a level of approximately 11.67 pathogenic bacteria per liter in the community distributed tap water. As

Table 3.14 For validation exercise 1: backwards calculation to estimate the level of infectious, pathogenic bacteria in the municipality's water, assuming that bacteria accounted for the risk not attributable to virus and protozoa

Estimated annual risk of bacterially induced illness per individual consumer	0.0420
Probability of illness (per incidence of bacterial infection)	÷ 0.4928
Probability of infection (per ingested pathogenic bacterium capable of exhibiting growth on culture medium)	÷ 0.00001
Days of water ingestion per year	÷ 365.25
Liters of water ingested per person per day	÷ 2.0
Estimated concentration of organisms (pathogenic bacteria capable of exhibiting growth on culture medium) per liter of water following municipal treatment of the water	= 11.66698

[a]This back calculation does not include an estimation for the reduction of risk affected by municipal treatment of the water, because no data were provided for the extent to which pathogenic bacteria were removed by the process used at the water treatment plants

suggested earlier in this chapter, some of these bacteria could represent organisms that have persisted through the processing regimen at the community drinking water treatment plants. A second source of bacteria in water distribution systems is accidental contamination, which can result from breaks in water distribution lines and also from inadvertant cross connections with pipes that carry sewage. A third source of bacterial contaminants is the biofilm that grows on the inside walls of water distribution pipes and water storage tanks.

3.4.4.2.2.2 *Second Risk Validation Exercise*

Several years later, after efforts to improve the efficiency of that community's drinking water treatment facilities, Payment and coworkers performed a second epidemiological study (Payment et al. 1997) during which they found that individuals ingesting conventionally treated, community-distributed tap water had an overall 0.66 annual risk for incidence of gastrointestinal illness. This contrasted with an annual illness risk of 0.58 among a control group of consumers who drank either bottled water, or drank tap water that originated from the same community distribution system but had gone through an in-home filtration system prior to ingestion. The difference in annual risk thereby attributed to microorganisms in the tap water during the second epidemiological study was 0.08. The validation exercise presented in this chapter for the second epidemiological study of Payment et al. (validation 2) used virus and protozoa data directly published as a part of that second study

Table 3.15 For validation exercise 2: calculating the estimation of risk due to infectious, pathogenic virus in the municipality's water

Concentration of organisms (virus that are infectious for mammalian cells, enumerated as infectious units) per liter of raw water	4.13
Liters of water ingested per person per day	× 2.0
Days of water ingestion per year	× 365.25
Probability of infection (per ingested virus capable of infecting mammalian cells)	× 0.5
Probability of illness (per incidence of viral infection)	× 0.5769
Estimate of residual virus-associated disease risk following municipal treatment of the water (this value is included because the data provided for viral levels were determined for the water prior to processing at the treatment plant), calculated as 1.000000 – 0.999921	× 0.000079
Estimated annual risk of virus-induced illness per individual consumer	= 0.068749

(Payment et al. 1997). The calculation presented in Table 3.15 suggests that of the 0.08 annual risk of illness, the amount of that which could have been due to virus in the drinking water, per individual consumer, was 0.068749 and this suggests that virus in the drinking water could have accounted for approximately 85.9% of the observed cases of illness. The calculations presented in Tables 3.16 and 3.17 suggest that protozoa of the genera *Cryptosporidium* and *Giardia* respectively could have accounted for approximately 1.5% and 4.2% of the observed illnesses. The data which were published by Payment et al. (1997) concerning bacterial concentrations in the tap water for this study represent indicator bacterial groups rather than known waterborne bacterial pathogens. As such, the only way to estimate the proportion of observed infectious disease which would have been attributable to bacterial pathogens in the water during the time of the epidemiological studies is by subtraction.

Knowing the total risk due to pathogens in the drinking water and then subtracting the estimates for risks attributable to virus and protozoa results in the estimated proportion of illnesses due to bacterial pathogens in the tap water shown in Table 3.18. The backwards calculation presented in Table 3.19 suggests that the remaining 0.0067 estimated annual risk of illness per individual consumer potentially attributed to bacterially induced illness could have been accounted for by an estimated average concentration of 1.86116 bacterial organisms (pathogenic bacteria capable of exhibiting growth on culture medium) per liter of water following municipal treatment of the water. This validation process reveals that for the two

Table 3.16 For validation exercise 2: calculating the estimation of risk due to *Cryptosporidium* in the municipality's water

Concentration of *Cryptosporidium* oocysts (measured by visual inspection following immunofluorescent staining) per liter of raw water	0.14
Liters of water ingested per person per day	× 2.0
Days of water ingestion per year	× 365.25
Probability of infection (per ingested *Cryptosporidium* oocyst)	× 0.033
Probability of illness (per incidence of cryptosporidial infection)	× 0.5538
Estimate of residual *Cryptosporidium*-associated disease risk following municipal treatment of the water (this value is included because the data provided for *Cryptosporidium* oocyst levels were determined for the water prior to processing at the treatment plant), calculated as 1.000000 – 0.999369	× 0.000631
Estimated annual risk of *Cryptosporidium*-induced illness per individual consumer	= 0.001179

Table 3.17 For validation exercise 2: calculating the estimation of risk due to *Giardia* in the municipality's water

Concentration of organisms (*Giardia* cysts, enumerated by visual inspection following immunofluorescent staining) per liter of raw water	2.00
Liters of water ingested per person per day	× 2.0
Days of water ingestion per year	× 365.25
Probability of infection (per ingested *Giardia* cyst)	× 0.1
Probability of illness (per incidence of infection by *Giardia*)	× 0.4617
Estimate of residual *Giardia*-associated disease risk following municipal treatment of the water (this value is included because the data provided for *Giardia* cyst levels were determined for the water prior to processing at the treatment plant), calculated as 1.000000 – 0.999950	× 0.000050
Estimated annual risk of *Giardia*-induced illness per individual consumer	= 0.003373

Table 3.18 For validation exercise 2: overall summary of risk estimations for the known levels of infectious, pathogenic virus and protozoans with the remaining risk assumed due to infectious, pathogenic bacteria

Total epidemiologically determined risk of illness per person per year associated with microbial contaminants in the municipally supplied tap water		0.08 (100.0%)
Estimated proportion of that risk attributable to virus in the municipally supplied tap water	0.06875 (85.9%)	
Estimated proportion of that risk attributable to *Cryptosporidium* in the municipally supplied tap water	0.00118 (1.5%)	
Estimated proportion of that risk attributable to *Giardia* in the municipally supplied tap water	0.00337 (4.2%)	
Total proportion of the epidemiologically determined risk which could be accounted for by the levels of virus, *Cryptosporidium*, and *Giardia* present in the municipally supplied tap water	0.07330 (91.6%)	0.07330 (91.6%)
Difference between the actual epidemiologically determined risk of illness and the level of risk predicted by this estimation technique using data for levels of virus and protozoa in the water		0.00670 (8.4%)

epidemiological studies published by Payment et al. (1991, 1997), bacterial pathogens could have accounted for approximately 16.1% of the observed cases of illness attributable to microorganisms in the community-distributed tap water for the first epidemiological study (Payment et al. 1991) and correspondingly could have accounted for 8.4% of the observed cases of illness for the second study (Payment et al. 1997).

Table 3.19 For validation exercise 2: backward calculation to estimate the level of infectious, pathogenic bacteria in the municipality's water, assuming that bacteria accounted for the risk not attributable to viruses and protozoa[a]

Estimated annual risk of bacterially induced illness per individual consumer	0.0067
Probability of illness (per incidence of bacterial infection)	÷ 0.4928
Probability of infection (per ingested pathogenic bacterium capable of exhibiting growth on culture medium)	÷ 0.00001
Days of water ingestion per year	÷ 365.25
Liters of water ingested per person per day	÷ 2.0
Estimated concentration of organisms (pathogenic bacteria capable of exhibiting growth on culture medium) per liter of water following municipal treatment of the water	= 1.86116

[a]This back calculation does not include an estimation for the reduction of risk affected by municipal treatment of the water, because no data were provided for the extent to which pathogenic bacteria were removed by the process used at the water treatment plants

3.5 Summary

Most disease transmission routes require that the causative microorganisms be exposed to the environment. The main purpose of this chapter was to demonstrate how mathematical models are used for describing the propagation of infectious diseases through populations, and to explain the various routes by which water associated infectious agents are transferred both to and between individuals. The major factor in determining whether an organism can be transmitted via a particular environmental route such as water, aerosols, or contact with the surface of objects, is whether or not that organism has evolved the ability to survive environmental exposure under the conditions it would encounter along that route. As an example, microorganisms which lack stability in environmental waters are unlikely to be transmitted by water. Some diseases can be transmitted by more than a single route. Tularemia, for example, can be acquired by ingesting contaminated food or water, inhalation of microbial aerosols, contact of bare skin with contaminated materials, and even acquired through insect bites.

Pathogenic microorganisms frequently can be found in source waters, and the ingestion of microbially contaminated water clearly has the potential for causing infectious disease. Microorganisms can escape even well operated, community based drinking water treatment processes. The results from this risk analysis technique indicate that direct water recycling may not be a good thing although indirect recycling occurs nearly everywhere.

This chapter described the means by which humans attempt to prevent the transmission of waterborne infectious agents of disease, and demonstrated two approaches which can be used to model the effectiveness of these prevention techniques. The first approach used historical data for typhoid to show how intercepting a primary route of disease transmission can dramatically reduce the incidence of that disease within a community even when there exist other, secondary routes for transmission of the disease. The use of a disease transmission model illustrated how this type of disease reduction can be accomplished. The second approach used a risk estimation technique.

Protozoan cysts and oocysts represent the microbial group most easily removed by filtration because of their relatively large size, although they are difficult to chemically disinfect. Virus are in particular difficult to remove by filtration because of their typically small size, although fortunately the virus seem inactivated with relative ease by using chlorine for chemical disinfection (Hurst 2001). The difficulty of virus removal by filtration, plus the fact that virus have a high probability of infection per ingested particle, likely are the reasons why virus seem to represent such a large amount of the risk from waterborne infections unless something such as a cholera outbreak is overwhelming the community.

Acknowledgements I thank my dear friends and colleagues Patricia A. Murphy, Robert M. Clark, and Stig E. Regli for their assistance with initial development of the concepts presented in this chapter.

Dedication I particularly wish to thank my very dear friend Gunther Franz Craun, a true Bohemian by birth, for his kindness. Gunther helped me to recognize that my studies regarding the presence and fate of microorganisms in the environment should include understanding the microbes relationship to human health. Thirty years ago he also encouraged me to begin editing books about microbiology, and in appreciation for his kind encouragement I dedicate this chapter to Gunther.

Gunther Franz Craun

Compliance with Ethical Standards

Conflict of Interest Christon J. Hurst declares that he has no conflict of interest.

Ethical Approval This article does not contain any studies with human participants or animals performed by the author.

References

Abraham R, Minor P, Dunn G et al (1993) Shedding of virulent poliovirus revertants during immunization with oral poliovirus vaccine after prior immunization with inactivated polio vaccine. J Infect Dis 168:1105–1109

Ahmad K, Jahan K, Huq I (1984) Decontamination of drinking water by alum for the preparation of oral rehydration solution. Food Nutr Bull 6(2):54–57

Ahmed A, Lee KS, Bukhsh A et al (2018) Outbreak of vaccine-preventable diseases in Muslim majority countries. J Infect Public Health 11(2):153–155

Amos GCA, Hawkey PM, Gaze WH et al (2014) Waste water effluent contributes to the dissemination of CTX-M-15 in the natural environment. J Antimicrob Chemother 69:1785–1791

Anders J, Shireley LA, Volmer L et al (1994) Rapid assessment of vectorborne diseases during the midwest flood – United States, 1993. Morb Mortal Wkly Rep 43:481–483

Assaad F, Borecka I (1977) Nine-year study of WHO virus reports on fatal virus infections. Bull World Health Organ 55:445–453

Atchison CG, Wintermeyer LA, Kelly JR et al (1993) Public health consequences of a flood disaster – Iowa, 1993. Morb Mortal Wkly Rep 42:653–656

Baker Z, Harrison RW, Miller BF (1941) Action of synthetic detergents on the metabolism of bacteria. J Exp Med 73:249–271

Baldursson S, Karanis P (2011) Waterborne transmission of protozoan parasites: review of worldwide outbreaks – an update 2004–2010. Water Res 45(20):6603–6614. https://doi.org/10.1016/j.watres.2011.10.013

Bernard RP (1965) The Zermatt typhoid outbreak in 1963. J Hyg (Cambridge) 63:537–563

Bernstein DI, Ziegler JM, Ward RL (1986) Rotavirus fecal IgA antibody response in adults challenged with human rotavirus. J Med Virol 20:297–304

Birkhead G, Vogt RL (1989) Epidemiologic surveillance for endemic *Giardia lamblia* infection in Vermont the roles of waterborne and person-to-person transmission. Am J Epidemiol 129:762–768

Birkhead GS, Morse DL, Levine WC et al (1993) Typhoid fever at a resort hotel in New York: a large outbreak with an unusual vehicle. J Infect Dis 167:1228–1232

Blake PA, Rosenberg ML, Costa JB et al (1977) Cholera in Portugal, 1974 I. Modes of transmission. Am J Epidemiol 105(4):337–343

Blake PA, Allegra DT, Snyder JD et al (1980) Cholera – a possible endemic focus in the United States. N Engl J Med 302:305–309

Bongard J, Savage R, Dern R et al (1994) Cryptosporidium infections associated with swimming pools – Dane County, Wisconsin, 1993. Morb Mortal Wkly Rep 43:561–572

Caprioli A, Luzzi I, Rosmini F et al (1994) Communitywide outbreak of hemolytic-uremic syndrome associated with non-0157 verocytotoxin-producing *Escherichia coli*. J Infect Dis 169:208–211

Cash RA, Music SI Libonati JP et al (1974) Response of man to infection with *Vibrio cholerae*. I. clinical, serologic, and bacteriologic responses to a known inoculum. J Infect Dis 129:45–52

Centers for Disease Control (1990) Protection against viral hepatitis: recommendations of the immunization practices advisory committee (ACIP). Morb Mortal Wkly Rep 39(RR-2):1–26

Centers for Disease Control (1991) Vaccinia (smallpox) vaccine: recommendations of the immunization practices advisory committee (ACIP). Morb Mortal Wkly Rep 40(RR-14):1–10
Centers for Disease Control and Prevention (1993) Update: Cholera – Western hemisphere, 1992. Morb Mortal Wkly Rep 42:89–91
Centers for Disease Control and Prevention (1994) Outbreak of *Shigella flexneri* 2a infections on a cruise ship. Morb Mortal Wkly Rep 43:657
Centers for Disease Control and Prevention (2018a) Incidence of typhoid fever, by year – United States, 1920–1960. https://www.cdc.gov/healthywater/observances/dww-graph.html. Accessed 9 March 2018
Centers for Disease Control and Prevention (2018b) Recommended vaccines by age. https://www.cdc.gov/vaccines/vpd/vaccines-age.html. Accessed 9 Mar 2018
Centers for Disease Control and Prevention (2018c) *Vibrio* species causing Vibriosis. https://www.cdc.gov/vibrio/index.html. Accessed 12 Mar 2018
Cholera Working Group, International Centre for Diarrhoeal Diseases Research, Bangladesh (1993) Large epidemic of cholera-like disease in Bangladesh caused by *Vibrio cholerae* 0139 synonym Bengal. Lancet 342:387–390
Clark HF, Offit PA, Dolan KT et al (1986) Response of adult human volunteers to oral administration of bovine and bovine/human reassortant rotaviruses. Vaccine 4(Mar):25–31
Clifford CP, Crook DWM, Conlon CP et al (1990) Impact of waterborne outbreak of cryptosporidiosis on AIDS and renal transplant patients. Lancet 335:1455–1456
Coronado VG, Beck-Sague CM, Hutton MD et al (1993) Transmission of multidrug-resistant Mycobacterium tuberculosis among persons with human immunodeficiency virus infection in an urban hospital: epidemiologic and restriction fragment length polymorphism analysis. J Infect Dis 168:1052–1055
Cotton P (1993) Health threat from mosquitos rises as flood of the century finally recedes. J Am Med Assoc 270(6):685–686
Couch RB, Cate TR, Gerone PJ et al (1965) Production of illness with a small-particle aerosol of Coxsackie A21. J Clin Invest 44:535–542
Craun GF (2012) The importance of waterborne disease outbreak surveillance in the United States. Ann Ist Super Sanita 48(4):447–459
Crawford FG, Vermund SH (1988) Human cryptosporidiosis. Crit Rev Microbiol 16:113–159
Cristino JAGM, Carvalho MIP, Salgado MJ (1988) An outbreak of cryptosporidiosis in a hospital day-care centre. Epidemiol Infect 101:355–359
Dahling DR, Safferman RS (1979) Survival of enteric viruses under natural conditions in a subarctic river. Appl Environ Microbiol 38:1103–1110
DuPont HL, Chappell CL, Sterling CR et al (1995) The infectivity of *Cryptosporidium parvum* in healthy volunteers. N Engl J Med 332(13):855–859
Editorial (1923) Typhoid fever death rates. Am J Public Health 13(8):660–661
Efstratiou A, Ongerth JE, Karanis P (2017) Waterborne transmission of protozoan parasites: review of worldwide outbreaks – an update 2011–2016. Water Res 114(May):14–22. https://doi.org/10.1016/j.watres.2017.01.036
Eichold BH, Williamson JR, Woernle CH et al (1993) Isolation of *Vibrio cholerae* 01 from oysters – Mobile bay, 1991–1992. Morb Mortal Wkly Rep 42:91–93
Fang ZY, Ye Q, Ho M-S et al (1989) Investigation of an outbreak of adult diarrhea rotavirus in China. J Infect Dis 160:948–953
Foster SO, Palmer EL, Gary GW Jr et al (1980) Gastroenteritis due to rotavirus in an isolated Pacific island group: an epidemic of 3,439 cases. J Infect Dis 141(1):32–39
Franklin B "Anonymous" (1735) Protection of Towns from Fire. Pennsylvania Gazette: 4 February. https://founders.archives.gov/documents/Franklin/01-02-02-0002. Accessed 5 March 2018
Geldreich EE, Fox KR, Goodrich JA et al (1992) Searching for a water supply connection in the Cabool, Missouri disease outbreak of *Escherichia coli* 0157:H7. Water Res 26:1127–1137
Gove S, Ali-Salad A, Farah MA et al (1987) Enterically transmitted non-A, non-B hepatitis – East Africa. Morb Mortal Wkly Rep 36:241–244

Grabowski DJ, Tiggs KJ, Senke HW et al (1989) Common-source outbreak of giardiasis – New Mexico. Morb Mortal Wkly Rep 38:405–407

Grimes DJ (1991) Ecology of estuarine bacteria capable of causing human disease: a review. Estuaries 14(4):345–360

Grohmann GS, Glass RI, Pereira HG et al (1993) Enteric viruses and diarrhea in HIV-infected patients. N Engl J Med 329:14–20

Grundmann H, Kropec A, Hartung D et al (1993) *Pseudomonas aeruginosa* in a neonatal intensive care unit: reservoirs and ecology of the nosocomial pathogen. J Infect Dis 168:943–947

Holmquist CA (1924) Relation of improvements in water supplies to typhoid fever and other intestinal diseases. J N Engl Water Works Assoc 38:237–247

Hornick RB, Greisman SE, Woodward TE et al (1970a) Typhoid fever: pathogenesis and immunologic control, first of two parts. N Engl J Med 283:686–691

Hornick RB, Greisman SE, Woodward TE et al (1970b) Typhoid fever: pathogenesis and immunologic control, second of two parts. N Engl J Med 283:739–746

Hoxie NJ, Davis JP, Vergeront JM et al (1997) Cryptosporidiosis-associated mortality following a massive waterborne outbreak in Milwaukee, Wisconsin. Am J Public Health 87:2032–2035

Hurst CJ (1991) Using linear and polynomial models to examine the environmental stability of viruses. In: Hurst CJ (ed) Modeling the environmental fate of microorganisms. American Society for Microbiology, Washington, pp 137–159

Hurst CJ (2001) Disinfection of water: drinking water, recreational water and wastewater. In: Block SS (ed) Disinfection, sterilization and preservation, 5th edn. Lippincott Williams & Wilkins, Philadelphia, pp 1023–1047

Hurst CJ (2003) Pathogens in the aquatic ecosystem: what do we really know? In Agua 2003: Seminario Internacional sobre Métodos Naturales para el Tratamiento de Aguas Residuales. Cartagena. Universidad del Valle/Instituto Cinara. p 208–212

Hurst CJ (2016a) Defining the concept of a species physiological boundaries and barriers. In: Hurst CJ (ed) Their World: a diversity of microbial environments. Advances in environmental microbiology, vol 1. Springer, Cham, pp 35–67

Hurst CJ (2016b) Opportunistic algae fungi and Ichthyosporea associated with mammalian livestock disease. In: Hurst CJ (ed) The Rasputin effect: when commensals and symbionts become parasitic. Advances in environmental microbiology, vol 3. Springer, Cham, pp 261–300

Hurst CJ (ed) (2017) Modeling the transmission and prevention of infectious disease. Advances in environmental microbiology, vol 4. Springer, Cham

Hurst CJ (2018) Opportunistic bacteria associated with mammalian livestock disease. In: Hurst CJ (ed) The connections between ecology and infectious disease. Advances in environmental microbiology, vol 5. Springer, Cham

Hurst CJ, Adcock NJ (2000) Relationship between humans and their viruses. In: Hurst CJ (ed) Viral ecology. Academic Press, San Diego, pp 519–548

Hurst CJ, Murphy PA (1996) Transmission and prevention of infectious disease. In: Hurst CJ (ed) Modeling disease transmission and its prevention by disinfection. Cambridge University Press, Cambridge, pp 3–54

Hurst CJ, Farrah SR, Gerba CP et al (1978) Development of quantitative methods for the detection of enteroviruses in sewage sludges during activation and following land disposal. Appl Environ Microbiol 36:81–89

Hurst CJ, McClellan KA, Benton WH (1988) Comparison of cytopathogenicity, immunofluorescence and in situ DNA hybridization as methods for the detection of adenoviruses. Water Res 22:1547–1552

Hurst CJ, Wild DK, Clark RM (1992) Comparing the accuracy of equation formats for modeling microbial population decay rates. In: Hurst CJ (ed) Modeling the metabolic and physiologic activities of microorganisms. Wiley, New York, pp 149–175

Hurst CJ, Clark RM, Regli SE (1996) Estimating the risk of acquiring infectious disease from ingestion of water. In: Hurst CJ (ed) Modeling disease transmission and its prevention by disinfection. Cambridge University Press, Cambridge, pp 99–139

Islam A (1990) Giardiasis in developing countries. In: Meyer EA (ed) Giardiasis. Elsevier, Amsterdam, pp 235–266
Johnson GA (1916) The typhoid toll. J Am Water Works Assoc 3(2):249–326
Kapikian AZ, Wyatt RG, Levine MM et al (1983) Oral administration of Human rotavirus to volunteers: induction of illness and correlates of resistance. J Infect Dis 147:95–106
Katz M, Plotkin SA (1967) Minimal infective dose of attenuated poliovirus for man. Am J Public Health 57:1837–1840
Kehr RW, Butterfield CT (1943) Notes on the relation between coliforms and enteric pathogens. Public Health Rep 58(15):589–607
Khuroo MS (1991) Hepatitis E: the enterically transmitted non-A, non-B hepatitis. Indian J Gastroenterol 10(3):96–100
Kim-Farley RJ, Rutherford G, Lichfield P et al (1984) Outbreak of paralytic poliomyelitis, Taiwan. Lancet ii:1322–1324
Koprowski H (1955) Living attenuated poliomyelitis virus as an immunizing agent of man. SA Med J 29:1134–1142
Koprowski H (1956) Immunization against poliomyelitis with living attenuated virus. Am J Trop Med Hyg 5:440–452
Langmuir AD, Nathanson N, Hall WJ (1956) Surveillance of poliomyelitis in the United States in 1955. Am J Public Health Nations Health 46:75–88
Leonard AFC, Zhang L, Balfour AJ et al (2015) Human recreational exposure to antibiotic resistant bacteria in coastal bathing waters. Environ Int 82(Sep):92–100. https://doi.org/10.1016/j.envint.2015.02.013
Leonard AFC, Singer A, Ukoumunne OC et al (2018a) Is it safe to go back into the water? A systematic review and meta-analysis of the risk of acquiring infections from recreational exposure to seawater. Int J Epidemiol 47(2):572–586. https://doi.org/10.1093/ije/dyx281
Leonard AFC, Yin XL, Zhang T et al (2018b) A coliform-targeted metagenomic method facilitating human exposure estimates to *Escherichia coli*-borne antibiotic resistance genes. FEMS Microbiol Ecol 94(3). https://doi.org/10.1093/femsec/fiy024
Leonard AFC, Zhang L, Balfour AJ et al (2018c) Exposure to and colonisation by antibiotic-resistant *E. coli* in UK coastal water users: Environmental surveillance, exposure assessment, and epidemiological study (Beach Bum Survey). Environ Int 114:326–333. pii:S0160-4120(17)31234-5. https://doi.org/10.1016/j.envint.2017.11.003
Levine MM, Black R, Clements ML (1984) Pathogenesis of enteric infections caused by Vibrio. In: Colwell R (ed) Vibrios in the environment. Wiley, New York, pp 109–122
Lopez CE, Dykes AC, Juranek DD et al (1980) Waterborne giardiasis: a communitywide outbreak of disease and a high rate of asymptomatic infection. Am J Epidemiol 112:495–507
Lowry PW, Blankenship RJ, Gridley W et al (1991) A cluster of legionella sternal-wound infections due to postoperative topical exposure to contaminated tap water. N Engl J Med 324(2):109–113
Lund E, Nissen B (1986) Low technology water purification by bentonite clay flocculation as performed in Sudanese villages virological examinations. Water Res 20:37–43
Lynch MF, Blanton EM, Bulens S et al (2009) Typhoid fever in the United States, 1999–2006. JAMA 302(8):859–865
Mac Kenzie WR, Hoxie NJ, Proctor ME et al (1994) A massive outbreak in Milwaukee of cryptosporidium infection transmitted through the public water supply. N Engl J Med 331:161–167
Mahmoudi M-R, Ongerth JE, Karanis P (2017) Cryptosporidium and cryptosporidiosis: the Asian perspective. Int J Hyg Environ Health 220:1098–1109
Marchione M (1994) Crypto report cites faulty equipment. Milwaukee J July 21, A1–9
Mata L, Simhon A, Padilla R et al (1983) Diarrhea associated with rotaviruses, enterotoxigenic *Escherichia coli*, Campylobacter, and other agents in Costa Rican children, 1976–1981. Am J Trop Med Hyg 32:146–153
Mbithi JN, Springthorpe VS, Sattar SA (1993) Comparative in vivo efficiencies of hand-washing agents against Hepatitis A virus (HM-175) and poliovirus type 1 (Sabin). Appl Environ Microbiol 59:3463–3469

McCarthy SA, Khambaty FM (1994) International dissemination of epidemic *Vibrio cholerae* by cargo ship ballast and other nonpotable waters. Appl Environ Microbiol 60:2597–2601

McGirk T (2015) Taliban assassins target Pakistan's Polio vaccinators. Natl Geogr. https://news.nationalgeographic.com/2015/03/150303-polio-pakistan-islamic-state-refugees-vaccination-health. Accessed 5 Mar 2018

Melnick JL (1947) Poliomyelitis virus in urban sewage in epidemic and in nonepidemic times. Am J Hyg 45:240–253

Melnick JL, Emmons J, Coffey JH et al (1954) Seasonal distribution of coxsackie viruses in urban sewage and flies. Am J Hyg 59:164–184

Minor TE, Allen CI, Tsiatis AA et al (1981) Human infective dose determinations for oral poliovirus type 1 vaccine in infants. J Clin Microbiol 13:388–389

Morris JG Jr, Takeda T, Tall BD et al (1990) Experimental non-O group 1 *Vibrio cholerae* gastroenteritis in humans. J Clin Invest 85:697–705

Mosley JW (1966) Transmission of viral diseases by drinking water. In: Berg G (ed) Transmission of viruses by the water route. Wiley, New York, pp 5–23

Nalin DR (1976) Cholera, copepods, and chitinase. Lancet ii:958

Nalin DR, Levine RJ, Levine MM et al (1978) Cholera, non-vibrio cholera, and stomach acid. Lancet ii:856–859

Nash TE, Herrington DA, Losonsky GA et al (1987) Experimental human infections with *Giardia lamblia*. J Infect Dis 156:974–984

Ndayimirije N, Maregeya M, Nshimirimana D et al (1993) Epidemic cholera – Burundi and Zimbabwe, 1992–1993. Morb Mortal Wkly Rep 42:407–416

NSF International (2011) Household germ study. https://www.nsf.org/newsroom_pdf/2011_NSF_Household_Germ_Study_exec-summary.pdf. Accessed 9 Mar 2018

Ortega YR, Sterling CR, Gilman RH et al (1993) Cyclospora species – a new protozoan pathogen of humans. N Engl J Med 328:1308–1312

Paul JR, Trask JD, Gard S (1940) II. Poliomyelitic virus in urban sewage. J Exp Med 71:765–777

Payment P, Franco E (1993) *Clostridium perfringens* and somatic coliphages as indicators of the efficiency of drinking water treatment for viruses and protozoan cysts. Appl Environ Microbiol 59:2418–2424

Payment P, Trudel M, Plante R (1985) Elimination of viruses and indicator bacteria at each step of treatment during preparation of drinking water at seven water treatment plants. Appl Environ Microbiol 49:1418–1428

Payment PL, Richardson J, Siemiatycki R et al (1991) A randomized trial to evaluate the risk of gastrointestinal disease due to consumption of drinking water meeting current microbiological standards. Am J Public Health 81:703–708

Payment PJ, Siemiatycki L, Richardson G et al (1997) A prospective epidemiological study of gastrointestinal health effects due to the consumption of drinking water. Int J Environ Health Res 7:5–31

Plutzer J, Lassen B, Jokelainen P et al (2018) Review of Cryptosporidium and Giardia in the eastern part of Europe, 2016. Euro Surveill 23(4):16-00825. https://doi.org/10.2807/1560-7917.ES.2018.23.4.16-00825

Ramalingaswami V, Purcell RH (1988) Waterborne non-A, non-B hepatitis. Lancet 1:571–573

Regli S, Rose JB, Haas CN et al (1991) Modeling the risk from giardia and viruses in drinking water. J Am Water Works Assoc 83:76–84

Rendtorff RC (1954) The experimental transmission of human intestinal protozoan parasites. II. *Giardia lamblia* cysts given in capsules. Am J Hyg 59:209–220

Rodriguez WJ, Kim HW, Brandt CD et al (1983) Use of electrophoresis of RNA from human rotavirus to establish the identity of strains involved in outbreaks in a tertiary care nursery. J Infect Dis 148(1):34–40

Rosado-García FM, Guerrero-Flóreza M, Karanis G et al (2017) Water-borne protozoa parasites: the Latin American perspective. Int J Hyg Environ Health 220(5):783–798. https://doi.org/10.1016/j.ijheh.2017.03.008

Saulsbury FT, Winkelstein JA, Yolken RH (1980) Chronic rotavirus infection in immunodeficiency. J Pediatr 97(1):61–65
Schiff GM, Stefanovic' GM, Young EC et al (1984) Studies of echovirus-12 in volunteers: determination of minimal infectious dose and the effect of previous infection on infectious dose. J Infect Dis 150:858–866
Schurmann W, Eggers HJ (1985) An experimental study on the epidemiology of enteroviruses: water and soap washing of poliovirus 1-contaminated hands, its effectiveness and kinetics. Med Microbiol Immunol 174:221–236
Simhon A, Chrystie IL, Totterdell BM et al (1981) Sequential rotavirus diarrhoea caused by virus of same subgroup. Lancet ii:1174
Sorvillo FJ, Fujioka K, Nahlen B et al (1992) Swimming-associated cryptosporidiosis. Am J Publ Health 82:742–744
Stewart SJ (1991) Francisella. In: Balows A, Hausler WJ Jr, Herrmann KL et al (eds) Manual of clinical microbiology, 5th edn. American Society for Microbiolgy, Washington, pp 454–456
Swerdlow DL, Ries AA (1992) Cholera in the Americas: guidelines for the clinician. J Am Med Assoc 267:1495–1499
Swerdlow DL, Ries AA (1993) *Vibrio cholerae* non-01 – the eighth pandemic? Lancet 342 (8868):382–283
Tambini G, Andrus JK, Marques E et al (1993) Direct detection of wild poliovirus circulation by stool surveys of healthy children and analysis of community wastewater. J Infect Dis 168:1510–1514
Tauxe RV, Hargrett-Bean N, Patton CM et al (1988) Campylobacter isolates in the United States, 1982–1986. Morb Mortal Wkly Rep 37(SS-2):1–13
Taylor FS (1942) The conquest of bacteria. Philosophical Library, New York
Taylor JP, Perdue JN, Dingley D (1985) Cryptosporidiosis outbreak in a day-care center. Am J Dis Child 139:1023–1025
Transnational Development Clinic (2013) Peacekeeping without accountability the United Nations' responsibility for the Haitian cholera epidemic. https://law.yale.edu/system/files/documents/pdf/Clinics/Haiti_TDC_Final_Report.pdf. Accessed 9 March 2018
Turney C, Green-Smith M, Shipp M et al (1994) *Escherichia coli* 0157:H7 outbreak linked to home-cooked hamburger – California, July 1993. Morb Mortal Wkly Rep 43:213–216
United States Environmental Protection Agency (1997) Exposure factors handbook. United States Government, Washington. https://cfpub.epa.gov/ncea/risk/recordisplay.cfm?deid=12464. Accessed 9 March 2018
van Lieverloo JH, Hoogenboezem W, Veenendaal G et al (2012) Variability of invertebrate abundance in drinking water distribution systems in the Netherlands in relation to biostability and sediment volumes. Water Res 46(16):4918–4932
van Wijngaarden JK, van Loon AM, Oostvogel P et al (1992) Update: poliomyelitis outbreak – Netherlands, 1992. Morb Mortal Wkly Rep 41:917–919
Viswanathan R (1957) Epidemiology. Indian J Med Res 45(Suppl):1–29
Ward RL, Bernstein DI, Young EC et al (1986) Human rotavirus studies in volunteers: determination of infectious dose and serological response to infection. J Infect Dis 154:871–880
Ward RL, Bernstein DI, Shukla R et al (1989) Effects of antibody to rotavirus on protection of adults challenged with a human rotavirus. J Infect Dis 159:79–88
Weissman JB, DeWitt WE, Thompson J et al (1974) A case of cholera in Texas, 1973. Am J Epidemiol 100:487–498

Chapter 4
The Ecology of Bacterial Agents of Foodborne Illness

Alexander Gill and John W. Austin

Abstract Eating and drinking are essential to human survival but come with an inherent level of risk. To intake nutrition, humans must introduce a foreign substance into their bodies, and this substance may be contaminated by either infectious microorganisms or microbial toxins, threats which are undetectable to human senses. Bacterial agents of foodborne illness can be divided into two groups, foodborne pathogens which infect humans and producers of foodborne toxins. For the microbiologist concerned with the bacterial agents of foodborne illness, an understanding of the ecology of these organisms is fundamental to addressing the practical issue of reducing human exposure to these organisms or their toxins. It is important to note that bacterial agents of foodborne illness are not obligate human pathogens. Humans are not a definitive, or even necessarily a desirable, host for these microorganisms, and the microorganisms may thrive in a variety of environments without ever coming into contact with human populations. Thus, the potential for these organisms to cause human illness is dependent on ecological conditions that bring the human and pathogen into contact.

4.1 Introduction

Though the nature of foodborne diseases and probability of risk vary regionally, depending upon the local hygiene infrastructure and food preparation customs, there is no region in which foodborne illness is not a significant contributor to morbidity and mortality. It has been estimated that, in the United States, 48 million people suffer from foodborne illness each year, with the economic cost exceeding $15.5 billion dollars annually (Hoffmann et al. 2015). In countries with weak public health and hygiene infrastructure, the impacts can be more severe. While the full burden of

A. Gill · J. W. Austin (✉)
Microbiology Research Division, Bureau of Microbial Hazards, Food Directorate, Health Products and Food Branch, Health Canada, Ottawa, ON, Canada
e-mail: john.austin@canada.ca

C. J. Hurst (ed.), *The Connections Between Ecology and Infectious Disease*, Advances in Environmental Microbiology 5,
https://doi.org/10.1007/978-3-319-92373-4_4

foodborne illness in low- and middle-income countries is not known (Grace 2015), the African, Southeast Asian, and Eastern Mediterranean regions have been reported to have the highest foodborne illness burden (Havelaar et al. 2015). In addition to the human cost of illness, food producers and distributors can suffer significant financial losses in responding to contamination events. The potential scale of such losses is illustrated by the proposal by European Union officials in 2011 to provide 150 million € in compensation to vegetable producers whose products were destroyed in the response to an outbreak of pathogenic *Escherichia coli*. In this chapter we discuss how an understanding of the ecology of the bacteria which cause foodborne illness can be used to identify potential food safety risks and strategies to prevent human exposure.

The term "oecologie" (ecology) was introduced to the scientific community by the German zoologist Ernst Haeckel in 1866 (Haeckel 1866). Based on the root of the Greek word *oikos* meaning house, he wrote: "By ecology we mean the body of knowledge concerning the economy of nature—the investigation of the total relations of the animal both to its inorganic and organic environment." Thus, ecology can be understood as the study of the interrelationships that occur between organisms and between organisms and their physicochemical environment.

To simplify our discussion of this topic, we have divided the ecology of foodborne bacterial pathogens into two domains. These domains are the ecology of the bacterial agent in the food production system and the ecology of the bacterial agent in the food product. The ecology of bacteria in the food production system includes identification of environments and hosts (plant and animal) which are potential sources of the bacterial agents, vectors for transmission onto food ingredients, and environmental factors which determine the potential for survival and replication of the organisms within the food processing environment. The ecology of bacterial agents in the food product is dependent upon those factors which determine the microbes' survival and replication in foods and includes factors intrinsic to food (pH, water activity, oxidation-reduction potential, nutrients, and antimicrobials), extrinsic to food (temperature, relative humidity, and gaseous environment), and interrelationships among microorganisms. The chapter concludes with several examples of foodborne pathogenic bacteria and their ecology in foods and food production systems.

4.2 The Ecology of Pathogens in the Food Production System

Bacteria may be introduced to foods at multiple points in the chain of production, processing, distribution, and consumer preparation. Unless controlled, once introduced bacteria may replicate on foods, resulting in an amplification of health risk and the potential for further distribution of the pathogen to other foods or food contact surfaces. Distribution and preparation are the stages in which the prevention of

Table 4.1 Minimum and optimum growth conditions of selected foodborne pathogens

Organism	Temperature (°C) (minimum/optimum)	pH (minimum/optimum)	Minimum water activity
Bacillus cereus	4/30–40	4.8–5.0/6.0–7.0	0.93
Campylobacter jejuni	30/42–43	4.8/6.5–7.5	0.987
Clostridium botulinum Group I	10/35–40	4.6/7	0.94
Clostridium botulinum Group II	3.0/28–30	5/7	0.97
Clostridium perfringens	4/43–47	5/7.2	0.93–0.97
Escherichia coli	7–8/35–40	4.4/6–7	0.95
Listeria monocytogenes	0/37	4.6–5.0/7.0	0.90–0.93
Salmonella	2–7/35–43	3.8/7.0–7.5	0.93–0.94
Staphylococcus aureus	6.7–7/37	4.0/6.0–7.0	0.83–0.85
Yersinia enterocolitica	−1.3/25–37	4.2/7.2	0.95

Minimum/optimum data from ICMSF (1996) and Stern et al. (1980)

contamination is in principle most easily achieved and safety of the product is largely dependent on appropriate packaging and handling. In contrast production and processing are the stages where initial contamination is most likely and in which process hygiene is most difficult to control.

4.2.1 Agricultural Environment

The agricultural environment is a common origin for foodborne pathogens. Infectious foodborne pathogens are adapted to colonize warm-blooded hosts, with optimum growth rates typically either at or near body temperature (Table 4.1). Host animals provide an opportunity for bacteria to replicate to high concentrations in a nutrient-rich environment. The pathogen may be dispersed between livestock, wild animals, and humans by direct contact or indirectly via contaminated water or feed. Consequently, though foodborne pathogens may be distributed throughout the environment, replication within animal hosts plays an important role in their survival and distribution.

4.2.1.1 Poultry and Red Meats

The muscle tissue of healthy animals is effectively sterile; thus, the presence of foodborne pathogens on meat is dependent upon transfer to exposed tissue surfaces during carcass dressing and processing. Pathogens transferred to meat during these processes typically originate from populations colonizing the outer surface (skin, hide, and feathers), the gastrointestinal tract, or the esophagus (Gill and Gill 2012).

The probability of such transfer is dependent upon the hygiene of carcass dressing and breaking. The degree of hygiene achievable with good manufacturing practices varies between animals and with the diligence of individual operators. Microflora transfer is easier to control with large ruminants such as cattle than with pigs or poultry. The expectations of process hygiene are reflected by pathogen testing standards for different meat products (Gill and Gill 2014). Exceptions to this rule are the obligate pathogens *Brucella* and *Mycobacterium* which infect the internal tissues, but these bacteria cannot replicate in the tissue of dead animals. Some specialized products are prepared from tissues which, unlike muscle tissue, have a significant native microflora (e.g., pig's head, chitterlings). The initial microflora of these products cannot be controlled by harvesting practice.

4.2.1.2 Seafood

As with red meats and poultry, the flesh of healthy finfish, molluscs, and crustaceans is essentially sterile, but bacterial pathogens may be present on the skin, gills, gastrointestinal tract, or other surfaces exposed to the environment. Processing may remove contaminated tissues, but typically no effort is made to prevent the transfer of microflora to other tissues during processing operations.

Of the foodborne pathogens associated with seafood, *Vibrio* spp. and *Clostridium botulinum* type E are native to aquatic environments. The presence of foodborne pathogens which cannot replicate but can survive for extended periods in natural waters (*Listeria monocytogenes*, *Salmonella*, and *Yersinia*) is indicative of contamination of water systems with agricultural runoff or sewage (Novoslavskij et al. 2016).

4.2.1.3 Other Foods of Animal Origin

Potentially any pathogens which colonize dairy livestock, or are present in the production environment, may contaminate milk. Milk is invariably contaminated with microflora present on the epidermis of the udder. This may include skin flora such as Staphylococci and enteric organisms such as *E. coli*, *Salmonella*, and *Campylobacter* (Oliver et al. 2005). Microflora may also be introduced from milking equipment or from infected udder tissues. These sources of contamination can be minimized, but not eliminated, by good hygienic practice and maintenance of animal health.

The shell of eggs may become contaminated after laying, with a wide variety of organisms from the production environment. The eggshell presents an exposed environment, unsuitable for the survival of most pathogens, and the shell can be decontaminated by chemical wash treatments. The primary pathogen associated with eggs is *Salmonella*, which can colonize the reproductive tract and be vertically transmitted during egg formation (Keller et al. 1995). *Salmonella* can also be internalized to the egg through pores in the shell, and this route has been proposed for other pathogens, including *Campylobacter* (Cox et al. 2012).

4.2.1.4 Fruits and Vegetables

The control of contamination during growth and harvest is a particular concern with fruits and vegetables which are consumed raw, without the removal of an external skin or rind. Options for postharvest decontamination are highly limited; at a commercial scale, the only widespread method is washing with chlorine (Erickson 2012). Washing with sanitizers is not particularly effective and may only serve to redistribute pathogens, spreading contamination from one or two plants to a whole production lot (Zhang et al. 2009; Danyluk and Schaffner 2011). Consequently, products such as lettuce, spinach, kale, herbs, tomatoes, cantaloupes, and watermelon are increasingly recognized as causes of bacterial foodborne illness (Erickson 2012). In the United States in 2012, plant products were identified as a vehicle for 33% of foodborne outbreaks, accounting for 42% of total cases of reported foodborne illness (Bennett et al. 2014).

Contamination of fruits and vegetables with bacterial pathogens may occur before or following harvest. Prior to harvest bacteria can be introduced from the soil; manure; irrigation water; inputs such as fungicides, insecticides, and fertilizers; fecal shedding by wild and domestic animals (including birds, reptiles, and insects); and human sewage (Olaimat and Holley 2012). During harvesting bacterial pathogens may be introduced from equipment and wash water and by workers with inadequate hygiene (Olaimat and Holley 2012).

The bacterial pathogens primarily associated with outbreaks involving fruits and vegetables are *Salmonella enterica*, *E. coli*, *L. monocytogenes*, *Shigella*, and *Listeria monocytogenes*, along with *Clostridium* and *Bacillus* spp. which are commonly part of the microflora of soil and plants. It should be noted that of these listed microorganisms, *Shigella* are host specialized for humans and are not well adapted for prolonged survival in external environments; therefore, the presence of *Shigella* is indicative of direct or indirect contamination with human sewage. Pathogenic *E. coli* and *Salmonella enterica* have a wide host range and can survive for prolonged periods outside of an animal host but are not thought to constitute part of the normal microflora of plants. The risk posed by these pathogens can be reduced by efforts to control exposure of crops to domestic and wild animals and by strict monitoring and control over the hygiene of irrigation and wash waters.

The risk of illness is not equal for all commercially exploited species of plants. Sprouts are generally recognized as posing a particular risk as the seeds are germinated and grown under conditions of temperature and humidity which will support the replication of enteric pathogens such as *Salmonella enterica* and *E. coli*. The largest reported foodborne outbreak of verotoxigenic *E. coli* occurred in Europe in 2011. In Germany alone there were over 3800 cases of serious illness and 53 deaths (Radosavljevic et al. 2015). The outbreak was eventually traced to fenugreek sprouts (Radosavljevic et al. 2015).

While small populations of endophytic microorganisms have been recovered from some plant species (Samish and Etinger-Tulczynska 1963), it was thought that the internal tissues of plants are essentially sterile, and spoilage and pathogenic bacteria were localized to the plant surface. More recently, microorganisms responsible for foodborne illness have been demonstrated to penetrate the internal tissues of a wide range of plants, including leafy greens, tomatoes, mung beans, pecans, almonds, and some fruits (Deering et al. 2012). Internalization may occur via leaf stomata or lateral root junctions, but only when high concentrations of enteric pathogens are present. Internalized bacterial cells are protected prior to harvest from UV light and desiccation and are also protected from postharvest decontamination efforts (Olaimat and Holley 2012). The significance of enteric pathogens internalized to plant tissues as a cause of foodborne outbreaks remains contentious.

The risk associated with fruits and vegetables may be on the increase due to changes in the way these products are processed and consumed. In some markets an increasing proportion of fruit and vegetable products are sold as prepared convenience foods, such as premade salads. These products are minimally processed with vegetables and fruits washed and precut. Tissue damage from cutting or peeling releases nutrients that can support bacterial replication (Harris et al. 2003). Packaging can maintain product humidity to reduce drying stress on the product, but this will simultaneously protect contaminating pathogens from dissection stress and possibly maintain water activity above the minimum threshold for growth (Harris et al. 2003). Under such conditions there is a significant risk of amplification of pathogen numbers on contaminated product during storage and distribution, particularly if storage temperature is not adequately controlled.

4.2.2 Processing Environment

The degree of processing that foods undergo prior to distribution to consumers is highly variable. The spectrum ranges from fresh fruits and vegetables that may simply have inedible material trimmed away and be washed to remove soil and dust to products composed of multiple ingredients which undergo complex multistep processing. An example of the latter is salami and other traditional fermented meats, which are composed of fat and muscle tissue from multiple animal species, starter culture, multiple flavoring spices, carbohydrate, and salts. Processing of these fermented meat products often consists of chopping raw meat, mixing ingredients, stuffing the resulting batter into casings (animal intestine or synthetic), fermentation, drying, and smoking. The final microflora of the product is dependent upon these processes. In the course of processing, there are multiple opportunities for microorganisms to be transferred between the food, ingredients, processing equipment, and environmental surfaces.

Though processing environments are complex and vary considerably, there are some principles to consider which determine the risk potential for pathogens, specifically whether the level of risk is constant, reduced, or amplified. A potential for risk amplification exists because the food itself provides nutrients to support bacterial replication. It is not uncommon that replication of bacteria can occur during processing. The source of such bacteria may be the food itself, or transferred from equipment or environmental surfaces, including the clothing and protective equipment of workers (Yang et al. 2015; Youssef et al. 2013).

Whether a pathogen can replicate during food processing depends upon the environmental conditions and the adaptability of the specific organism. Pathogens may colonize the processing environment and replicate in transient communities or form persistent communities which can serve as an ongoing source of contamination. Pathogens commonly associated with this behavior include *E. coli* (Yang et al. 2015), *Salmonella enterica* (Podolak et al. 2010), and *L. monocytogenes* (Malley et al. 2015).

Bacteria transferred onto equipment surfaces in the processing environment will initially attach as single cells. Such single cells are relatively vulnerable to desiccation and removal by sanitation processes, but if growth of the bacteria occurs, a bacterial biofilm can form. Biofilms are complex communities of bacterial cells embedded in an extracellular matrix which form at the interface between two phases (solid-liquid, solid-air, liquid-air) (Stoodley et al. 2002; Costerton et al. 1995). Attachment of *Salmonella* to food processing surfaces and to other *Salmonella* may occur via amyloid fibers produced on the cell surface of *Salmonella* and other enteric bacteria that facilitate adherence to biotic and abiotic surfaces, leading to biofilm development (Austin et al. 1998b). Foodborne pathogens may form biofilms alone or be members of mixed species biofilms, in either case the cells which compose the biofilm typically display far greater resistance to stresses such as antimicrobials, desiccation, oxygen levels, and physical removal (Giaouris et al. 2015).

Control of pathogens in the food processing environment can be achieved through controlling the hygiene of the food production system. The specific details of a successful hygiene control system must match the specific physical site, production inputs, and the needs of the processor. It should be developed and applied in a systematic manner, such as Hazard Analysis and Critical Control Point (Hulebak and Schlosser 2002). However, four basic principles can be identified: prevention or minimization of bacterial transfer, prevention of bacterial replication, decontamination of the food, and sanitation of the production environment, including equipment and surfaces.

Bacteria can be transferred by any physical contact between solid surfaces, fluids, and aerosols generated by processing operations. The transfer of bacteria may introduce pathogens to foods or surfaces in the processing environment from which pathogens can circulate back to the food. Transfer can be minimized by the use of physical barriers (separation of operations, closed pipes), minimizing processing operations, and minimizing the distance food must travel through the processing environment.

There are limited means to control the growth potential of bacteria on the food itself during processing. The availability of free water can be minimized, by keeping processing as dry as possible. Free water may be present as a component of the raw food, as an exudate, added as an ingredient or used to rinse the food and production surfaces. Temperature may be reduced to control growth potential but must also be high enough to provide a tolerable working environment.

Food products can be decontaminated to reduce pathogen numbers present. Decontamination is not possible with all foods as there are limitations on the effective treatments available due to the necessity to preserve sensory qualities or to conform to with regulatory requirements. To be successful, decontamination must be combined with prevention of bacterial transfer to prevent recontamination. An example of this approach is milk pasteurization. Milk cannot be harvested aseptically and will inevitably be contaminated with microflora. The raw milk is pumped through a pasteurizer, which applies a thermal treatment which is typically sufficient to kill pathogens associated with milk. The pasteurized milk is then transferred through a closed system to directly fill containers for commercial distribution.

The purpose of the fourth principle, sanitation of processing surfaces, is commonly thought of as killing microorganisms, but the primary role of sanitation is to deny habitat to bacteria in the processing environment. Ideally, equipment and surfaces should be designed for "cleanability," which is a combination ensuring exposure to sanitation treatments and physical removal of soil and attached cells. The removal of organic material reduces the potential for bacterial attachment and biofilm formation. Cell attachment is promoted by a coating of organic material, and nutrients are required for microbial cell replication and biofilm formation. Similarly, water is essential for the survival and replication of bacteria, and so drying of surfaces can effectively control pathogen colonization. Depending upon the type of equipment, it either may be cleaned in place (CIP) or dismantled for cleaning. Available CIP regimes vary in effectiveness in eliminating adherent bacteria (Austin and Bergeron 1995; Bremer et al. 2006), depending upon the cleaning agent used, contact time, cleaning temperature, and design of the processing equipment. Though less commonly used, equipment surfaces can also be decontaminated by physical processes such as heat (dry or steam) or UV light.

4.3 Survival and Replication of Bacteria in Foods

The study of microbial ecology of foods began in the 1930s. Haines (Haines 1933; Haines and Elliot 1944) and later Ingram (1949) and Westerdijk (1949) were the pioneers in microbial ecology of foods. Many of the original studies of the primary factors affecting food spoilage, such as temperature, pH, water activity, redox potential, and gaseous environment, were carried out in the two decades following the Second World War (Barnes 1994). Prior to this pioneering work on microbial ecology of foods, it was thought that most species of microorganisms could be isolated from nearly any food. This belief was based upon the isolation of microorganisms, in the absence of information on their relative abundance (Mossel et al. 1995). Westerdijk (1949), while examining spoilage fungi on decomposing organic substrates, was the first to apply the idea of "association," or a typical microflora associated with specific substrates.

4.3.1 Properties Intrinsic to Foods

Characteristics of foods play an important role in controlling bacterial growth. These characteristics are intrinsic to the food and include pH of the food, availability of water, oxidation-reduction potential of the food, the presence of naturally occurring antimicrobials, and interactions between microorganisms within the food. Foodborne microorganisms require specific conditions for growth (Table 4.1). Control of the growth of these microorganisms is often achieved by a combination of factors rather than a single factor. The use of combinations of inhibitory factors, such as refrigeration temperature, reduced water activity, lowered pH, and low redox potential to control growth of foodborne microorganisms, is referred to as "hurdle technology" (Leistner and Gould 1994).

4.3.1.1 pH

Acidification of food by the addition of organic acids is the primary means of preventing the growth of human pathogens in a wide range of fermented and acidified ready-to-eat foods. Most microorganisms common to foods grow optimally at neutral pH (pH values from 6.6 to 7.5). The minimum pH limits for the growth of selected food pathogenic bacteria are listed in Table 4.1.

Acid tolerance of microorganisms is dependent upon several factors including the nutrient content and physical nature of the food itself, temperature, nature of the acidulent, the presence of preservatives, water activity, and redox potential. Some foodborne bacteria exhibit an acid tolerance response that allows them to survive typically lethal levels of acidity (Rowbury and Goodson 1998; Alvarez-Ordóñez et al. 2015). The minimum pH allowing growth of bacteria is usually dependent upon the type of acid used to adjust the pH. Hydrochloric acid and citric acid often allow growth at lower pH than acetic acid or lactic acid (Smelt et al. 1982; Álvarez-Ordóñez et al. 2010).

4.3.1.2 Water Activity

Water activity (a_w) may be thought of as the free water within a food which is available to microorganisms. Within foods, water is present in both bound and free forms. The bound water is involved in the hydration of molecules and dissolution of ions in foods and is not available for microbial growth. The free water is not involved in interactions with the food and its constituents and is available for microbial growth. Water activity is defined as the ratio of the water vapor pressure of the food (p) relative to the vapor pressure of pure water ($p0$) at the same temperature, i.e., $a_w = p/p0$. Values for water activity range from 0 to 1 (Hurst et al. 1979).

The water activity of foods can be reduced by reducing the moisture content (drying) or by the addition of solutes. In traditional food preservation, the solutes most commonly used are sodium chloride or sugar, the latter often in the form of either sucrose or honey. The solute used to control a_w may influence the lower a_w limits for growth. In the case of *C. botulinum*, NaCl, KCl, glucose, and sucrose show similar effects, while glycerol allows growth at lower a_w (Sperber 1983). The minimum a_w required for growth may be raised significantly by other factors, such as increased acidity or the use of preservatives (Baird-Parker and Freame 1967).

Decreasing water activity is an effective means of preventing the growth of spoilage organisms and bacterial pathogens on many processed or prepared foods. The water activity of most fresh foods including meat, fish, fruits, vegetables, milk, and eggs is 0.98–0.99. Breads, fruit juices, salted fish, sausages, and cheeses have water activities in the range of 0.93–0.98. Fermented or dry-cured meats have water activity in the range of 0.85–0.93. Fruit preserves typically have water activities of 0.75–0.80, and honey, dried egg, and chocolate have water activities below 0.60. The limiting water activity for most spoilage bacteria is 0.91, and most foodborne pathogens require a minimum water activity for growth of 0.93 or higher (Table 4.1). Exceptions are *Staphylococcus aureus*, which may grow at water activities as low as 0.83, and *Listeria monocytogenes* which may grow at water activities as low as 0.90. However, low water activity alone cannot ensure safety as some foodborne pathogens such as *Cronobacter*, *Salmonella enterica*, and *Listeria monocytogenes* are desiccation resistant and can survive for prolonged periods in foods with water activities well below the minimum required for growth.

4.3.1.3 Oxidation-Reduction (Redox) Potential

Oxidation-reduction potential is the ease with which a substrate loses or gains electrons. When a substrate loses electrons, it becomes oxidized, while if a substrate gains electrons, it is reduced. Oxidation-reduction reactions are coupled reactions in which one substrate is oxidized and a second substrate is simultaneously reduced. The electron donor, since it reduces an oxidized substrate, is a reducing agent. The electron acceptor, since it oxidizes a reduced substrate, is an oxidizing agent. The redox potential, termed the "Eh," is measured in millivolts (mV). Oxidized substrates have larger mV values (+mV), while reduced substrates have lower values (−mV).

The redox potential of a food is dependent upon the chemical composition of the food and the storage atmosphere. Fresh foods undergoing respiration will be in the reduced state. The presence of reducing substances such as ascorbic acid, reducing sugars, and the sulfhydryl group of amino acids also contributes to a reduced state. When respiration stops, and oxygen diffuses in, fresh foods reach an oxidized state. Foods stored in air will have a larger Eh (+mV) than foods stored either under vacuum or in a modified atmosphere containing elevated levels of either CO_2 or N_2 or reduced O_2.

While it is commonly assumed that *C. botulinum* cannot grow in foods exposed to O_2, the Eh of most foods is usually low enough to allow its growth (Kim and Foegeding 1993). *Clostridium botulinum* grows optimally at an Eh of −350 mV, but its growth initiation may occur in the range of +30 to +250 mV (Kim and Foegeding 1993).

4.3.1.4 Natural Antimicrobials in Foods

Foods are highly complex in their chemical composition and may contain compounds with antimicrobial activity at sufficient concentration to impact the growth and survival of bacteria. The list of natural antimicrobials in foods is extensive, and a variety of naturally occurring antimicrobial systems include those derived from both animal and plant tissues.

Agents considered to be natural antimicrobials are diverse and include enzymes such as lysozyme and lactoperoxidase; iron-binding proteins such as lactoferrin, ovotransferrin, and lactoferricin; low molecular weight hydrophobic molecules and phenolics from plants such as eugenol, carvacrol and thymol, and phytoalexins; and peptides produced by bacteria such as nisin (Beuchat and Golden 1989; Juneja et al. 2012; Holley and Patel 2005). Some of these agents are established in commercial use, such as lysozyme and nisin which are both used to control growth of Gram-positive bacteria. Others are part of the normal composition of foods and may impact the growth and survival of microflora, including pathogens. Examples of this class are lactoferrin in milk (Masson and Heremans 1971; Rainard 1986), ovotransferrin in eggs (Wu and Acero-Lopez 2012), and the plant oil phenolics (Holley and Patel 2005).

4.3.1.5 Interactions Between Microorganisms

Foods typically have a complex microflora which can significantly impact the survival and replication of foodborne pathogens by either inhibiting or enhancing growth.

Fermentation is one of the oldest forms of food preservation. Fermented foods such as bread, cheese, and wine have been produced for thousands of years and are strongly linked to culture and tradition. Lactic acid inhibits bacterial growth by lowering the pH of food. Lactic acid bacteria (LAB), those which produce lactic acid, include the genera *Lactobacillus*, *Lactococcus*, *Pediococcus*, and *Leuconostoc*. The various species of LAB produce lactic acid by utilization of hexoses via the Embden-Meyerhof pathway. The fermentative processes involved in the manufacture of sauerkraut, olives, yoghurt, and sausage all produce lactic acid as the primary acid (Blom and Mortvedt 1991). In addition to lactic acid, antimicrobial metabolic products produced during fermentations include propionic and acetic acids, diacetyl, hydrogen peroxide, fatty acids, aldehydes, and other materials such as bacteriocins.

Destruction of competing flora may lead to unrestricted growth of a pathogen. "Pasteurization" of fish by irradiation enhances the growth of *C. botulinum* (Cann et al. 1966), probably by reduction of the normal spoilage flora which would compete with *C. botulinum*. Microorganisms with high metabolic activity may consume nutrients making them unavailable to less active organisms. Streptococci may inhibit *Staphylococcus* by consuming the B vitamins nicotinamide, niacin, and biotin (Iandolo et al. 1965; Haines and Harmon 1973).

In some cases, microorganisms may facilitate the survival or growth of foodborne pathogenic bacteria. For example, some species of free-living protozoans—such as *Acanthamoeba castellanii*, *A. polyphaga*, and *Tetrahymena pyriformis*—can act as hosts of pathogenic bacteria (Vaerewijck et al. 2014). The simultaneous finding of free-living protozoans and several enteric pathogenic bacteria on dishcloths has been used to suggest the possibility that the contaminating bacteria may have survived within the protozoans (Chavatte et al. 2014).

Metabolic activity by other flora can result in changes to the physicochemical properties of food that in turn will promote growth of the pathogen. The pH of acidic foods can be increased by the metabolic activity of some microflora, raising the pH above the level which restricts pathogen growth. For example, the growth of molds has been reported to increase the pH of tomato products sufficiently to permit the growth of *C. botulinum* (Huhtanen et al. 1976). A similar metabiotic relationship has been demonstrated where growth of proteolytic molds in tomatoes (Wade and Beuchat 2003; Wade et al. 2003) or cantaloupe (Richards and Beuchat 2005) increases the pH and promotes the growth of *Salmonella enterica*.

4.3.2 Properties Extrinsic to Foods

4.3.2.1 Temperature

All microorganisms have temperature-defined growth limits. Cell replication ceases below a minimum temperature, maximum growth rate is observed at an optimum temperature, and there is a maximum growth temperature above which replication ceases and thermal inactivation of the microbial cells occurs. Foodborne bacterial pathogens are almost exclusively mesophilic, with their optimum growth temperatures being between 30 and 40 °C and their minimum growth temperatures typically around 5 °C (Table 4.1).

Maintaining either food or the processing environment below the minimum growth temperature by refrigeration or freezing is an effective method of preventing pathogen replication and toxin production. To be an effective control on pathogens, temperature must be maintained below the growth minimum throughout food processing and storage. Given the complexity of food production and distribution, this temperature requirement can be challenging, and significant growth or toxin production can occur within even short periods of intermittent temperature increase. Similarly, in processing environments where the ambient temperature is maintained below the growth minimum, microbial replication can occur on equipment or surfaces that are heated by either friction or heat conduction.

Thermal treatments such as cooking and pasteurization are among the most common means used to decontaminate foods. The spores of *Bacillus* and *Clostridium* are considerably more heat resistant than either their own vegetative cells or the vegetative cells of food associated flora. Consequently, there is a potential for these spores to survive heat treatment and then outgrow in foods once the food has cooled (Gilbert et al. 1974).

4.3.2.2 Gaseous Environment

Modified atmosphere packaging (MAP) is increasingly used to extend the shelf life and improve the quality of foods. MAP systems include vacuum packaging and packaging under defined gas mixtures utilizing oxygen, carbon dioxide, nitrogen, and argon (Gill and Gill 2010). It is known that MAP achieves shelf life extension primarily by reducing the availability of oxygen to aerobic microorganisms, thereby inhibiting the growth of spoilage bacteria. Carbon dioxide is inhibitory to the growth of bacteria by dissolving in water to form carbonic acid, lowering the pH (Gill and Gill 2005). Nitrogen and argon are inert gasses, which are used to flush out oxygen and provide a headspace (Gill and Gill 2005; Spencer 2005).

The bacterial pathogens of concern with MAP foods, particularly ready-to-eat foods, are *C. botulinum* and *L. monocytogenes*. Replication of *Enterobacteriaceae*, including pathogens such as *E. coli*, *Salmonella enterica*, *Shigella*, and *Yersinia*, is inhibited at pH 5.8 or below under anaerobic conditions (Gill and Gill 2005). A pH below 5.8 is typical of fresh red and fermented meats.

MAP ready-to-eat foods present a risk for botulism, as the low-oxygen environment of MAP allows the growth of *C. botulinum* and inhibits the growth of spoilage organisms. Thus, the consumer cannot rely on spoilage to indicate the food is not safe for consumption. Ready-to-eat foods, by definition, do not receive a heating process prior to consumption. Any neurotoxin that may be formed during storage of the food is not inactivated by cooking. The high incidence of type E spores in fish (Dodds and Austin 1997) has made the safety of MAP smoked fish a particular concern. *Clostridium botulinum* can grow and produce neurotoxin in MAP fish, and, depending on conditions, neurotoxin may be present before the fish is considered spoiled (Dufresne et al. 2000).

Table 4.2 Incidence rate of illness from selected foodborne pathogens, per 100,000

Pathogen	USA (2014)	Canada (2013)	European Union (2013)	Australia (2011)
Brucella	NA	0.05	0.11	NA
Campylobacter	13.45	29.14	64.8	115.8
Listeria	0.24	0.36	0.44	0.3
Salmonella	15.45	17.57	20.4	54.3
Shigella	5.81	1.94	NA	2.2
Verotoxigenic *Escherichia coli*	2.35	1.8	1.59	0.4
Vibrio	0.45	NA	NA	NA
Yersinia	0.28	NA	1.94	NA

NA: Not available

USA 2014—CDC Food, Foodborne Diseases Active Surveillance Network (FoodNet), United States http://www.cdc.gov/foodnet/trends/tables-2014.html

Canada 2013—Public Health Agency of Canada—Notifiable Diseases On-Line—http://www.cdc.gov/foodnet/trends/tables-2014.html

European Union 2013. European Food Safety Authority. The European Union summary report on trends and sources of zoonoses, zoonotic agents and food-borne outbreaks in 2013. EFSA Journal 2015;13(1):3991. doi: https://doi.org/10.2903/j.efsa.2015.3991

Australia 2011. The OzFoodNet Working Group. 2015. Monitoring the incidence and causes of diseases potentially transmitted by food in Australia: Annual report of the OzFoodNet network, 2011. Commun Dis Intell 2015;39(2):E236–E264

4.4 Case Studies

Both the reported incidence rate (Table 4.2) of specific foodborne diseases and their association with foods (Table 4.3) vary between pathogen species. Below is presented a series of case studies in which the relationship between selected foodborne pathogens and the foods they are associated with is discussed. These case studies are intended to illustrate how the hazards associated with particular combinations of pathogen and food can be understood, including why some combinations can be effectively controlled, whereas others remain a challenge.

In some of the case studies below, a value for infectious dose is presented as an indicator of the relative risk of infectivity by the pathogens. Infectious dose is not the minimum number of pathogen cells necessary to cause illness but is the number of pathogen cells required to result in a high probability (50–100%) of infecting the host (Hurst et al. 1996). Infection in volunteer exposure studies may be determined by an increase in antibody titer or shedding of the pathogen. Estimates may also be made from outbreak investigations, but these results are not consistent with volunteer studies and often indicate lower infective doses (Hurst et al. 1996; Todd et al. 2008).

The reciprocal of the infectious dose can be viewed as the probability that a single cell can establish an infection (Hurst et al. 1996). For example, if a potential host is exposed to a single cell of a pathogen with an infectious dose of 1000 cells, the probability of infection is 0.001. Variables known to affect the probability of

Table 4.3 Identified vehicles of foodborne disease outbreaks reported to the US-CDC (1998–2013) (http://wwwn.cdc.gov/foodborneoutbreaks/)

			Meat (%)	Fresh fruit and vegetables (%)	Dairy (%)	Seafood (%)	Eggs (%)	Cold prepared [a] (%)	Notable associations (%)
Infectious pathogens									
Campylobacter	Outbreaks	245	30.6	9.0	51.0	2.9	0	4.1	Poultry (21.2)
	Cases	6101	15.3	11.6	59.3	8.2	0	4.2	Poultry (7.9)
Pathogenic *E. coli*	Outbreaks	326	48.2	27.9	8.3	1.2	0	3.7	Beef (39.0)
	Cases	9127	29.6	39.8	6.9	0.8	0	12.0	Beef (20.4)
Listeria	Outbreaks	35	20.8	10.4	33.3	4.2	0	2.1	
	Cases	648	31.1	26.2	21.3	1.0	0	8.0	
Salmonella	Outbreaks	1265	41.3	17.4	7.0	5.1	10.0	2.1	Poultry (23.0)
	Cases	44,852	33.1	29.3	5.6	2.9	11.1	1.5	Poultry (18.6)
Shigella	Outbreaks	155	28.2	28.2	2.8	8.5	0	23.9	
	Cases	7278	15.4	46.6	2.7	1.8	0	31.9	
Vibrio	Outbreaks	126	0.8	0	0	98.4	0	0	Raw seafood (52.4)
	Cases	1617	2.9	0	0	96.8	0	0	Raw seafood (48.6)
Yersinia enterocolitica	Outbreaks	11	90.9	0	9.1	0	0	0	Pork (81.8)
	Cases	103	84.5	0	15.5	0	0	0	Pork (75.7)
Foodborne intoxication									
Bacillus cereus	Outbreaks	440	35.9	6.6	2.5	3.9	0	6.4	Rice (27.7)
	Cases	5706	44.6	4.1	3.7	4.8	0	4.5	Rice (23.8)
Clostridium botulinum[b]	Outbreaks	47	8.5	2.1	0	38.3	4.3	0	Preserved vegetables, fish, traditional northern foods, prison-prepared alcohol (pruno)
	Cases	166	7.2	1.2	0	30.7	3.0	0	
Clostridium perfringens	Outbreaks	729	67.8	3.7	1.5	2.2	0	1.5	Sauces, beef, poultry
	Cases	26,122	72.1	1.7	2.3	2.2	0	1.8	
Staphylococcus aureus	Outbreaks	499	59.1	7.0	2.0	6.8	1.6	1.8	Dairy, pork, poultry, egg products
	Cases	8179	69.4	11.2	2.9	2.8	1.9	4.0	

Percentages do not total to 100%, since outbreak reports may list multiple food vehicles or involve foods not included in this table

[a]Cold prepared foods include pasta salad, potato salad, dips, and salsa

[b]A single case of botulism is considered to be an outbreak

infection include the specific strain of the pathogen; the food matrix, which may protect against digestive processes; and the health status of the exposed individual (Hurst et al. 1996; Todd et al. 2008). Consequently, infectious dose estimates should be considered conservatively high and that food contaminated below the infectious dose represents a reduced but potentially still significant risk to the consumer.

4.4.1 *Brucella*

The bacteria of the genus *Brucella* are zoonotic pathogens which are known to infect a number of mammalian hosts including humans, cattle, goats, sheep, dogs, rodents, and marine mammals. Ten species of *Brucella* have been identified, and three of these species which infect domestic livestock are of particular concern as a cause of human illness, *Brucella melitensis* which primarily infects sheep and goats, *Brucella abortus* which infects cattle, and *Brucella suis* which infects swine (Godfroid 2014).

In humans *Brucella* infections cause brucellosis, a disease which is rarely fatal but can cause miscarriage and prolonged debilitation, with a wide range of symptoms depending on the tissues infected (Franco et al. 2007). Worldwide, brucellosis is one of the most common bacterial zoonoses, with an estimated 500,000 new cases each year and prevalence rates greater than 10 per 100,000 in some countries (Pappas et al. 2006). However, in some jurisdictions, brucellosis has effectively been eliminated as a disease by routine pasteurization of milk and the establishment of *Brucella*-free herds.

Brucella are Gram-negative bacteria forming small cocci or short rods. The members of this group are considered aerobic, although optimal growth for some strains is under microaerobic conditions. The optimal pH for growth is 6.6–7.4, and replication can occur between 20 and 40 °C, with 37 °C as the optimum (Corbel and Banai 2005). *Brucella* cells can survive for a period in water and soil but do not replicate in these environments. They are unusual among foodborne pathogens, as they are an obligate parasite that naturally replicates following invasive infection of phagocytic host cells (Atluri et al. 2011).

The primary mode of *Brucella* transmission to humans is as a foodborne disease, usually through consumption of unpasteurized dairy products (Godfroid 2014). Infection through consumption of either raw or undercooked offal of infected animals may also occur but is much less common (Chan et al. 1989; Yoo et al. 2015). Humans can become infected with *Brucella* organisms through either ingestion, inhalation, or contact with the conjunctiva of the eye and skin abrasions (Franco et al. 2007). Infection may result from close contact with infected animals or, very rarely, human-to-human contact (Franco et al. 2007; Palanduz et al. 2000).

Differences in the potential for milk and meat to serve as vehicles for the transmission of *Brucella* can be explained by the mechanism of infection. *Brucella* infection in female mammals leads to infection of the mammary glands and shedding in the milk and other mammary secretions (Meador et al. 1989; Xavier et al. 2009).

Transmission of *Brucella* in mother's milk appears to be an important mechanism for the intergenerational transmission of this parasite. Standard dairy pasteurization conditions kill *Brucella*, and the adoption of routine pasteurization effectively eliminates milk products as a vehicle for this pathogen. In raw milk, no reduction in *Brucella* was observed during 4 days of storage at 4 °C (Falenski et al. 2011). In fermented dairy products, *Brucella* numbers have been observed to decline during manufacture and storage at 4 and 24 °C, from a combination of low pH, low water activity, and interactions with starter cultures (Falenski et al. 2011; Santiago-Rodríguez et al. 2015; El-daher et al. 1990; Méndez-González et al. 2011). Individual studies have reported *Brucella* surviving at levels exceeding 100 CFU/g at up to 18–50 days of storage, making it impossible to establish manufacturing conditions or a storage period that would ensure the safety of *Brucella*-contaminated dairy products.

Brucella can be isolated in tissues throughout animal carcasses but is primarily concentrated in the placenta, mammary glands, lymph nodes, spleen, or localized lesions (Hutchings et al. 1951; Xavier et al. 2009; Carvalho-Neta et al. 2010). This means the probability of exposure through normal muscle tissue is low, and, unlike enteric bacteria, contamination from the gastrointestinal tract or skin during slaughter is highly unlikely. Additionally, the requirement for temperatures to be above 20 °C for *Brucella* replication means that no amplification of risk within food items can occur unless significant temperature abuse occurs. However, viable cells of *Brucella* have been reported to persist in refrigerated, frozen, or salted meat for weeks or months (Hutchings et al. 1951; Ivanova 1960). Thus meat contaminated with *Brucella* remains a hazard if consumed raw or undercooked (Kara et al. 2012).

The *Brucella* species that are a primary risk to humans have been effectively eliminated in some jurisdictions through control programs that utilize vaccination, testing, and culling to reduce the reservoir hosts of these pathogens (Moreno 2014). This strategy has been successful due to the restricted host range and limited capacity for environmental persistence of *Brucella*. Success is also dependent upon the potential for wild herds to serve as a reservoir and the willingness of authorities to commit the necessary resources for ongoing control programs (Moreno 2014).

4.4.2 Campylobacter

Campylobacter is a commonly reported cause of bacterial foodborne infection in Canada (29.1 per 100,000), the EU (64.8 per 100,000), Australia (115.8 per 100,000), and the United States (13.5 per 100,000) (Table 4.2). Compared to other major bacterial pathogens, the great majority of *Campylobacter* illnesses are sporadic cases, with disproportionally few cases associated with an outbreak (Table 4.3). Currently, 17 clinically relevant species of *Campylobacter* have been reported to be associated with human illness (Kaakoush et al. 2015); this section will discuss only two of these species, *C. coli* and *C. jejuni*, as these are primarily reported as a cause of foodborne illness.

Campylobacter are Gram-negative microaerophilic bacteria, with a spiral or curved rod morphology. They require 3–15% oxygen and 3–5% CO_2 for growth and temperatures above 29 °C with an optimum of 42 °C (Vandamme et al. 2005). Rather than generating energy by the fermentation of carbohydrates, *Campylobacter* utilize either tricarboxylic acid cycle (TCA) intermediates or amino acids as substrates for respiratory energy generation, with *C. coli* and *C. jejuni* using oxygen as a terminal electron acceptor though atmospheric oxygen concentrations are typically lethal (Vandamme et al. 2005). *Campylobacter* can be isolated from a wide range of mammals and birds, including agricultural and companion species (Horrocks et al. 2009). The physiological characteristics of *Campylobacter* indicate specialization for existence as an intracellular parasite in animal hosts, as this environment is associated with the availability of TCA cycle intermediates, low oxygen levels, and the required growth temperatures.

The infectious dose of *C. coli* is unknown, while that of *C. jejuni* has been determined as approximately 500 cells by volunteer studies (Todd et al. 2008). Both *C. jejuni* and *C. coli* infections have an onset of 24–72 h, resulting in acute gastroenteritis, with symptoms including watery or bloody diarrhea, fever, and cramps (Kaakoush et al. 2015). Gastroenteritis caused by either *C. jejuni* or *C. coli* is generally self-limiting and typically resolves within 6 days. Infection with *C. jejuni* can trigger a range of autoimmune disorders, including Guillain-Barré syndrome and reactive arthritis, as some strains produce cell surface antigens (lipo-oligosaccharides) that mimic the gangliosides on human nervous tissue (Godschalk et al. 2004).

Since replication of *Campylobacter* is inhibited below 30 °C and the physicochemical conditions found in many foods are either inhibitory or lethal for this bacterial genus, the food safety risk associated with *Campylobacter* is dependent upon the ability of the bacteria to persist. Many *Campylobacter* strains are sensitive to oxygen and will die if exposed to atmospheric levels, but a significant minority (35.7%) of *C. jejuni* strains can survive 24 h of aerobic incubation in broth media (Oh et al. 2015). *Campylobacter* are also highly sensitive to desiccation; cells suspended in blood die as blood drops dry on an exposed surface but persist in liquid blood (Humphrey et al. 1995). The persistence of *Campylobacter* on foods and in the environment can be enhanced by their entry into a viable but non-culturable (VBNC) physiological state (Jackson et al. 2009). The VNBC state can be triggered by exposure to refrigeration temperatures (Chaisowwong et al. 2012) and pH stress (Chaveerach et al. 2003). Alternately, there is evidence that the survival of *Campylobacter* is enhanced by participation in biofilm communities and parasitism of amoebae (Indikova et al. 2015).

Both *C. jejuni* and *C. coli* are carried and shed by a wide range of domestic animals, including cattle, pigs, and sheep, but the meat primarily associated with *Campylobacter* outbreaks is poultry (Table 4.3). The prevalence of *Campylobacter* in poultry, especially broiler chickens, can be very high, and they colonize the crop and cecum at high concentrations (Musgrove et al. 2001). Cattle and other meat animals also carry and shed *Campylobacter* (Horrocks et al. 2009), so rather than the risk associated with poultry being due primarily to the presence or absence of the pathogen, a possible explanation can be found in the differences in the processing of poultry versus larger meat animals. The infectious dose of *Campylobacter* is relatively high compared to verotoxigenic *Escherichia coli* (VTEC) and *Salmonella enterica*, and typical poultry slaughter and processing operations will result in higher loads of fecal organisms including *Campylobacter* compared to red meats (Luber and Bartelt 2007). The smaller relative size of poultry makes it more difficult to remove the gastrointestinal tract hygienically. Similarly, the transfer of fecal flora from the hides of larger animals to the carcass is easier to prevent during dehiding as the skin may be removed as a single piece, whereas poultry must be plucked and then the skin removed or retained. Additionally *Campylobacter* are highly vulnerable to desiccation, and the carcass surfaces of mammals are typically dried during cooling, while poultry is kept wet. Consequently, *Campylobacter* are more likely to be present and will be present in higher numbers and have a far better probability of survival on poultry than on red meat.

A number of strategies for the control of *Campylobacter* during poultry production have been tried. Unlike *Salmonella* which can colonize eggs to spread vertically between generations, *Campylobacter* is dependent upon horizontal transmission by host animals (Callicott et al. 2006). This creates a potential for the use of rigorous biocontrol to prevent flock colonization (Guerin et al. 2007). Changes in the processing of poultry may also result in significant reductions in *Campylobacter*, for example, freezing chicken meat for 24 h will reduce *Campylobacter* numbers by 1 $\log_{10}$ unit (Sampers et al. 2010). Such control methods can be effective, as demonstrated by the experience of New Zealand, which in 2006 introduced a multifaceted program to reduce the risk of campylobacteriosis from chicken. The success of this program was indicated by a 54% reduction in the nationwide incidence of campylobacteriosis within 2 years of implementation (Sears et al. 2011). The New Zealand program included biosecurity to reduce flock contamination, changes to processing practices and regulatory targets for poultry carcasses, freezing of chicken meat prior to retail sale, the promotion of leak proof packaging at retail, and consumer education in safe handling practices (Sears et al. 2011).

4.4.3 *Enterobacteriaceae*

The *Enterobacteriaceae* are a large family of rod-shaped Gram-negative bacteria. They are facultative anaerobes, with both fermentative and respiratory metabolism, and demonstration of motility by peritrichous flagella is common. Individual species or strains of this bacterial family exhibit varying degrees of host or environmental specialization, but they are generally adaptable organisms, and representatives of this genus can be isolated from water, soil, and a wide variety of plants and animals. They may be pathogens of specific plants or animals, but many other members of the genus are apparent symbionts (Brenner and Farmer 2005). Five members of the *Enterobacteriaceae* are of significance as causes of human foodborne illness,

Cronobacter sp., *Escherichia coli*, *Shigella* sp., *Salmonella* sp., and *Yersinia enterocolitica.*

4.4.3.1 *Cronobacter*

The genus *Cronobacter* is a member of the *Enterobacteriaceae* and was originally classified as *Enterobacter sakazakii.* Subsequent molecular typing led to the reclassification of these organisms as seven species within a new genus, *Cronobacter* (Iversen et al. 2008). Currently, ten species and three subspecies of *Cronobacter* are recognized (LSPN 2017). Members of the genus *Cronobacter* are not obligate human pathogens or symbionts, and, like the closely related *Enterobacter*, they may be isolated from a wide range of environmental sources (Grimont and Grimont 2005) and foods including herbs, spices, salads, grains, meat, dairy, and seafood (Jaradat et al. 2014). Thus, exposure to this organism is probably not uncommon, yet illness is associated with infants and the adults over 70 (Patrick et al. 2014). In adults, *Cronobacter* has been isolated from a wide variety of infections (Farmer 2015), most often in patients who are elderly, immunocompromised, or suffering some other underlying health condition. These adult infections appear to be the consequence of environmental exposure.

The status of *Cronobacter* as a foodborne pathogen is related to outbreaks of neonatal meningitis, bacteremia, and necrotizing enterocolitis associated with consumption of powdered infant formula contaminated with *Cronobacter sakazakii* (Farmer 2015). Though outbreaks are a rare occurrence, the results of illness can be devastating, with *C. sakazakii* neonatal meningitis having a fatality rate of 41.9%, with a high rate of neurological complications in survivors (Friedemann 2009).

The association of this opportunistic pathogen with disease in neonates can be considered to be a consequence in changes in human behavior which have created exposure routes and vulnerable populations. These changes include the adoption of powdered infant formula as a convenient source of nutrition for infants and the increased survival rate of preterm neonates, which is linked to advances in medical care (Goldenberg et al. 2008). Preterm neonates are more vulnerable to opportunistic pathogens and face barriers to breastfeeding due to their immaturity and delays in milk expression by their mothers and thus are more likely to receive artificial nutrition, including powdered infant formula (Ziegler 2011).

Why *Cronobacter sakazakii* is associated with powdered infant formula and neonatal disease is not fully clear. The thermotolerance of *C. sakazakii* is similar to other *Enterobacteriaceae*, and cells should be killed by pasteurization (Iversen et al. 2004). The presence of *C. sakazakii* in powdered infant milk formula is likely due to contamination after pasteurization during the manufacturing process (Shaker et al. 2008). The explanation seems to lie in the unusually high resistance to desiccation of *Cronobacter* species (Osaili and Forsythe 2009); however, the specific processing failures that cause contamination remain obscure (Norberg et al. 2012).

4.4.3.2 Pathogenic *Escherichia coli*

Escherichia coli colonizes the gut of most mammalian and avian species, including humans, typically as a commensal (Scheutz and Strockbine 2005). The maximum growth rate for *E. coli* occurs under conditions found in the gastrointestinal tract of its hosts, 37 °C, pH 6–7, an environment rich in complex nutrients. However, *E. coli* is well adapted for growth and survival under a wider range of conditions, from 7 to 46 °C, pH 4.4 to 9, and water activity above 0.95. If conditions are not permissive of growth, then *E. coli* can persist for prolonged periods, with survival for months or years in water, soil, sediment, or plants (Sjogren 1994; Van Elsas et al. 2011; Pachepsky and Shelton 2011).

While the majority of *E. coli* strains are harmless to humans, some strains possess virulence factors which make them enteric pathogens. Currently, seven enteric pathotypes of *E. coli* are distinguished on the basis of pathology and the possession of specific virulence factors. These *E. coli* pathotypes have been designated as enterotoxigenic (ETEC), enteropathogenic (EPEC), diffusely adherent (DAEC), enteroaggregative (EAEC), enteroinvasive (EIEC), adherent invasive (AIEC), and verotoxigenic (VTEC) *E. coli* (Croxen et al. 2013).

The relative prevalence of the different *E. coli* pathotypes in foods is difficult to assess due to the absence of comprehensive surveys. As a cause of foodborne illness, VTEC (also known as Shiga toxin-producing *E. coli* or enterohemorrhagic *E. coli*) is the pathotype most commonly reported as a cause of outbreaks in industrialized nations. In the United States, from 1998 to 2013, there were 485 outbreaks of *E. coli* (11,600 cases) (CDC 2015). The VTEC were involved in 91.6% of reported *E. coli* outbreaks and 76.6% of reported cases of *E. coli* illnesses (CDC 2015).

The predominance of VTEC as a cause of reported outbreaks results from the low infectious dose of some strains of these pathogens, <100 cells (Todd et al. 2008), and the risk of severe patient outcomes (Karch et al. 2005; Bennish 1991). The infectious dose of EIEC can be as low as 100 cells, but the infectious dose established for other pathotypes is much greater, EPEC 10^6, ETEC 10^7, and EAEC 10^8 (Todd et al. 2008). Although pathotypes other than VTEC and EIEC are associated with uncomplicated diarrhea, their impact on human health should not be underestimated. These microorganisms are a significant cause of illness and mortality, especially among infants and young children in areas where public health and sanitation infrastructure is poor, and the simple but effective treatments for diarrheal illness such as oral rehydration solutions made with potable quality water are not available (Das et al. 2014).

The foods associated with outbreaks of *E. coli* are very diverse including fresh meats, processed foods, fresh fruits and vegetables, fruit juice, sprouts, seafood, and dairy. Outbreaks of foodborne *E. coli* in the United States have predominately involved meat (48.2%) and fresh fruits and vegetables (27.9%) (CDC 2015). As a common commensal, *E. coli* is routinely transferred onto meat surfaces during slaughter and carcass breaking operations for both mammalian and avian species. Fresh fruits and vegetables are a major source of outbreaks of pathogenic *E. coli*, and

US data indicates that these food products result in a larger number of illnesses per outbreak (CDC 2015). This is attributable to the raw consumption of these food products, whereas fresh meat is typically cooked prior to consumption. Probable routes for the contamination of fresh fruits and vegetables by *E. coli* are irrigation water, processing water, harvesting techniques and equipment, and contamination from processing workers in addition to defecation onto crops by either wild or domestic animals.

Ready-to-eat meat and fermented dairy products do not typically provide environments suitable for the replication of *E. coli*. Replication of *E. coli* may be inhibited by a combination of physicochemical parameters such as pH, water activity, and organic acids and by interactions with starter cultures (Lücke 2000). Although *E. coli* often declines in these products during typical production and storage conditions, infectious doses of this pathogen can persist for months of storage (McQuestin et al. 2009; Peng et al. 2011).

4.4.3.3 *Shigella*

Shigella spp. were historically distinguished from *E. coli* on the basis of pathology and biochemical reactions, but molecular taxonomy confirms that they have a very close relationship with *E. coli*, particularly enteroinvasive *E. coli* (EIEC) which have a very similar pathology (Strockbine and Maurelli 2005). Whether *Shigella* represents a pathotype of *E. coli* (Van Den Beld and Reubsaet 2012) or a separate species within the genus *Escherichia* (Zuo et al. 2013) remains disputed. However, to understand the role of *Shigella* in foodborne illness, it is useful to view *Shigella* as a pathotype of *E. coli* composed of strains that have become specialized for humans as hosts (Van Den Beld and Reubsaet 2012).

As with VTEC, *Shigella* has a low infectious dose, <100 cells (Todd et al. 2008), a risk of severe patient outcomes (Karch et al. 2005; Bennish 1991) and the associated outbreaks can involve a similar diversity of foods. However, due to the host specialization of *Shigella*, there are important differences in the transmission of these pathogens and the foods associated with outbreaks. In the United States, from 1998 to 2013, there were 155 foodborne outbreaks of *Shigella* (7278 cases) (Table 4.3) (CDC 2015). Unlike *E. coli* where a clear majority of outbreaks and cases involved two product type meats and fresh fruits and vegetables, *Shigella* outbreaks were divided closely between meats (28.2%), fresh fruits and vegetables (28.2%), and unheated prepared foods, such as potato or pasta salad, dips, and salsa (23.9%) (Table 4.3). The greatest proportion of cases of illness were associated with fresh fruits and vegetables (46.6%) followed by unheated prepared foods (31.9%) and meats (15.4%) (Table 4.3). Since *Shigella* is adapted for a human host, contamination of food occurs either as a consequence of poor hygiene by infected food production and service workers or the use of water for irrigation or food preparation from a source contaminated with human feces. In areas where water treatment and distribution infrastructure are relatively good, infected workers are most likely the source. This explains the association of outbreaks with prepared foods which are consumed uncooked. The lower association of *Shigella* with meats and dairy compared to pathogenic *E. coli* (Table 4.3) probably reflects that *Shigella* is not carried by domestic animals, and the incident represents the probability of transmission to foods post cooking.

4.4.3.4 *Salmonella enterica*: Typhoidal and Non-typhoidal

The genus *Salmonella* contains two species, *Salmonella enterica* and *S. bongori*, of which *S. enterica* is a human pathogen. *Salmonella enterica* is subdivided into six subspecies, which are further subdivided by serology into more than 2000 serotypes or serovars (Popoff and Le Minor 2005). Individual serotypes of *S. enterica* vary in host specificity, with some serotypes more adapted to humans and more commonly associated with human illness. Serotypes that cause human illness are distinguished by clinical pathology into two groups, typhoidal and non-typhoidal *Salmonella* (Gal-Mor et al. 2014). Infection with non-typhoidal *Salmonella* typically results in self-limiting acute gastroenteritis and watery diarrhea, with onset of symptoms within 6–12 h of ingestion and a 10-day course of illness (Gal-Mor et al. 2014). Infection with typhoidal *Salmonella* has an average incubation period of up 14 days and 21 days of illness, with symptoms of sustained fever, chills, abdominal pain, swelling of the liver and spleen, rash, nausea, diarrhea or constipation, and coughing (Gal-Mor et al. 2014). The difference in the course of illness is related to the tissues infected; both typhoidal and non-typhoidal *Salmonella* initially invade the intestinal epithelium of the small intestine (Liu et al. 1988), but non-typhoidal *Salmonella* provoke a strong inflammatory response which aids in clearing of the microorganism, while the inflammatory response to typhoidal strains is relatively weak (Gal-Mor et al. 2014). Consequently, infection of immunocompetent hosts by non-typhoidal *Salmonella* is limited to the intestinal epithelium, whereas typhoidal strains are able to cross the intestine and spread to the mesenteric lymph nodes, liver, spleen, bone marrow, and gall bladder (Gordon 2008).

Typhoidal *Salmonella* serotypes such as Typhi, Sendai, and Paratyphi are specialized to human hosts (Gal-Mor et al. 2014). Like other specialized human pathogens, their ability to spread between hosts and cause disease is associated with the failure or absence of systematic hygiene systems or personal hygiene; thus, the burden of disease associated with these pathogens is borne largely by nonindustrialized countries. Typhoidal *Salmonella* have the ability to asymptomatically infect hosts, who become “carriers” and shed the bacterium in their feces. If employed in food production or preparation, such individuals can be a reoccurring source of foodborne outbreaks (Marineli et al. 2013).

Non-typhoidal serotypes typically have broader host specificity than do typhoidal strains. The health risk associated with individual non-typhoidal serotypes may vary greatly; this is reflected in the widely ranging estimates of the infectious dose, in individual outbreaks from 10 to 10^{11} cells (Todd et al. 2008).

A significant proportion of *Salmonella* outbreaks in the United States (41.3%) are associated with meats, particularly poultry (23.0%) (Table 4.3). *Salmonella* can colonize practically all domestic animals and as with other enteric organisms can be transferred to meat surfaces during carcass breaking. Like *Campylobacter*, the relatively high proportion of *Salmonella* outbreaks associated with poultry probably reflects the hygiene challenges inherent in poultry slaughter and processing.

The second most common food associated with *Salmonella* outbreaks are fresh fruits and vegetables (Table 4.3). If inoculated onto plant surfaces, through contaminated water or defecation by humans or animals, the ability of *Salmonella* to survive at low water activities and its physiological adaptability can allow it to survive in the environment and colonize plant tissues. Successful colonization of plant tissues by *Salmonella* may also involve fimbriae and cellulose production for attachment and interactions with plant-associated microflora (Brandl et al. 2013). The cutting or breaking of plant surfaces intentionally or unintentionally during processing and transport may increase the availability of nutrients and moisture which in turn facilitate bacterial replication, and under such conditions temperature remains the only limit on *Salmonella* replication.

Of the foodborne bacterial pathogens, *Salmonella* has the greatest association with outbreaks involving eggs (Table 4.3). *Salmonella* are capable of colonizing the reproductive tract of birds and can internalize into the egg during the process of egg formation (Keller et al. 1995; Gantois et al. 2008). Alternatively, the shell of the egg can also become contaminated by *Salmonella* from the gastrointestinal tract of birds either during passage of the egg through the oviduct or after laying, and these subsequently contaminating bacterial cells can then penetrate the shell (De Reu et al. 2006). Once internalized into the egg, *Salmonella* may replicate in the albumen or the yolk (Humphrey and Whitehead 1993). *Salmonella* on the surface of eggs can be removed by appropriate washing methods, but internalized *Salmonella* are protected from decontamination treatments.

Low-moisture foods are generally considered safe, as the low water activity inhibits replication, but occasionally outbreaks involving *Salmonella* in low-moisture foods are reported. Products typically involved in these outbreaks include spices, chocolate (Werber et al. 2005), tree nuts, peanut butter (Viazis et al. 2015), tea, and powdered infant formula (Finn et al. 2013). Though *Salmonella* require a water activity above 0.93 for replication, similar to other *Enterobacteriaceae*, *Salmonella* can persist for prolonged periods in foods with water activities below 0.85 (Finn et al. 2013). The risk possessed by *Salmonella* in foods with low water activity may be enhanced by physiological adaptation, including cell filament formation, resulting in increased resistance to other physicochemical stresses, such as heat, pH, and sanitizers (Finn et al. 2013; Mattick et al. 2000). Alternatively, outbreaks can occur with high-moisture foods which are prepared with contaminated low-moisture ingredients and thus subsequently allow replication of *Salmonella*. High-fat, low-moisture foods may also be protective of cells during digestion following consumption, increasing the probability of infection (Aviles et al. 2013).

4.4.3.5 *Yersinia enterocolitica*

Yersinia enterocolitica causes acute, self-limiting gastrointestinal illness, which typically resolves in 1–3 days. Importantly, *Y. enterocolitica* is notable for its cold tolerance; it can grow at 0 to 40 °C (Bottone 1997). Also notable is the fact that *Y. enterocolitica* can be isolated from a wide range of environmental sources and animal hosts, but foodborne infection is primarily associated with pork (Drummond et al. 2012). Of the 11 outbreaks with identified causes reported to the CDC from 1998 to 2013, nine (81.8%) involved pork, one outbreak involved an unspecified "deli meat," and one involved a dairy product (Table 4.3).

The intestinal contents and tonsils of pigs are important sources of *Y. enterocolitica* contamination of pork during slaughter and carcass breaking. The primacy of pork association with *Y. enterocolitica* infection may reflect that those biotypes of these bacterial species which are potential human pathogens are more common in pigs than in other hosts (Laukkanen-Ninios et al. 2014). Illness attributable to *Y. enterocolitica* associated with other foods does occur, though is far less common than pork. The capacity of *Y. enterocolitica* to survive and grow in a wide range of conditions means that, like other *Enterobacteriaceae*, contamination of plants in the field, shedding by food preparers, or cross contamination in the kitchen remain alternate routes of transmission.

4.4.4 Listeria monocytogenes

Listeria monocytogenes was first identified by E. G. D. Murray in 1924, as a Gram-positive bacillus responsible for mononucleosis in laboratory animals (Murray et al. 1926). *Listeria monocytogenes* is a nonspore-forming, facultative anaerobe, and its replication can occur from 0 to 45 °C. The species *L. monocytogenes* is ubiquitously distributed in the environment, and there is little evidence of host specificity. It has been isolated from soil, water and vegetation, silage, and both human and animal hosts (Fenlon 1985, 1989; Dowe et al. 1997). It appears that *L. monocytogenes* is a highly adaptable organism capable of life as an intracellular parasite in a wide range of eukaryotic hosts.

Of the major foodborne bacterial pathogens, *L. monocytogenes* has one of the lowest incidence rates (Table 4.2) in developed nations, but it has the highest case fatality rates, 21% in the United States in 2009–2011 (Silk et al. 2013). Patient outcomes from *L. monocytogenes* infection are highly dependent upon the underlying health status of the patient. In otherwise healthy people, listeriosis is characterized by self-limiting fever, aches, fatigue, nausea, and diarrhea (Ooi and Lorber 2005). However, in susceptible individuals (older adults, immunocompromised persons, newborns, and pregnant women), listeriosis can take the form of invasive infection, resulting in septicemia, meningitis, meningoencephalitis, or abortion (Doganay 2003).

The majority of human listeriosis cases are foodborne, and the most commonly implicated foods are ready-to-eat foods (Lianou and Sofos 2007). Due to its ubiquity in food production environments, it is not surprising that *L. monocytogenes* has been isolated from a variety of foods including milk and dairy products, meat products, egg products, seafood, vegetables, and other ready-to-eat foods (Ferreira et al. 2014; Fenlon et al. 1996; Dowe et al. 1997; Macgowan et al. 1994). The infective dose of *L. monocytogenes* is >1000 cells (Todd et al. 2008), which is relatively high compared to pathogens such as VTEC and *Salmonella.*

Outbreaks of *L. monocytogenes* are regularly traced back to the processing plant. Due to its ubiquity, *L. monocytogenes* is easily introduced into processing plants, and once introduced if sanitation and other control measures are inadequate, equipment and other environmental surfaces can support an effectively permanent population of *L. monocytogenes* (McCollum et al. 2013; Currie et al. 2015). Biofilm formation is an important factor in the persistence of such populations, and once established they may serve as a reoccurring source of outbreaks for months or years (Thimothe et al. 2004; Ferreira et al. 2014; Carpentier and Cerf 2011; Ortiz et al. 2010). This is illustrated by a human listeriosis outbreak that occurred in 2000 in the United States and was caused by a *L. monocytogenes* strain that appears to have persisted in the same food processing plant over at least 12 years, causing a sporadic case of listeriosis in 1988 and a subsequent outbreak in 2000 (Orsi 2008).

From these observations, it is clear that the prevention of *L. monocytogenes* outbreaks is the most easily controlled at the processing plant and is dependent upon establishing and maintaining an effective and rigorous sanitation plan. Due to the risk of colonization of the processing environment by *L. monocytogenes*, it is desirable to assess the effectiveness of the sanitation plan with microbiological testing which allows differentiation between transient and persistent populations of *L. monocytogenes.* This can be achieved by molecular subtyping of isolates. If the presence of a persistent population is indicated, then the sanitation plan must be thoroughly revised to address the deficiency that allows the persistent population to exist.

4.4.5 Mycobacterium

The members of the genus *Mycobacterium* are slightly curved or straight rods which possess a cell wall structure that is not typical of either Gram-positive or Gram-negative bacteria. Instead the cell wall contains large quantities of unique lipids. They are aerobic or microaerobic, requiring complex nutrition, and are very slow growing even under optimal conditions (Magee and Ward 2012).

Infection of humans with specific species of *Mycobacterium* can cause the diseases tuberculosis and leprosy. Tuberculosis may result from infection by either inhalation or ingestion and is caused by a number of closely related *Mycobacterium* species (*M. tuberculosis*, *M. bovis*, *M. caprae*, and *M. avium*) that may infect domestic animals or humans (Magee and Ward 2012). As a foodborne disease, milk once was a significant vector, and foodborne tuberculosis is primarily associated with the consumption of milk contaminated with *M. bovis*. Foodborne transmission of tuberculosis is virtually unknown where the control measures of milk pasteurization and establishment of tuberculosis-free herds have been employed (Thoen et al. 2006).

In contrast to milk, the consumption of beef from infected cattle is not associated with the transmission of *M. bovis*. Three factors which limit the potential for transmission of *M. bovis* by meat have been identified (de la Rua-Domenech 2006). Firstly, *M. bovis* grows slowly even under optimal conditions and will not replicate on postmortem tissue. Secondly, the bacterium is localized in lesions which only form in the skeletal muscle of animals with advanced infection. Thirdly, *M. bovis* is not notably resistant to thermal treatment and is destroyed by normal cooking conditions.

It has been proposed that Crohn's disease, a form of chronic inflammatory bowel disease, is caused by *M. avium paratuberculosis* and that foodborne transmission plays an important role in exposure. However, a causative role for *M. avium paratuberculosis* in Crohn's disease has not been proven (Liverani et al. 2014), and though the organism has been recovered from beef and dairy products, evidence for significant foodborne exposure is scant (Gill et al. 2011).

4.4.6 Vibrio

Among members of the family *Vibrionaceae*, the genus *Vibrio* is composed of facultative anaerobic, Gram-negative, straight, or slightly curved rods. This genus contains a diverse range of species which are found worldwide as an essential component of aquatic ecosystems. All species are halophiles, with optimum growth occurring at sodium concentrations from 5 to 700 mM. All species are able to replicate at 20 °C, but the temperature range for individual species may range from as low as 4 °C to as high as 37 °C. The distribution and range of individual species of *Vibrio* appear to reflect specialized adaptions for specific ecological niches and localities. There is evidence that a diverse range of abiotic and biotic factors influence the proliferation of specific *Vibrio* species, and it is unclear to what extent *Vibrio* live as planktonic organisms or are dependent on symbiotic or parasitic relationships with plants, algae, animals, and protozoa (Takemura et al. 2014). Conflicting results of individual studies may reflect ecological specialization of strains within species. What is clear is that water temperature and salinity play a major role in determining the geographical range of species (Takemura et al. 2014).

Among the 70 species of *Vibrio* that have been identified, 11 species have been reported to cause human illness. Of these pathogens, three species are the most relevant for consideration as a cause of foodborne illness, *V. cholerae*, *V. parahaemolyticus*, and *V. vulnificus*. These three species are genetically and phenotypically diverse with virulent and avirulent strains reported (Sanjuan et al. 2009; Ronholm et al. 2015).

Those *V. cholerae* strains capable of producing cholera toxin are the most notorious of the *Vibrio*, as they can cause large-scale outbreaks under conditions of poor water sanitation. Poor sanitation is a necessity for cholera outbreaks as the infectious dose is high (10^6–10^9 cells). Infection is most often asymptomatic, but of the patients who develop diarrhea, approximately 20% suffer acute diarrheal illness which can kill with hours of onset (WHO 2014), if treatment and supportive medical care are unavailable.

It is known that *V. cholerae* prefers water temperatures from 10 to 30 °C and 1–10 ppt saline, with the environmental presence of these organisms generally decreasing as salinity increases (Takemura et al. 2014). Crustaceans serve as a reservoir for *V. cholerae* in marine environments; the association of *V. cholerae* with the chitin exoskeleton of crustaceans has provided the microorganism with a number of advantages, including food availability, adaptation to environmental nutrient gradients, tolerance to stress, and protection from predators (Pruzzo et al. 2008). Chitinase produced by *V. cholerae* allows the microorganism to derive carbon, nitrogen, and energy from crustacean exoskeleton (Meibom et al. 2004). However, outbreaks are associated with fecal contamination of fresh water. Consequently, though cholera is endemic in some regions, the disease is practically unknown where water and sewage sanitation are well established. Foodborne transmission may occur when food or beverages are prepared with contaminated water or by infected individuals.

In America, *V. parahaemolyticus* and *V. vulnificus* are the *Vibrio* species primarily reported as a cause of foodborne disease and are almost exclusively associated with seafood, particularly uncooked products such as raw crustaceans and bivalve shellfish (Table 4.3). The two species *V. parahaemolyticus* and *V. vulnificus* differ significantly in the risks they pose. *Vibrio parahaemolyticus* has a high infectious dose of 10^5–10^7 cells (Todd et al. 2008), with an incubation period of 4–90 h, with most patients experiencing disease symptoms within 24 h. The associated illness usually involves 3 days of self-limiting diarrhea, accompanied by symptoms of headache, vomiting, nausea, abdominal cramps, and low fever, though life-threating septicemia can occur in patients with underlying health conditions (Su and Liu 2007).

Vibrio vulnificus is a much more dangerous pathogen; it has an infectious dose estimated at less than 10 cells (Todd et al. 2008); the incubation period is short, being 26 h on average; and this organism causes septicemia with symptoms of fever, chills, nausea, and abdominal pain (Oliver 2013). In the United States, *V. vulnificus* is responsible for 95% of seafood-related deaths and has a 50% case fatality rate (Oliver 2013). The population vulnerable to *V. vulnificus* is very well defined; cases almost always (94%) involve an underlying health condition (liver disease, diabetes, immunocompromising illness), and it is most common in males over 40. In experiments with rats, the vulnerability of males has been demonstrated to be due to a protective effect by estrogen against the activity of *V. vulnificus* endotoxin (lipopolysaccharide) (Oliver 2013).

Vibrio parahaemolyticus is associated with water temperatures greater than 20 °C, preferring 25–35 °C, and within that temperature range, it tolerates salinity levels from 3 to 35 ppt (Takemura et al. 2014). Cell numbers in the water column increase as temperatures increase above 15 °C, with cell density in water declining below that 15 °C, though *V. parahaemolyticus* can be isolated from marine sediments year-round irrespective of temperature (Kaneko and Colwell 1973). *Vibrio vulnificus* is associated with a wide temperature range, 10–32 °C, at 5–10 ppt salinity, but prefers water above 20 °C when salinity exceeds 10 ppt (Takemura et al. 2014). The preference of both *Vibrio* species for warmer water temperatures is reflected in the greater risk associated with the ingestion of raw oysters in summer (Cook et al. 2002). Raw oysters and other filter feeding molluscs represent an increased risk as filter feeding can very effectively concentrate *Vibrio* from the environment into the animal's tissues and result in the tissue containing concentrations several orders of magnitude higher than the surrounding waters (Takemura et al. 2014).

While there is seasonality in pathogen prevalence and numbers, the risks associated with *Vibrio* on raw seafood cannot be managed by limiting harvesting seasons. Many jurisdictions conduct bacteriological testing of harvested molluscs and either limit or expand fisheries on that basis, though this approach cannot ensure safety. Once fish or shellfish have been harvested, growth of *V. parahaemolyticus* is inhibited below 13 °C and *V. vulnificus* below 11 °C (Kim et al. 2012). Neither of these bacterial species display significant heat resistance, and so the associated illnesses are commonly linked to raw or undercooked foods.

4.4.7 Sporeformers

Spore-forming bacteria can persist as spores in the environment for years or decades. This persistence makes them common contaminants of foods, especially those foods in close contact with soils. Three species of sporeformers, *C. botulinum*, *C. perfringens*, and *B. cereus*, produce toxins that can cause illness in humans and animals, and many species of sporeformers cause spoilage of foods (Setlow and Johnson 2007). Spore-forming bacteria present a special problem to food processors, as endospores can resist extreme conditions of heat, pressure, desiccation, radiation, and chemical sterilants. Endospores are metabolically inactive yet are capable of sensing their environment and will germinate into vegetative cells given the presence of appropriate conditions of moisture, temperature, and nutrients.

4.4.7.1 *Bacillus cereus*

Bacillus cereus sensu lato is a group of species of Gram-positive, facultative anaerobic, spore-forming rods. The taxonomy of *B. cereus* sensu lato is complex and changing. To date, seven members have been described: *Bacillus cereus* sensu stricto, *Bacillus mycoides*, *Bacillus pseudomycoides*, *Bacillus thuringiensis*, *Bacillus weihenstephanensis*, *Bacillus anthracis*, and *Bacillus cytotoxicus* (Guinebretiere et al. 2013). Unlike the closely related species *B. anthracis* and *B. thuringiensis*, *B. cereus* does not enter a pathogenic life cycle in the natural environment. *Bacillus cereus* has a saprophytic lifestyle, where spores germinate when exposed to moisture and nutrients, after which the vegetative cells multiply and then sporulate when nutrients are exhausted (Stenfors Arnesen et al. 2008). It has been suggested that *B. cereus* may be the most common aerobic sporeformer and is widely distributed in soils, sediments, dust, the rhizosphere of some plants (Halverson et al. 1993; Young et al. 1995), rain and groundwater (Brillard et al. 2015), and the intestinal tract of invertebrates (Jensen et al. 2003; Margulis et al. 1998).

Several members of *B. cereus* sensu lato are commonly isolated from foods and are associated with two food poisoning syndromes in humans—emetic (vomiting) syndrome and diarrheal syndrome (Stenfors Arnesen et al. 2008). Both syndromes are typically mild, with patients often recovering within 24 h. However, lethal cases of emetic intoxications, associated with the consumption of rice or pasta dishes, have been reported after the ingestion of food contaminated with high amounts of emetic toxin (Dierick et al. 2005; Naranjo et al. 2011; Mahler et al. 1997; Shiota et al. 2010).

The bacterium is so commonly isolated that it is often simply declared as a "contaminant" when isolated from clinical specimens and is given little concern (Bottone 2010). As a result of the ubiquitous distribution of *B. cereus* in foods, the bacterium is ingested in small numbers and becomes part of the transitory human intestinal flora (Jensen et al. 2003; Turnbull 1985). The spores of *B. cereus* are particularly a concern in the food industry because these spores are not inactivated either by pasteurization (Carlin et al. 2000; Choma et al. 2000; Crielly et al. 1994) or gamma radiation (Mtenga et al. 2012, 2013; Thayer and Boyd 1994), and their hydrophobic nature allows them to adhere to food processing surfaces (Faille et al. 2002; Tauveron et al. 2006).

The emetic syndrome is caused by the cyclic dodecadepsipeptide cereulide, which is produced in food before ingestion (Agata et al. 1995), causing a foodborne intoxication. Emetic food poisoning by *B. cereus* is most often associated with consumption of contaminated cooked rice and pasta dishes that were not properly cooled. Some strains have the ability to grow at low temperatures and can be regarded as psychrotrophs with a minimal growth temperature of 4 °C (van Netten et al. 1990). Rice can become contaminated with microorganisms from either soil, irrigation water, animal feces, or insects (Kim et al. 2014; Haque and Russell 2005; Choi et al. 2014), with *Bacillus* spp. being one of the primary contaminants (Cottyn et al. 2001). In a recent study (Kim et al. 2014), *B. cereus* spores were isolated from 96.8%, 80.6%, and 57.8% of rough, brown, and white rice samples, respectively. Levels of *B. cereus* in unhusked rice have been estimated to be as high as 1.2 to 3.5×10^3 cfu g^{-1} (Sarrías et al. 2002). Spores prepared from *B. cereus* strains isolated from rice are heat resistant, having D_{90}-values ranging from 3.23 to 23.26 min (Sarrías et al. 2002). Thus spores survive the cooking process, to germinate, grow, and produce cereulide in the cooked product.

Bacillus cereus can also cause a diarrheal syndrome due to release of enterotoxins produced during growth in the small intestine after ingestion of high numbers of viable cells or spores (Clavel et al. 2004; Granum et al. 1993). Diarrheal syndrome is most often associated with sauces and vegetables, meat products, soups, puddings, and milk products (Kotiranta et al. 2000; Stenfors Arnesen et al. 2008).

4.4.7.2 *Clostridium botulinum*

Popoff (2015) has described diseases due to toxigenic clostridia as "an accidental connection between a host and an environmental bacterium which synthesizes a specific toxic compound, rather than resulting from bacterial strategy to invade an organism and subsequently survive in this new environment sheltered from host defenses."

Since the sole virulence factor of *C. botulinum* is botulinum neurotoxin, neurotoxigenic clostridia are characterized into seven different types, A through G, based on the serological specificity of the neurotoxin produced. Human botulism, including foodborne, wound, and infant and adult colonization botulism, is associated with types A, B, E, and, very rarely, F. Neurotoxigenic clostridia are also divided into six groups based on phenotype, and this division is supported by genomic analysis. Group I includes BoNT/A-, BoNT/B-, and BoNT/F-producing *C. botulinum*. These strains are proteolytic, grow at a minimum temperature of 10 °C, and produce spores of high heat resistance. Group II includes BoNT/B-, BoNT/E-, and BoNT/F-producing *C. botulinum*. These strains are nonproteolytic, grow at refrigeration temperatures, and produce spores of low heat resistance. Group III includes BoNT/C- or BoNT/D-producing *C. botulinum*. These strains cause botulism in animals, but not in humans. Group IV includes BoNT/G-producing *Clostridium argentinense*. Group V includes neurotoxigenic BoNT/F-producing *Clostridium baratii*; and Group VI includes neurotoxigenic BoNT/E-producing *Clostridium butyricum* (Smith et al. 2015). The genes encoding botulinum toxin types B, E, and F are not limited to strains with similar genomic backgrounds but have crossed species. Type B toxin is produced by Groups I and II *C. botulinum*; type E toxin is produced by Group II *C. botulinum* and neurotoxigenic strains of *C. butyricum*, while type F toxin is the most promiscuous toxin being produced by both Group I and II *C. botulinum* and by neurotoxigenic strains of *C. baratii*.

Spores of *C. botulinum* are ubiquitously distributed in soils and in marine and freshwater sediments. They are commonly isolated from many animals, including animals processed for food. Global distribution of *C. botulinum* roughly indicates *C. botulinum* type A spores predominate in soils in the western United States, China, Brazil, and Argentina and type B spores in the eastern United States, the United Kingdom, and much of continental Europe. Of the type B spores, most American type B strains are proteolytic (Group I), while most European strains are nonproteolytic (Group II). The majority of botulism cases caused in northern latitudes are a result of *C. botulinum* type E, indicating the predominance of type E strains in northern regions and in most temperate aquatic regions and their surroundings. Types C and D are found more frequently in warmer environments. The reasons for this distribution pattern are not understood well. Type A appears to be favored by neutral to alkaline soil with low organic content; this is consistent with its virtual absence in the highly cultivated soils of the eastern United States and Europe. Type E is psychrotolerant, which undoubtedly plays a role in its prevalence in the North and in many aquatic environments. The high rate of botulism in Inuit communities of southern Ungava Bay correlates with the high prevalence of *C. botulinum* type E found in that environment (Leclair et al. 2013).

One of the primary hurdles to inhibit the growth of *C. botulinum* is low pH. The minimum pH allowing growth of *C. botulinum* group I is 4.6; for group II, it is approximately pH 5.0. Many fruits are sufficiently acidic to inhibit *C. botulinum* by their pH alone, while acidulents frequently are used to preserve other products. The rate of acidification of foods can be important in inhibition of microorganisms, and notably *C. botulinum* can grow in some acidified foods if excessively slow pH equilibration occurs.

Importantly, *C. botulinum* is not considered to be a successful "competitor" in terms of growing in foods with mixed bacterial flora. Lactic acid bacteria, including *Lactobacillus*, *Pediococcus*, and *Streptococcus*, can inhibit growth of *C. botulinum*, largely by reducing the pH, but also by the production of bacteriocins (Rodgers et al. 2003; Okereke and Montville 1991). An example of this is the "Wisconsin process" which uses *Pediococcus acidilactici* and a fermentable carbohydrate for producing bacon with 40 or 80 ppm sodium nitrite rather than conventional bacon made with 120 ppm nitrite in the United States (Tanaka et al. 1985).

The growth of other microorganisms may also protect consumers by causing spoilage that would make a toxic product less likely to be consumed. For example, *C. botulinum* was observed to grow and produced toxin in ready-to-eat bagged salads, but only after the salads were visibly spoiled (Austin et al. 1998a). In the United States and Canada, smoked fish is only permitted for sale when packed in a film permeable to oxygen, ensuring growth of aerobic spoilage flora and hence spoilage of the product when stored for an extended time at abusive temperatures (Dufresne et al. 2000).

Traditional northern foods in Canada, Alaska, and Greenland are common causes of botulism. Meat and fat from seal, whale, and walrus are aged without control of temperature, followed by consumption without cooking. This tradition has led to high rates of botulism in Arctic, and near-Arctic, communities. While this form of food preparation has been incorrectly referred to as "fermentation," the lack of fermentable carbohydrates in muscle and fat tissue results in the pH of these traditional foods remaining near neutral pH, with the resultant growth of *C. botulinum* type E.

Pruno, an alcoholic beverage prepared by prison inmates, has caused several botulism outbreaks in US prisons over the past several years (Adams et al. 2015; Vugia et al. 2009; Walters et al. 2015; Williams et al. 2014; Yasmin et al. 2015). Pruno is prepared from food scraps, typically fruits, sugar, and water. Notably, the pruno batches associated with botulism all had potatoes added to them (Walters et al. 2015). Potatoes were reported to produce higher levels of alcohol in a shorter fermentation time. Addition of potatoes, a food that has a history of causing botulism (Angulo et al. 1998; Bhutani et al. 2005; Seals et al. 1981), may have introduced spores to the pruno formula and would not contribute to a decrease in pH as addition of fruits would.

4.4.7.3 *Clostridium perfringens*

Similar to the other spore-forming bacteria, *C. perfringens* is widely spread in the environment. The species *C. perfringens* is classified into one of the five types (A–E) based on the production of four major toxins (alpha, beta, epsilon, and iota). This species is frequently present in the normal intestinal microbiota of humans and animals. Food poisoning caused by *C. perfringens* is attributed to a subset of type A strains which produce *C. perfringens* enterotoxin (CPE); CPE is a single polypeptide with a molecular weight of 3.5 kDa and is produced during sporulation of vegetative cells after large numbers have been ingested ($>10^6$ bacteria g^{-1}) which requires growth in temperature-abused foods.

Clostridium perfringens was the second most common cause of bacterial foodborne disease outbreaks in the United States during 1998–2010 (Grass et al. 2013) and the most common bacterial source of foodborne illness in Canada during 2000–2010 (Thomas et al. 2013).

The species *C. perfringens* has an extensive history of use as an indicator for fecal contamination, based on its association with the gastrointestinal tract of humans and other animals and the presence of its spores in sewage (Mueller-Spitz et al. 2010). Specifically, *C. perfringens* has been shown to be a conservative indicator for fecal excreta from carnivorous wildlife and human-associated sewage (Vierheilig et al. 2013). Important for its use as an indicator is the fact that *C. perfringens* does not reproduce in aquatic systems (Wright 1989; Desmarais et al. 2002). The spores may be detected long after a pollution event has occurred and far from the source (Vierheilig et al. 2013).

4.5 Conclusions

Exposure to bacterial foodborne pathogens and their potential to cause disease are not random, though these organisms are widely dispersed in the environment and part of the common microflora of agricultural environments. Even the most ubiquitous and adaptable of foodborne bacterial pathogens are associated with specific foods. These associations can be understood as arising from the ecological relationship of the individual pathogen with the food production environment, the food product itself, and its associated microflora.

Knowledge of these relationships is invaluable in informing the identification of potential control strategies and excluding control strategies which will be ineffective. For example, *Mycobacterium* and *Brucella* have been effectively eliminated as foodborne diseases in some jurisdictions by reduction in the availability of susceptible animal hosts through such measures as immunization and testing, as well as culling of herds to eliminate the presence of infected animals. But the success of these strategies is dependent upon the limited host range and poor survival of these pathogens outside a host. Consequently, it can be predicted that this approach is unlikely to be successful with pathogens such as *E. coli* and *Salmonella* which can survive and replicate in a far wider range of potential hosts and environmental niches. Similarly, some ready-to-eat foods, such as cheeses and fermented meats, have a surprisingly good history of safety with traditional production methods, which do not include a decontamination step, as the food itself is an unfavorable environment for pathogen survival and growth.

In general, the hazard potential for infectious foodborne pathogens depends on the probability of exposure of the consumer to an infectious dose of the organism. This in turn is dependent upon the probability of inoculation of the food with the pathogen and its potential for survival on the product throughout the shelf life of the product. Replication of the pathogen either on or within the food amplifies the risk but is not required for illness to occur. The key strategies for the control of infectious foodborne pathogens can be understood to be decontamination and the prevention of contamination, and decontamination is not always available for ready-to-eat products. In contrast, bacterial agents of intoxication present a risk not simply through their presence but through bacterial growth and toxin production. For these pathogens, food safety can be established by formulating, storing, and packaging food to prevent toxin production, even if the pathogen is present.

The value of the examples presented in this chapter is in illustrating the relationship between pathogens and food commodities and the factors which determine these relationships. The ecology of foodborne pathogens is not static but should be understood as an evolving system. New pathogens and new food pathogen associations may emerge with changes to food production and distribution. The potential causes of these changes are as diverse as the food production system. The locations in which foods are grown or harvested may alter due to changes in the environment or economic opportunity. The geographic range of pathogens may change in response to changes in host range or climatic conditions. Food processing, packaging, and distribution may be changed by industries seeking new opportunities. Fashions in food consumption and preparation regularly change among consumers. The principles presented in this chapter will hopefully be of value to microbiologists attempting to address the food safety challenges of the future.

Compliance with Ethical Standards

Conflict of Interest Alexander Gill declares that he has no conflict of interest. John W. Austin declares that he has no conflict of interest.

Ethical Approval This article does not contain any studies with human participants or animals performed by either of the authors.

References

Adams LE, Yasmin S, Briggs G, Redden K, Silvas S, Anderson S, Weiss J, Tsang CA, Henke E, Francies J, Herrick K, Lira R, Livar E, Thompson G, Sunenshine R, Robinson BF, Bisgard KM, Komatsu KK (2015) Alcohol production, prevention strategies, and inmate knowledge about the risk for botulism from pruno consumption in a correctional facility--Arizona, 2013. J Correct Health Care 21(4):335–342. https://doi.org/10.1177/1078345815599763

Agata N, Ohta M, Mori M, Isobe M (1995) A novel dodecadepsipeptide, cereulide, is an emetic toxin of *Bacillus cereus*. FEMS Microbiol Lett 129(1):17–19

Álvarez-Ordóñez A, Fernández A, Bernardo A, López M (2010) Acid tolerance in *Salmonella typhimurium* induced by culturing in the presence of organic acids at different growth temperatures. Food Microbiol 27(1):44–49. https://doi.org/10.1016/j.fm.2009.07.015

Alvarez-Ordóñez A, Broussolle V, Colin P, Nguyen-The C, Prieto M (2015) The adaptive response of bacterial foodborne pathogens in the environment, host and food: implications for food safety. Int J Food Microbiol 213:99–109. https://doi.org/10.1016/j.ijfoodmicro.2015.06.004

Angulo FJ, Getz J, Taylor JP, Hendricks KA, Hatheway CL, Barth SS, Solomon HM, Larson AE, Johnson EA, Nickey LN, Ries AA (1998) A large outbreak of botulism: the hazardous baked potato. J Infect Dis 178(1):172–177

Atluri VL, Xavier MN, De Jong MF, Den Hartigh AB, Tsolis RM (2011) Interactions of the human pathogenic *Brucella* species with their hosts. Annu Rev Microbiol 65:523–541. https://doi.org/10.1146/annurev-micro-090110-102905

Austin JW, Bergeron G (1995) Development of bacterial biofilms in dairy processing lines. J Dairy Res 62(3):509–519

Austin JW, Dodds KL, Blanchfield B, Farber JM (1998a) Growth and toxin production by *Clostridium botulinum* on inoculated fresh-cut packaged vegetables. J Food Prot 61(3):324–328

Austin JW, Sanders G, Kay WW, Collinson SK (1998b) Thin aggregative fimbriae enhance *Salmonella enteritidis* biofilm formation. FEMS Microbiol Lett 162(2):295–301

Aviles B, Klotz C, Smith T, Williams R, Ponder M (2013) Survival of *Salmonella enterica* serotype Tennessee during simulated gastric passage is improved by low water activity and high fat content. J Food Prot 76:333–337. https://doi.org/10.4315/0362-028X.JFP-12-280

Baird-Parker AC, Freame B (1967) Combined effect of water activity, pH and temperature on the growth of *Clostridium botulinum* from spore and vegetative cell inocula. J Appl Bacteriol 30 (3):420–429

Barnes EM (1994) Personal recollections of developments in food microbiology. J Appl Bacteriol 76:5–12

Bennett SD, Manikonda K, Mungai E, Dewey-Mattia D, Gould LH (2014) Surveillance for foodborne disease outbreaks, United States, 2012, Annual Report. US Department of Health and Human Services, Centers for Disease Control and Prevention, Atlanta, GA

Bennish ML (1991) Potentially lethal complications of shigellosis. Rev Infect Dis 13(Suppl 4): S319–S324

Beuchat LR, Golden DA (1989) Antimicrobials occurring naturally in foods. Food Technol 43 (1):134–142

Bhutani M, Ralph E, Sharpe MD (2005) Acute paralysis following "a bad potato": a case of botulism. Can J Anaesth = Journal canadien d'anesthesie 52(4):433–436. https://doi.org/10.1007/bf03016290

Blom H, Mortvedt C (1991) Antimicrobial substances produced by food associated microorganisms. Biochem Soc Trans 19(3):694–698

Bottone EJ (1997) *Yersinia enterocolitica*: the charisma continues. Clin Microbiol Rev 10 (2):257–276

Bottone EJ (2010) *Bacillus cereus*, a volatile human pathogen. Clin Microbiol Rev 23(2):382–398. https://doi.org/10.1128/cmr.00073-09

Brandl M, Cox C, Teplitski M (2013) *Salmonella* interactions with plants and their associated microbiota. Phytopathology 103(4):316–325. https://doi.org/10.1094/PHYTO-11-12-0295-RVW

Bremer PJ, Fillery S, McQuillan AJ (2006) Laboratory scale clean-in-place (CIP) studies on the effectiveness of different caustic and acid wash steps on the removal of dairy biofilms. Int J Food Microbiol 106(3):254–262

Brenner DJ, Farmer JJ (2005) Enterobacteriaceae. In: Brenner DJ, Krieg NR, Staley JT (eds) Bergey's manual of systematic bacteriology, vol 2B, 2nd edn. Springer and Bergey's Manual Trust, Michigan, pp 587–607

Brillard J, Dupont CM, Berge O, Dargaignaratz C, Oriol-Gagnier S, Doussan C, Broussolle V, Gillon M, Clavel T, Berard A (2015) The water cycle, a potential source of the bacterial pathogen *Bacillus cereus*. Biomed Res Int 2015:15. https://doi.org/10.1155/2015/356928

Callicott KA, Friethriksdottir V, Reiersen J, Lowman R, Bisaillon JR, Gunnarsson E, Berndtson E, Hiett KL, Needleman DS, Stern NJ (2006) Lack of evidence for vertical transmission of *Campylobacter* spp. in chickens. Appl Environ Microbiol 72(9):5794–5798. https://doi.org/10.1128/AEM.02991-05

Cann DC, Wilson BB, Shewan JM, Roberts TA, Rhodes DN (1966) A comparison of toxin production by *Clostridium botulinum* type E in irradiated and unirradiated vacuum packed fish. J Appl Bacteriol 29(3):540–548

Carlin F, Guinebretiere M-H, Choma C, Pasqualini R, Braconnier A, Nguyen-the C (2000) Spore-forming bacteria in commercial cooked, pasteurised and chilled vegetable purées. Food Microbiol 17(2):153–165. https://doi.org/10.1006/fmic.1999.0299

Carpentier B, Cerf O (2011) Review: Persistence of *Listeria monocytogenes* in food industry equipment and premises. Int J Food Microbiol 145(1):1–8

Carvalho-Neta AV, Mol JPS, Xavier MN, Paixão TA, Lage AP, Santos RL (2010) Pathogenesis of bovine brucellosis. Vet J 184(2):146–155. https://doi.org/10.1016/j.tvjl.2009.04.010

CDC (2015) Foodborne outbreak online database. http://wwwn.cdc.gov/foodborneoutbreaks/Default.aspx. Accessed 9 July 2015

Chaisowwong W, Kusumoto A, Hashimoto M, Harada T, Maklon K, Kawamoto K (2012) Physiological characterization of *Campylobacter jejuni* under cold stresses conditions: its potential for public threat. J Vet Med Sci 74(1):43–50

Chan J, Baxter C, Wenman WM (1989) Brucellosis in an Inuit child, probably related to caribou meat consumption. Scand J Infect Dis 21(3):337–338. https://doi.org/10.3109/00365548909035706

Chavatte N, Baré J, Lambrecht E, Van Damme I, Vaerewijck M, Sabbe K, Houf K (2014) Co-occurrence of free-living protozoa and foodborne pathogens on dishcloths: implications for food safety. Int J Food Microbiol 191:89–96. https://doi.org/10.1016/j.ijfoodmicro.2014.08.030

Chaveerach P, ter Huurne AAHM, Lipman LJA, van Knapen F (2003) Survival and resuscitation of ten strains of *Campylobacter jejuni* and *Campylobacter coli* under acid conditions. Appl Environ Microbiol 69(1):711–714

Choi S, Kim H, Kim Y, Kim BS, Beuchat LR, Ryu JH (2014) Fate of *Bacillus cereus* and naturally occurring microbiota on milled rice as affected by temperature and relative humidity. Food Microbiol 38:122–127. https://doi.org/10.1016/j.fm.2013.08.016

Choma C, Guinebretiere MH, Carlin F, Schmitt P, Velge P, Granum PE, Nguyen-The C (2000) Prevalence, characterization and growth of *Bacillus cereus* in commercial cooked chilled foods containing vegetables. J Appl Microbiol 88(4):617–625

Clavel T, Carlin F, Lairon D, Nguyen-The C, Schmitt P (2004) Survival of *Bacillus cereus* spores and vegetative cells in acid media simulating human stomach. J Appl Microbiol 97(1):214–219

Cook DW, Oleary P, Hunsucker JC, Sloan EM, Bowers JC, Blodgett RJ, Depaola A (2002) *Vibrio vulnificus* and *Vibrio parahaemolyticus* in U.S. retail shell oysters: a national survey from June 1998 to July 1999. J Food Prot 65(1):79–87

Corbel MJ, Banai M (2005) *Brucella*. In: Brenner DJ, Krieg NR, Staley JT (eds) Bergey's manual of systematic bacteriology, vol 2C, 2nd edn. Springer and Bergey's Manual Trust, Michigan, pp 370–386

Costerton JW, Lewandowski Z, Caldwell DE, Korber DR, Lappinscott HM (1995) Microbial biofilms. Annu Rev Microbiol 49:711–745

Cottyn B, Regalado E, Lanoot B, De Cleene M, Mew TW, Swings J (2001) Bacterial populations associated with rice seed in the tropical environment. Phytopathology 91(3):282–292

Cox NA, Richardson LJ, Maurer JJ, Berrang ME, Fedorka-Cray PJ, Buhr RJ, Byrd JA, Lee MD, Hofacre CL, O'Kane PM, Lammerding AM, Clark AG, Thayer SG, Doyle MP (2012) Evidence for horizontal and vertical transmission in *Campylobacter* passage from hen to her progeny. J Food Prot 75(10):1896–1902. https://doi.org/10.4315/0362-028.jfp-11-322

Crielly EM, Logan NA, Anderton A (1994) Studies on the *Bacillus* flora of milk and milk products. J Appl Bacteriol 77(3):256–263

Croxen MA, Law RJ, Scholz R, Keeney KM, Wlodarska M, Finlay BB (2013) Recent advances in understanding enteric pathogenic *Escherichia coli*. Clin Microbiol Rev 26(4):822–880. https://doi.org/10.1128/cmr.00022-13

Currie A, Farber JM, Nadon C, Sharma D, Whitfield Y, Gaulin C, Galanis E, Bekal S, Flint J, Tschetter L, Pagotto F, Lee B, Jamieson F, Badiani T, MacDonald D, Ellis A, May-Hadford J, McCormick R, Savelli C, Middleton D, Allen V, Tremblay FW, MacDougall L, Hoang L, Shyng S, Everett D, Chui L, Louie M, Bangura H, Levett PN, Wilkinson K, Wylie J, Reid J, Major B, Engel D, Douey D, Huszczynski G, Di Lecci J, Strazds J, Rousseau J, Ma K, Isaac L, Sierpinska U (2015) Multi-province listeriosis outbreak linked to contaminated deli meat consumed primarily in institutional settings, Canada, 2008. Foodborne Pathog Dis 12 (8):645–652. https://doi.org/10.1089/fpd.2015.1939

Danyluk MD, Schaffner DW (2011) Quantitative assessment of the microbial risk of leafy greens from farm to consumption: preliminary framework, data, and risk estimates. J Food Prot 74 (5):700–708. https://doi.org/10.4315/0362-028x.jfp-10-373

Das JK, Salam RA, Bhutta ZA (2014) Global burden of childhood diarrhea and interventions. Curr Opin Infect Dis 27(5):451–458. https://doi.org/10.1097/qco.0000000000000096

de la Rua-Domenech R (2006) Human *Mycobacterium bovis* infection in the United Kingdom: incidence, risks, control measures and review of the zoonotic aspects of bovine tuberculosis. Tuberculosis (Edinb) 86(2):77–109. https://doi.org/10.1016/j.tube.2005.05.002

De Reu K, Grijspeerdt K, Messens W, Heyndrickx M, Uyttendaele M, Debevere J, Herman L (2006) Eggshell factors influencing eggshell penetration and whole egg contamination by different bacteria, including *Salmonella enteritidis*. Int J Food Microbiol 112(3):253–260. https://doi.org/10.1016/j.ijfoodmicro.2006.04.011

Deering AJ, Mauer LJ, Pruitt RE (2012) Internalization of *E. coli* O157:H7 and *Salmonella* spp. in plants: a review. Food Res Int 45(2):567–575. https://doi.org/10.1016/j.foodres.2011.06.058

Desmarais TR, Solo-Gabriele HM, Palmer CJ (2002) Influence of soil on fecal indicator organisms in a tidally influenced subtropical environment. Appl Environ Microbiol 68(3):1165–1172. https://doi.org/10.1128/aem.68.3.1165-1172.2002

Dierick K, Van Coillie E, Swiecicka I, Meyfroidt G, Devlieger H, Meulemans A, Hoedemaekers G, Fourie L, Heyndrickx M, Mahillon J (2005) Fatal family outbreak of *Bacillus cereus*-associated food poisoning. J Clin Microbiol 43(8):4277–4279. https://doi.org/10.1128/jcm.43.8.4277-4279.2005

Dodds KL, Austin JW (1997) *Clostridium botulinum*. In: Doyle MP, Beuchat LR, Montville TJ (eds) Food microbiology. Fundamentals and frontiers, 1st edn. ASM Press, Washington, DC, pp 288–304

Doganay M (2003) Listeriosis: clinical presentation. FEMS Immunol Med Microbiol 35 (3):173–175

Dowe MJ, Jackson ED, Mori JG, Bell CR (1997) *Listeria monocytogenes* survival in soil and incidence in agricultural soils. J Food Prot 60(10):1201–1207

Drummond N, Murphy BP, Ringwood T, Prentice MB, Buckley JF, Fanning S (2012) *Yersinia enterocolitica*: a brief review of the issues relating to the zoonotic pathogen, public health challenges, and the pork production chain. Foodborne Pathog Dis 9(3):179–189. https://doi.org/10.1089/fpd.2011.0938

Dufresne I, Smith JP, Liu JN, Tarte I, Blanchfield B, Austin JW (2000) Effect of films of different oxygen transmission rate on toxin production by *Clostridium botulinum* type E in vacuum packaged cold and hot smoked trout fillets. J Food Saf 20(4):251–268

El-daher N, Na'was T, Al-qaderi S (1990) The effect of the pH of various dairy products on the survival and growth of *Brucella melitensis*. Ann Trop Med Parasitol 84(5):523–528

Erickson MC (2012) Microbial ecology. In: Decontamination of fresh and minimally processed produce. Wiley-Blackwell, New York, pp 1–41. https://doi.org/10.1002/9781118229187.ch1

Faille C, Jullien C, Fontaine F, Bellon Fontaine MN, Slomianny C, Benezech T (2002) Adhesion of *Bacillus* spores and *Escherichia coli* cells to inert surfaces: role of surface hydrophobicity. Can J Microbiol 48(8):728–738

Falenski A, Mayer-Scholl A, Filter M, Göllner C, Appel B, Nöckler K (2011) Survival of *Brucella* spp. in mineral water, milk and yogurt. Int J Food Microbiol 145(1):326–330. https://doi.org/10.1016/j.ijfoodmicro.2010.11.033

Farmer JJ 3rd (2015) My 40-year history with *Cronobacter/Enterobacter sakazakii* - lessons learned, myths debunked, and recommendations. Front Pediatr 3:84. https://doi.org/10.3389/fped.2015.00084

Fenlon DR (1985) Wild birds and silage as reservoirs of *Listeria* in the agricultural environment. J Appl Bacteriol 59(6):537–543

Fenlon DR (1989) Silage as a potential source of *Listeria monocytogenes* in the food chain. Microbiol Aliments Nutr 7(2):171–173

Fenlon DR, Wilson J, Donachie W (1996) The incidence and level of *Listeria monocytogenes* contamination of food sources at primary production and initial processing. J Appl Bacteriol 81 (6):641–650

Ferreira V, Wiedmann M, Teixeira P, Stasiewicz MJ (2014) *Listeria monocytogenes* persistence in food-associated environments: epidemiology, strain characteristics, and implications for public health. J Food Prot 77(1):150–170. https://doi.org/10.4315/0362-028x.jfp-13-150

Finn S, Condell O, McClure P, Amezquita A, Fanning S (2013) Mechanisms of survival, responses and sources of *Salmonella* in low-moisture environments. Front Microbiol 4:331. https://doi.org/10.3389/fmicb.2013.00331

Franco MP, Mulder M, Gilman RH, Smits HL (2007) Human brucellosis. Lancet Infect Dis 7 (12):775–786. https://doi.org/10.1016/s1473-3099(07)70286-4

Friedemann M (2009) Epidemiology of invasive neonatal *Cronobacter* (*Enterobacter sakazakii*) infections. Eur J Clin Microbiol Infect Dis 28(11):1297–1304. https://doi.org/10.1007/s10096-009-0779-4

Gal-Mor O, Boyle EC, Grassl GA (2014) Same species, different diseases: how and why typhoidal and non-typhoidal *Salmonella enterica* serovars differ. Front Microbiol 5:391. https://doi.org/10.3389/fmicb.2014.00391

Gantois I, Eeckhaut V, Pasmans F, Haesebrouck F, Ducatelle R, Van Immerseel F (2008) A comparative study on the pathogenesis of egg contamination by different serotypes of *Salmonella*. Avian Pathol 37(4):399–406. https://doi.org/10.1080/03079450802216611

Giaouris E, Heir E, Desvaux M, Hebraud M, Moretro T, Langsrud S, Doulgeraki A, Nychas GJ, Kacaniova M, Czaczyk K, Olmez H, Simoes M (2015) Intra- and inter-species interactions within biofilms of important foodborne bacterial pathogens. Front Microbiol 6:841. https://doi.org/10.3389/fmicb.2015.00841

Gilbert RJ, Stringer MF, Peace TC (1974) The survival and growth of *Bacillus cereus* in boiled and fried rice in relation to outbreaks of food poisoning. J Hyg (Lond) 73(3):433–444

Gill AO, Gill CO (2005) Preservative packaging for fresh meats, poultry and fin fish. In: Han JH (ed) Innovation in food packaging. Elsevier Academic, Amsterdam, pp 204–226

Gill AO, Gill CO (2010) Packaging and the shelf life of fresh red and poultry meats. Chapter 14. In: Robertson GL (ed) Food packaging and shelf life: a practical guide. Taylor & Francis Group, LLC, Boca Raton, FL, pp 259–277

Gill AO, Gill CO (2012) Microbial decontamination in the food industry: novel methods and applications. In: Demirci A, Ngadi MO (eds) Decontamination of fresh and processed meats. Woodhead, Sawston, pp 30–59

Gill AO, Gill CO (2014) Developments in sampling and test methods for pathogens in fresh meat. Adv Microb Food Saf 2:157–280. https://doi.org/10.1533/9781782421153.3.257

Gill CO, Saucier L, Meadus WJ (2011) *Mycobacterium avium* subsp. paratuberculosis in dairy products, meat, and drinking water. J Food Prot 74(3):480–499. https://doi.org/10.4315/0362-028X.JFP-10-301

Godfroid J (2014) Brucellosis. In: Tham W, Danielsson-Tham M-L (eds) Food associated pathogens. CRC, Boca Raton, FL, pp 33–49

Godschalk PC, Heikema AP, Gilbert M, Komagamine T, Ang CW, Glerum J, Brochu D, Li J, Yuki N, Jacobs BC, van Belkum A, Endtz HP (2004) The crucial role of *Campylobacter jejuni* genes in anti-ganglioside antibody induction in Guillain-Barre syndrome. J Clin Invest 114 (11):1659–1665. https://doi.org/10.1172/JCI15707

Goldenberg RL, Culhane JF, Iams JD, Romero R (2008) Epidemiology and causes of preterm birth. Lancet (London, England) 371(9606):75–84. https://doi.org/10.1016/s0140-6736(08)60074-4

Gordon MA (2008) *Salmonella* infections in immunocompromised adults. J Infect 56(6):413–422. https://doi.org/10.1016/j.jinf.2008.03.012

Grace D (2015) Food safety in low and middle income countries. Int J Environ Res Public Health 12 (9):10490–10507. https://doi.org/10.3390/ijerph120910490

Granum PE, Brynestad S, Osullivan K, Nissen H (1993) Enterotoxin from *Bacillus cereus* – production and biochemical characterization. Netherlands Milk Dairy J 47(2):63–70

Grass JE, Gould LH, Mahon BE (2013) Epidemiology of foodborne disease outbreaks caused by *Clostridium perfringens*, United States, 1998-2010. Foodborne Pathog Dis 10(2):131–136. https://doi.org/10.1089/fpd.2012.1316

Grimont PAD, Grimont F (2005) Enterobacter. In: Brenner DJ, Krieg NR, Staley JT (eds) Bergey's manual of systematic bacteriology, vol 2B, 2nd edn. Springer and Bergey's Manual Trust, Michigan, pp 661–670

Guerin MT, Martin W, Reiersen J, Berke O, McEwen SA, Bisaillon JR, Lowman R (2007) A farm-level study of risk factors associated with the colonization of broiler flocks with *Campylobacter* spp. in Iceland, 2001-2004. Acta Vet Scand 49:18. https://doi.org/10.1186/1751-0147-49-18

Guinebretiere MH, Auger S, Galleron N, Contzen M, De Sarrau B, De Buyser ML, Lamberet G, Fagerlund A, Granum PE, Lereclus D, De Vos P, Nguyen-The C, Sorokin A (2013) *Bacillus cytotoxicus* sp. nov. is a novel thermotolerant species of the *Bacillus cereus* Group occasionally associated with food poisoning. Int J Syst Evol Microbiol 63(Pt 1):31–40. https://doi.org/10.1099/ijs.0.030627-0

Haeckel E (1866) Die Generelle Morphologie der Organismen (The general morphology of organisms). Reimer, Berlin

Haines RB (1933) The bacterial flora developing on stored lean meat, especially with regard to 'slimy' meat. J Hygiene 33:175–182

Haines RB, Elliot EML (1944) Some bacteriological aspects of dehydrated foods. J Hyg 43:370–381

Haines WC, Harmon LG (1973) Effect of selected lactic acid bacteria on growth of *Staphylococcus aureus* and production of enterotoxin. Appl Microbiol 25(3):436–441

Halverson LJ, Clayton MK, Handelsman J (1993) Population biology of *Bacillus cereus* UW85 in the rhizosphere of field-grown soybeans. Soil Biol Biochem 25(4):485–493. https://doi.org/10.1016/0038-0717(93)90074-I

Haque A, Russell NJ (2005) Phenotypic and genotypic characterisation of *Bacillus cereus* isolates from Bangladeshi rice. Int J Food Microbiol 98(1):23–34. https://doi.org/10.1016/j.ijfoodmicro.2004.04.025

Harris LJ, Farber JN, Beuchat LR, Parish ME, Suslow TV, Garrett EH, Busta FF (2003) Outbreaks associated with fresh produce: incidence, growth, and survival of pathogens in fresh and fresh-cut produce. Compr Rev Food Sci Food Saf 2(1 Suppl):78–141

Havelaar AH, Kirk MD, Torgerson PR, Gibb HJ, Hald T, Lake RJ, Praet N, Bellinger DC, de Silva NR, Gargouri N, Speybroeck N, Cawthorne A, Mathers C, Stein C, Angulo FJ, Devleesschauwer B (2015) World Health Organization global estimates and regional comparisons of the burden of foodborne disease in 2010. PLoS Med 12(12):e1001923. https://doi.org/10.1371/journal.pmed.1001923

Hoffmann S, Maculloch B, Batz M (2015) Economic burden of major foodborne illnesses acquired in the United States. U.S. Department of Agriculture, Economic Research Service, Washington, DC

Holley RA, Patel D (2005) Improvement in shelf-life and safety of perishable foods by plant essential oils and smoke antimicrobials. Food Microbiol 22(4):273–292. https://doi.org/10.1016/j.fm.2004.08.006

Horrocks SM, Anderson RC, Nisbet DJ, Ricke SC (2009) Incidence and ecology of *Campylobacter jejuni* and coli in animals. Anaerobe 15(1-2):18–25. https://doi.org/10.1016/j.anaerobe.2008.09.001

Huhtanen CN, Naghski J, Custer CS, Russell RW (1976) Growth and toxin production by *Clostridium botulinum* in moldy tomato juice. Appl Environ Microbiol 32(5):711–715

Hulebak KL, Schlosser W (2002) Hazard Analysis and Critical Control Point (HACCP) history and conceptual overview. Risk Anal 22(3):547–552. https://doi.org/10.1111/0272-4332.00038

Humphrey TJ, Whitehead A (1993) Egg age and the growth of *Salmonella enteritidis* PT4 in egg contents. Epidemiol Infect 111(2):209–219

Humphrey T, Mason M, Martin K (1995) The isolation of *Campylobacter jejuni* from contaminated surfaces and its survival in diluents. Int J Food Microbiol 26(3):295–303

Hurst A, Collins-Thompson DL, Alexander M (1979) Food as a bacterial habitat. Springer, Berlin, pp 79–134

Hurst CJ, Clark RM, Regli SE (1996) Estimating the risk of acquiring infectious disease from ingestion of water. In: Hurst CJ (ed) Modeling disease transmission and its prevention by disinfection. Cambridge University Press, Cambridge, pp 99–139

Hutchings LM, McCullough NB, Donham CR, Eisele CE, Bunnell DE (1951) The viability of *Brucella melitensis* in naturally infected cured hams. Public Health Rep 66(43):1402–1408

Iandolo JJ, Clark CW, Bluhm L, Ordal ZJ (1965) Repression of *Staphylococcus aureus* in associative culture. Appl Microbiol 13(5):646–649

ICMSF (1996) Characteristics of microbial pathogens. Microorganisms in foods. Springer, London

Indikova I, Humphrey TJ, Hilbert F (2015) Survival with a helping hand: *Campylobacter* and microbiota. Front Microbiol 6:1266. https://doi.org/10.3389/fmicb.2015.01266

Ingram M (1949) Science in the imported meat industry. Hygiene and storage. J R Sanit Inst Lond 69:39–47

Ivanova VI (1960) Survival of Brucella in salted and frozen meat. Труды Всесоюзного Института Ветеринарной Санитарии Proc All-Union Inst Vet Sanit 17:76–81

Iversen C, Lane M, Forsythe SJ (2004) The growth profile, thermotolerance and biofilm formation of *Enterobacter sakazakii* grown in infant formula milk. Lett Appl Microbiol 38(5):378–382. https://doi.org/10.1111/j.1472-765X.2004.01507.x

Iversen C, Mullane N, McCardell B, Tall BD, Lehner A, Fanning S, Stephan R, Joosten H (2008) *Cronobacter* gen. nov., a new genus to accommodate the biogroups of *Enterobacter sakazakii*, and proposal of *Cronobacter sakazakii* gen. nov., comb. nov., *Cronobacter malonaticus* sp. nov., *Cronobacter turicensis* sp. nov., *Cronobacter muytjensii* sp. nov., *Cronobacter dublinensis* sp. nov., *Cronobacter* genomospecies 1, and of three subspecies, *Cronobacter dublinensis* subsp. *dublinensis* subsp. nov., *Cronobacter dublinensis* subsp. *lausannensis* subsp. nov. and *Cronobacter dublinensis* subsp. *lactaridi* subsp. nov. Int J Syst Evol Microbiol 58(Pt 6):1442–1447. https://doi.org/10.1099/ijs.0.65577-0

Jackson DN, Davis B, Tirado SM, Duggal M, van Frankenhuyzen JK, Deaville D, Wijesinghe MAK, Tessaro M, Trevors JT (2009) Survival mechanisms and culturability of *Campylobacter jejuni* under stress conditions. Antonie Van Leeuwenhoek 96(4):377–394

Jaradat ZW, Al Mousa W, Elbetieha A, Al Nabulsi A, Tall BD (2014) *Cronobacter* spp. – opportunistic food-borne pathogens. A review of their virulence and environmental-adaptive traits. J Med Microbiol 63(Pt 8):1023–1037. https://doi.org/10.1099/jmm.0.073742-0

Jensen GB, Hansen BM, Eilenberg J, Mahillon J (2003) The hidden lifestyles of *Bacillus cereus* and relatives. Environ Microbiol 5(8):631–640. https://doi.org/10.1046/j.1462-2920.2003.00461.x

Juneja VK, Dwivedi HP, Yan X (2012) Novel natural food antimicrobials. Annu Rev Food Sci Technol 3(1):381–403. https://doi.org/10.1146/annurev-food-022811-101241

Kaakoush NO, Castano-Rodriguez N, Mitchell HM, Man SM (2015) Global epidemiology of *Campylobacter* infection. Clin Microbiol Rev 28(3):687–720. https://doi.org/10.1128/CMR.00006-15

Kaneko T, Colwell RR (1973) Ecology of *Vibrio parahaemolyticus* in Chesapeake Bay. J Bacteriol 113(1):24–32

Kara R, Gök V, Kenar B, Akkaya L (2012) Behavior of *Brucella abortus* and *Brucella melitensis* in raw meatball (Cig Kofte). J Food Res 1(2):274. https://doi.org/10.5539/jfr.v1n2p274

Karch H, Tarr PI, Bielaszewska M (2005) Enterohaemorrhagic *Escherichia coli* in human medicine. Int J Med Microbiol 295(6–7):405–418. https://doi.org/10.1016/j.ijmm.2005.06.009

Keller LH, Benson CE, Krotec K, Eckroade RJ (1995) *Salmonella enteritidis* colonization of the reproductive tract and forming and freshly laid eggs of chickens. Infect Immunity 63 (7):2443–2449

Kim J, Foegeding PM (1993) Principles of control. In: Hauschild AHW, Dodds KL (eds) *Clostridium botulinum.* Ecology and control in foods. Marcel Dekker, New York, pp 121–176

Kim YW, Lee SH, Hwang IG, Yoon KS (2012) Effect of temperature on growth of *Vibrio parahaemolyticus* and *Vibrio vulnificus* in flounder, salmon sashimi and oyster meat. Int J Environ Res Public Health 9(12):4662–4675

Kim B, Bang J, Kim H, Kim Y, B-s K, Beuchat LR, Ryu J-H (2014) *Bacillus cereus* and *Bacillus thuringiensis* spores in Korean rice: prevalence and toxin production as affected by production area and degree of milling. Food Microbiol 42(0):89–94. https://doi.org/10.1016/j.fm.2014.02.021

Kotiranta A, Lounatmaa K, Haapasalo M (2000) Epidemiology and pathogenesis of *Bacillus cereus* infections. Microb Infect 2(2):189–198

Laukkanen-Ninios R, Fredriksson-Ahomaa M, Korkeala H (2014) Enteropathogenic *Yersinia* in the pork production chain: challenges for control. Compr Rev Food Sci Food Saf 13(6):1165–1191. https://doi.org/10.1111/1541-4337.12108

Leclair D, Farber JM, Doidge B, Blanchfield B, Suppa S, Pagotto F, Austin JW (2013) Distribution of *Clostridium botulinum* type E strains in Nunavik, Northern Quebec, Canada. Appl Environ Microbiol 79(2):646–654. https://doi.org/10.1128/AEM.05999-11

Leistner L, Gould GW (1994) Principles and applications of hurdle technology. Blackie Academic and Professional, Glasgow, pp 1–21

Lianou A, Sofos JN (2007) A review of the incidence and transmission of *Listeria monocytogenes* in ready-to-eat products in retail and food service environments. J Food Prot 70(9):2172–2198

Liu SL, Ezaki T, Miura H, Matsui K, Yabuuchi E (1988) Intact motility as a *Salmonella typhi* invasion-related factor. Infect Immunity 56(8):1967–1973

Liverani E, Scaioli E, Cardamone C, Dal Monte P, Belluzzi A (2014) *Mycobacterium avium* subspecies paratuberculosis in the etiology of Crohn's disease, cause or epiphenomenon? World J Gastroenterol 20(36):13060–13070. https://doi.org/10.3748/wjg.v20.i36.13060

LSPN (2017) List of prokaryotic names with standing in nomenclature. http://www.bacterio.net/index.html. Accessed 24 Apr 2017

Luber P, Bartelt E (2007) Enumeration of *Campylobacter* spp. on the surface and within chicken breast fillets. J Appl Microbiol 102(2):313–318

Lücke FK (2000) Utilization of microbes to process and preserve meat. Meat Sci 56(2):105–115

Macgowan AP, Bowker K, McLauchlin J, Bennett PM, Reeves DS (1994) The occurrence and seasonal changes in the isolation of *Listeria* spp. in shop bought food stuffs, human faeces, sewage and soil from urban sources. Int J Food Microbiol 21(4):325–334

Magee JG, Ward AC (2012) *Mycobacterium.* In: Brenner DJ, Krieg NR, Staley JT (eds) Bergey's manual of systematic bacteriology, vol 5A, 2nd edn. Springer and Bergey's Manual Trust, Michigan, p 312

Mahler H, Pasi A, Kramer JM, Schulte P, Scoging AC, Bar W, Krahenbuhl S (1997) Fulminant liver failure in association with the emetic toxin of *Bacillus cereus*. N Engl J Med 336 (16):1142–1148. https://doi.org/10.1056/nejm199704173361604

Malley TJ, Butts J, Wiedmann M (2015) Seek and destroy process: *Listeria monocytogenes* process controls in the ready-to-eat meat and poultry industry. J Food Prot 78(2):436–445. https://doi.org/10.4315/0362-028x.jfp-13-507

Margulis L, Jorgensen JZ, Dolan S, Kolchinsky R, Rainey FA, Lo SC (1998) The *Arthromitus* stage of *Bacillus cereus*: intestinal symbionts of animals. Proc Natl Acad Sci USA 95(3):1236–1241

Marineli F, Tsoucalas G, Karamanou M, Androutsos G (2013) Mary Mallon (1869-1938) and the history of typhoid fever. Ann Gastroenterol 26(2):132–134

Mattick KL, Jorgensen F, Legan JD, Lappin-Scott HM, Humphrey TJ (2000) Habituation of *Salmonella* spp. at reduced water activity and its effect on heat tolerance. Appl Environ Microbiol 66(11):4921–4925

Masson PL, Heremans JF (1971) Lactoferrin in milk from different species. Comp Biochem Physiol B 39(1):119–129

McCollum JT, Cronquist AB, Silk BJ, Jackson KA, O'Connor KA, Cosgrove S, Gossack JP, Parachini SS, Jain NS, Ettestad P, Ibraheem M, Cantu V, Joshi M, DuVernoy T, Fogg NW, Gorny JR, Mogen KM, Spires C, Teitell P, Joseph LA, Tarr CL, Imanishi M, Neil KP, Tauxe RV, Mahon BE (2013) Multistate outbreak of listeriosis associated with cantaloupe. N Engl J Med 369 (10):944–953. https://doi.org/10.1056/NEJMoa1215837

McQuestin OJ, Shadbolt CT, Ross T (2009) Quantification of the relative effects of temperature, pH, and water activity on inactivation of *Escherichia coli* in fermented meat by meta-analysis. Appl Environ Microbiol 75(22):6963–6972. https://doi.org/10.1128/aem.00291-09

Meador VP, Deyoe BL, Cheville NF (1989) Pathogenesis of *Brucella abortus* infection of the mammary gland and supramammary lymph node of the goat. Vet Pathol 26(5):357–368

Meibom KL, Li XB, Nielsen AT, Wu CY, Roseman S, Schoolnik GK (2004) The *Vibrio cholerae* chitin utilization program. Proc Natl Acad Sci USA 101(8):2524–2529

Méndez-González KY, Hernández-Castro R, Carrillo-Casas EM, Monroy JF, López-Merino A, Suárez-Güemes F (2011) *Brucella melitensis* survival during manufacture of ripened goat cheese at two temperatures. Foodborne Pathog Dis 8(12):1257–1261. https://doi.org/10.1089/fpd.2011.0887

Moreno E (2014) Retrospective and prospective perspectives on zoonotic brucellosis. Front Microbiol 5 (May). doi:https://doi.org/10.3389/fmicb.2014.00213

Mossel DAA, Corry JEL, Stuijk CB, Baird RM (1995) Essentials of the microbiology of foods. Wiley, Chichester, England

Mtenga AB, Kassim N, Lee WG, Shim WB, Yoon Y, Chung DH (2012) Resistance of *Bacillus cereus* and its enterotoxin genes in ready-to-eat foods to γ-irradiation. Food Sci Biotechnol 21 (2):443–452

Mtenga AB, Kassim N, Shim WB, Chung DH (2013) Efficiency of fractionated γ-irradiation doses to eliminate vegetative cells and spores of *Bacillus cereus* from raw rice. Food Sci Biotechnol 22 (2):577–584. https://doi.org/10.1007/s10068-013-0117-6

Mueller-Spitz SR, Stewart LB, Val Klump J, McLellan SL (2010) Freshwater suspended sediments and sewage are reservoirs for enterotoxin-positive *Clostridium perfringens*. Appl Environ Microbiol 76(16):5556–5562. https://doi.org/10.1128/aem.01702-09

Murray EGD, Webb RA, Swann MBR (1926) A disease of rabbits characterised by a large mononuclear leucocytosis, caused by a hitherto undescribed bacillus *Bacterium monocytogenes* (n.sp.). J Pathol Bacteriol 29(4):407–439. https://doi.org/10.1002/path.1700290409

Musgrove MT, Berrang ME, Byrd JA, Stern NJ, Cox NA (2001) Detection of *Campylobacter* spp. in ceca and crops with and without enrichment. Poult Sci 80(6):825–828

Naranjo M, Denayer S, Botteldoorn N, Delbrassinne L, Veys J, Waegenaere J, Sirtaine N, Driesen RB, Sipido KR, Mahillon J, Dierick K (2011) Sudden death of a young adult associated with *Bacillus cereus* food poisoning. J Clin Microbiol 49(12):4379–4381. https://doi.org/10.1128/jcm.05129-11

Norberg S, Stanton C, Ross RP, Hill C, Fitzgerald GF, Cotter PD (2012) *Cronobacter* spp. in powdered infant formula. J Food Prot 75(3):607–620. https://doi.org/10.4315/0362-028X.JFP-11-285

Novoslavskij A, Terentjeva M, Eizenberga I, Valciņa O, Bartkevičs V, Bērziņš A (2016) Major foodborne pathogens in fish and fish products: a review. Ann Microbiol 66(1):1–15. https://doi.org/10.1007/s13213-015-1102-5

Oh E, McMullen L, Jeon B (2015) Impact of oxidative stress defense on bacterial survival and morphological change in *Campylobacter jejuni* under aerobic conditions. Front Microbiol 6:295

Okereke A, Montville TJ (1991) Bacteriocin inhibition of *Clostridium botulinum* spores by lactic acid bacteria. J Food Prot 54(5):349–353, 356

Olaimat AN, Holley RA (2012) Factors influencing the microbial safety of fresh produce: a review. Food Microbiol 32(1):1–19. https://doi.org/10.1016/j.fm.2012.04.016

Oliver JD (2013) *Vibrio vulnificus*: death on the half shell. A personal journey with the pathogen and its ecology. Microb Ecol 65(4):793–799. https://doi.org/10.1007/s00248-012-0140-9

Oliver SP, Jayarao BM, Almeida RA (2005) Foodborne pathogens in milk and the dairy farm environment: food safety and public health implications. Foodborne Pathog Dis 2(2):115–129. https://doi.org/10.1089/fpd.2005.2.115

Ooi ST, Lorber B (2005) Gastroenteritis due to *Listeria monocytogenes*. Clin Infect Dis 40 (9):1327–1332. https://doi.org/10.1086/429324

Orsi RH (2008) Short-term genome evolution of *Listeria monocytogenes* in a non-controlled environment. BMC Genomics 9(1):539–539. https://doi.org/10.1186/1471-2164-9-539

Ortiz S, López V, Villatoro D, López P, Dávila JC, Martínez-Suárez JV (2010) A 3-Year surveillance of the genetic diversity and persistence of *Listeria monocytogenes* in an Iberian pig slaughterhouse and processing plant. Foodborne Pathog Dis 7(10):1177–1184. https://doi.org/10.1089/fpd.2010.0535

Osaili T, Forsythe S (2009) Desiccation resistance and persistence of *Cronobacter* species in infant formula. Int J Food Microbiol 136(2):214–220. https://doi.org/10.1016/j.ijfoodmicro.2009.08.006

Pachepsky YA, Shelton DR (2011) *Escherichia coli* and fecal coliforms in freshwater and estuarine sediments. Crit Rev Environ Sci Technol 41(12):1067–1110. https://doi.org/10.1080/10643380903392718

Palanduz A, Palanduz Ş, Güler K, Güler N (2000) Brucellosis in a mother and her young infant: probable transmission by breast milk. Int J Infect Dis 4(1):55–56

Pappas G, Papadimitriou P, Akritidis N, Christou L, Tsianos EV (2006) The new global map of human brucellosis. Lancet Infect Dis 6(2):91–99. https://doi.org/10.1016/s1473-3099(06)70382-6

Patrick ME, Mahon BE, Greene SA, Rounds J, Cronquist A, Wymore K, Boothe E, Lathrop S, Palmer A, Bowen A (2014) Incidence of *Cronobacter* spp. infections, United States, 2003–2009. Emerg Infect Dis 20(9):1520–1523. https://doi.org/10.3201/eid2009.140545

Peng S, Tasara T, Hummerjohann J, Stephan R (2011) An overview of molecular stress response mechanisms in *Escherichia coli* contributing to survival of shiga toxin-producing *Escherichia coli* during raw milk cheese production. J Food Prot 74(5):849–864. https://doi.org/10.4315/0362-028x.jfp-10-469

Podolak R, Enache E, Stone W, Black DG, Elliott PH (2010) Sources and risk factors for contamination, survival, persistence, and heat resistance of *Salmonella* in low-moisture foods. J Food Prot 73(10):1919–1936

Popoff MR (2015) From saprophytic to toxigenic Clostridia, a complex evolution based on multiple diverse genetic transfers and/or rearrangements. Res Microbiol 166(4):221–224. https://doi.org/10.1016/j.resmic.2015.02.008

Popoff MY, Le Minor L (2005) Salmonella. In: Brenner DJ, Krieg NR, Staley JT (eds) Bergey's manual of systematic bacteriology, vol 2B, 2nd edn. Springer and Bergey's Manual Trust, Michigan, pp 764–799

Pruzzo C, Vezzulli L, Colwell RR (2008) Global impact of *Vibrio cholerae* interactions with chitin. Environ Microbiol 10(6):1400–1410. https://doi.org/10.1111/j.1462-2920.2007.01559.x

Radosavljevic V, Finke E-J, Belojevic G (2015) *Escherichia coli* O104:H4 outbreak in Germany—clarification of the origin of the epidemic. Eur J Public Health 25(1):125–129. https://doi.org/10.1093/eurpub/cku048

Rainard P (1986) Bacteriostatic activity of bovine milk lactoferrin against mastitic bacteria. Vet Microbiol 11(4):387–392

Richards GM, Beuchat LR (2005) Metabiotic associations of molds and *Salmonella* Poona on intact and wounded cantaloupe rind. Int J Food Microbiol 97(3):327–339

Rodgers S, Peiris P, Casadei G (2003) Inhibition of nonproteolytic *Clostridium botulinum* with lactic acid bacteria and their bacteriocins at refrigeration temperatures. J Food Prot 66 (4):674–678

Ronholm J, Petronella N, Chew Leung C, Pightling AW, Banerjee SK (2015) Genomic features of environmental and clinical *Vibrio parahaemolyticus* isolates lacking recognized virulence factors are dissimilar. Appl Environ Microbiol 82(4):1102–1113. https://doi.org/10.1128/AEM.03465-15

Rowbury RJ, Goodson M (1998) Induction of acid tolerance at neutral pH in log-phase *Escherichia coli* by medium filtrates from organisms grown at acidic pH. Lett Appl Microbiol 26 (6):447–451

Samish Z, Etinger-Tulczynska R (1963) Distribution of bacteria within the tissue of healthy tomatoes. Appl Microbiol 11:7–10

Sampers I, Habib I, De Zutter L, Dumoulin A, Uyttendaele M (2010) Survival of *Campylobacter* spp. in poultry meat preparations subjected to freezing, refrigeration, minor salt concentration, and heat treatment. Int J Food Microbiol 137(2–3):147–153. https://doi.org/10.1016/j.ijfoodmicro.2009.11.013

Sanjuan E, Fouz B, Oliver JD, Amaro C (2009) Evaluation of genotypic and phenotypic methods to distinguish clinical from environmental *Vibrio vulnificus* strains. Appl Environ Microbiol 75 (6):1604–1613. https://doi.org/10.1128/AEM.01594-08

Santiago-Rodríguez MDR, Díaz-Aparicio E, Arellano-Reynoso B, García-Lobo JM, Gimeno M, Palomares-Reséndiz EG, Hernández-Castro R (2015) Survival of *Brucella abortus* aqpX mutant in fresh and ripened cheeses. Foodborne Pathog Dis 12(2):170–175. https://doi.org/10.1089/fpd.2014.1823

Sarrías JA, Valero M, Salmerón MC (2002) Enumeration, isolation and characterization of *Bacillus cereus* strains from Spanish raw rice. Food Microbiol 19(6):589–595

Scheutz F, Strockbine N (2005) Genus I. *Escherichia*. In: Garrity GM, Brenner DJ, Krieg NR, Staley JT (eds) Bergey's manual of systematic bacteriology, vol 2B. Springer, New York, pp 607–627

Seals JE, Snyder JD, Edell TA, Hatheway CL, Johnson CJ, Swanson RC, Hughes JM (1981) Restaurant-associated type A botulism: transmission by potato salad. Am J Epidemiol 113 (4):436–444

Sears A, Baker MG, Wilson N, Marshall J, Muellner P, Campbell DM, Lake RJ, French NP (2011) Marked campylobacteriosis decline after interventions aimed at poultry, New Zealand. Emerg Infect Dis 17(6):1007–1015

Setlow P, Johnson EA (2007) Spores and their significance. In: Doyle MP, Beuchat LR (eds) Food microbiology. Fundamentals and frontiers. ASM Press, Washington, DC, pp 35–68

Shaker RR, Osaili TM, Abu Al-Hasan AS, Ayyash MM, Forsythe SJ (2008) Effect of desiccation, starvation, heat, and cold stresses on the thermal resistance of *Enterobacter sakazakii* in rehydrated infant milk formula. J Food Sci 73(7):M354–M359. https://doi.org/10.1111/j.1750-3841.2008.00880.x

Shiota M, Saitou K, Mizumoto H, Matsusaka M, Agata N, Nakayama M, Kage M, Tatsumi S, Okamoto A, Yamaguchi S, Ohta M, Hata D (2010) Rapid detoxification of cereulide in *Bacillus cereus* food poisoning. Pediatrics 125(4):e951–e955. https://doi.org/10.1542/peds.2009-2319

Silk BJ, Mahon BE, Griffin PM, Hannah Gould L, Tauxe RV, Crim SM, Jackson KA, Gerner-Smidt P, Herman KM, Henao OL (2013) Vital signs: Listeria illnesses, deaths, and outbreaks – United States, 2009-2011. Morb Mortal Wkly Rep 62(22):448–452

Sjogren RE (1994) Prolonged survival of an environmental *Escherichia coli* in laboratory soil microcosms. Water Air Soil Pollut 75(3-4):389–403. https://doi.org/10.1007/bf00482948

Smelt J, Raatjes GJM, Crowther JS, Verrips CT (1982) Growth and toxin formation by *Clostridium botulinum* at low pH values. J Appl Bacteriol 52(1):75–82 23 ref

Smith TJ, Hill KK, Xie G, Foley BT, Williamson CH, Foster JT, Johnson SL, Chertkov O, Teshima H, Gibbons HS, Johnsky LA, Karavis MA, Smith LA (2015) Genomic sequences of six botulinum neurotoxin-producing strains representing three clostridial species illustrate the mobility and diversity of botulinum neurotoxin genes. Infect Genet Evol 30:102–113. https://doi.org/10.1016/j.meegid.2014.12.002

Spencer K (2005) Modified atmosphere packaging if ready-to-eat foods. In: Innovation in food packaging. Elsevier Academic, Amsterdam, pp 185–203

Sperber WH (1983) Influence of water activity on foodborne bacteria - a review. J Food Prot 46 (2):142–150

Stenfors Arnesen LP, Fagerlund A, Granum PE (2008) From soil to gut: *Bacillus cereus* and its food poisoning toxins. FEMS Microbiol Rev 32(4):579–606

Stern NJ, Pierson MD, Kotula AW (1980) Effects of pH and sodium chloride on *Yersinia enterocolitica* growth at room and refrigeration temperatures. J Food Sci 45:64–67

Stoodley P, Sauer K, Davies DG, Costerton JW (2002) Biofilms as complex differentiated communities. Annu Rev Microbiol 56:187–209

Strockbine NA, Maurelli AT (2005) Genus XXXV. *Shigella*. In: Garrity GM, Brenner DJ, Krieg NR, Staley JT (eds) Bergey's manual of systematic bacteriology. Springer, New York, pp 811–823

Su YC, Liu C (2007) *Vibrio parahaemolyticus*: a concern of seafood safety. Food Microbiol 24 (6):549–558. https://doi.org/10.1016/j.fm.2007.01.005

Takemura AF, Chien DM, Polz MF (2014) Associations and dynamics of *Vibrionaceae* in the environment, from the genus to the population level. Front Microbiol 5:38. https://doi.org/10.3389/fmicb.2014.00038

Tanaka N, Meske L, Doyle MP, Traisman E, Thayer DW, Johnston RW (1985) Plant trials of bacon made with lactic acid bacteria, sucrose and lowered sodium nitrite. J Food Prot 48(8):679–686

Tauveron G, Slomianny C, Henry C, Faille C (2006) Variability among *Bacillus cereus* strains in spore surface properties and influence on their ability to contaminate food surface equipment. Int J Food Microbiol 110(3):254–262

Thayer DW, Boyd G (1994) Control of enterotoxic *Bacillus cereus* on poultry or red meats and in beef gravy by gamma irradiation. J Food Prot 57(9):758–764

Thimothe J, Nightingale KK, Gall K, Scott VN, Wiedmann M (2004) Tracking of *Listeria monocytogenes* in smoked fish processing plants. J Food Prot 67(2):328–341

Thoen C, Lobue P, de Kantor I (2006) The importance of *Mycobacterium bovis* as a zoonosis. Vet Microbiol 112(2–4):339–345. https://doi.org/10.1016/j.vetmic.2005.11.047

Thomas MK, Murray R, Flockhart L, Pintar K, Pollari F, Fazil A, Nesbitt A, Marshall B (2013) Estimates of the burden of foodborne illness in Canada for 30 specified pathogens and unspecified agents, Circa 2006. Foodborne Pathog Dis 10(7):639–648

Todd ECD, Greig JD, Bartleson CA, Michaels BS (2008) Outbreaks where food workers have been implicated in the spread of foodborne disease. Part 4. Infective doses and pathogen carriage. J Food Prot 71(11):2339–2373

Turnbull PCB (1985) Intestinal carriage of *Bacillus cereus*: faecal isolation studies in three population groups. J Hyg 95(3):629–638

Vaerewijck MJM, Baré J, Lambrecht E, Sabbe K, Houf K (2014) Interactions of foodborne pathogens with free-living protozoa: potential consequences for food safety. Compr Rev Food Sci Food Saf 13(5):924–944. https://doi.org/10.1111/1541-4337.12100

Van Den Beld MJC, Reubsaet FAG (2012) Differentiation between *Shigella*, enteroinvasive *Escherichia coli* (EIEC) and noninvasive *Escherichia coli*. Eur J Clin Microbiol Infect Dis 31 (6):899–904. https://doi.org/10.1007/s10096-011-1395-7

Van Elsas JD, Semenov AV, Costa R, Trevors JT (2011) Survival of *Escherichia coli* in the environment: fundamental and public health aspects. ISME J 5(2):173–183. https://doi.org/10.1038/ismej.2010.80

van Netten P, van De Moosdijk A, van Hoensel P, Mossel DA, Perales I (1990) Psychrotrophic strains of *Bacillus cereus* producing enterotoxin. J Appl Bacteriol 69(1):73–79

Vandamme P, Dewhirst FE, Paster BJ, On SLW (2005) Campylobacteraceae. In: Garrity GM, Brenner DJ, Krieg NR, Staley JT (eds) Bergey's manual of systematic bacteriology, vol 2. Springer Science. New York, NY, pp 1147–1160

Viazis S, Beal JK, Monahan C, Lanier WA, Kreil KR, Melka DC, Boden WD, Dion JL, Miller ZA, Nguyen TA, Gieraltowski LB, Zink DL (2015) Laboratory, environmental, and epidemiologic investigation and regulatory enforcement actions in response to an outbreak of *Salmonella* Bredeney infections linked to peanut butter. Open Forum Infect Dis 2(3):ofv114. https://doi.org/10.1093/ofid/ofv114

Vierheilig J, Frick C, Mayer RE, Kirschner AKT, Reischer GH, Derx J, Mach RL, Sommer R, Farnleitner AH (2013) *Clostridium perfringens* is not suitable for the indication of fecal pollution from ruminant wildlife but is associated with excreta from nonherbivorous animals and human sewage. Appl Environ Microbiol 79(16):5089–5092. https://doi.org/10.1128/aem.01396-13

Vugia DJ, Mase SR, Cole B, Stiles J, Rosenberg J, Velasquez L, Radner A, Inami G (2009) Botulism from drinking pruno. Emerg Infect Dis 15(1):69–71. https://doi.org/10.3201/eid1501.081024

Wade WN, Beuchat LR (2003) Metabiosis of proteolytic moulds and *Salmonella* in raw, ripe tomatoes. J Appl Microbiol 95(3):437–450

Wade WN, Vasdinnyei R, Deak T, Beuchat LR (2003) Proteolytic yeasts isolated from raw, ripe tomatoes and metabiotic association of *Geotrichum candidum* with *Salmonella*. Int J Food Microbiol 86(1-2):101–111

Walters MS, Sreenivasan N, Person B, Shew M, Wheeler D, Hall J, Bogdanow L, Leniek K, Rao A (2015) A qualitative inquiry about pruno, an illicit alcoholic beverage linked to botulism outbreaks in United States prisons. Am J Public Health 105(11):2256–2261. https://doi.org/10.2105/ajph.2015.302774

Werber D, Dreesman J, Feil F, van Treeck U, Fell G, Ethelberg S, Hauri AM, Roggentin P, Prager R, Fisher IS, Behnke SC, Bartelt E, Weise E, Ellis A, Siitonen A, Andersson Y, Tschape H, Kramer MH, Ammon A (2005) International outbreak of *Salmonella* Oranienburg due to German chocolate. BMC Infect Dis 5:7. https://doi.org/10.1186/1471-2334-5-7

Westerdijk J (1949) The concept 'association' in mycology. Antonie van Leeuwenhoek 15:187–189

WHO (2014) Cholera. WHO fact sheet N°107. WHO, Geneva

Williams BT, Schlein SM, Caravati EM, Ledyard H, Fix ML (2014) Emergency department identification and critical care management of a Utah prison botulism outbreak. Ann Emerg Med 64(1):26–31. https://doi.org/10.1016/j.annemergmed.2013.10.027

Wright RC (1989) The survival patterns of selected faecal bacteria in tropical fresh waters. Epidemiol Infect 103(3):603–611

Wu J, Acero-Lopez A (2012) Ovotransferrin: structure, bioactivities, and preparation. Food Research Int 46(2):480–487. https://doi.org/10.1016/j.foodres.2011.07.012

Xavier MN, Paixão TA, Poester FP, Lage AP, Santos RL (2009) Pathological, immunohistochemical and bacteriological study of tissues and milk of cows and fetuses experimentally infected with *Brucella abortus*. J Comp Pathol 140(2-3):149–157. https://doi.org/10.1016/j.jcpa.2008.10.004

Yang X, Tran F, Youssef MK, Gill CO (2015) Determination of sources of *Escherichia coli* on beef by multiple-locus variable-number tandem repeat analysis. J Food Prot 78(7):1296–1302

Yasmin S, Adams L, Briggs G, Weiss J, Bisgard K, Anderson S, Tsang C, Henke E, Vasiq M, Komatsu K (2015) Outbreak of botulism after consumption of illicit prison-brewed alcohol in a maximum security prison--Arizona, 2012. J Correct Health Care 21(4):327–334. https://doi.org/10.1177/1078345815604752

Yoo JR, Heo ST, Lee KH, Kim YR, Yoo SJ (2015) Foodborne outbreak of human brucellosis caused by ingested raw materials of fetal calf on Jeju Island. Am J Trop Med Hyg 92(2):267–269. https://doi.org/10.4269/ajtmh.14-0399

Young CS, Lethbridge G, Shaw LJ, Burns RG (1995) Survival of inoculated *Bacillus cereus* spores and vegetative cells in non-planted and rhizosphere soil. Soil Biol Biochem 27(8):1017–1026

Youssef MK, Badoni M, Yang X, Gill CO (2013) Sources of *Escherichia coli* deposited on beef during breaking of carcasses carrying few *E. coli* at two packing plants. Food Control 31 (1):166–171. https://doi.org/10.1016/j.foodcont.2012.09.045

Zhang G, Ma L, Phelan VH, Doyle MP (2009) Efficacy of antimicrobial agents in lettuce leaf processing water for control of *Escherichia coli* O157:H7. J Food Prot 72(7):1392–1397

Ziegler EE (2011) Meeting the nutritional needs of the low-birth-weight infant. Ann Nutr Metab 58 (Suppl 1):8–18. https://doi.org/10.1159/000323381

Zuo G, Xu Z, Hao B (2013) *Shigella* strains are not clones of *Escherichia coli* but sister species in the Genus *Escherichia*. Genomics Proteomics Bioinformatics 11(1):61–65. https://doi.org/10.1016/j.gpb.2012.11.002

Chapter 5
Impact of Mycobacterial Biofilms on Public Health

Anil K. Ojha

Abstract The genus *Mycobacterium* represents over 150 bacterial species of the actinomycetes family, inhabiting a wide range of ecological niches—from soil and aquatic environments to intracellular phagosomes in human bodies. *Mycobacterium tuberculosis (Mtb)*, the etiological agent of tuberculosis in human, is the predominant mycobacterial pathogen that resides in an estimated one-third of the world's human population, causing disease in hundreds of millions and killing over a million people every year. In addition, several environmental mycobacteria including *Mycobacterium avium* and *Mycobacterium abscessus* are opportunistic pathogens that can establish infection in immunocompromised individuals. A common characteristic of all mycobacterial infections, regardless of the species, is their extraordinary recalcitrance to antibiotic regimens, although the underlying basis of drug resistance remains unclear. Recent studies suggest a possible linkage between mycobacterial tolerance to antibiotics and their propensity to grow as organized multicellular aggregates, called biofilms. This chapter describes the linkage and its implication in controlling mycobacterial infections in humans.

5.1 Introduction

Homeostasis in the interfaces with commensal polymicrobial communities is a critical aspect of human life, perturbation of which is both a cause and a consequence of infection by pathogenic microbes. The defining feature of microbial pathogens is their ability to colonize the human body and inflict tissue damage, typically associated with the microbes either inducing inflammation, or producing toxins, or both. However, the line of separation between infectious and commensal microbes is blurred by those microbial species that can successfully establish infections inside

A. K. Ojha (✉)
Wadsworth Center, NY State Department of Health, New York, NY, USA

Department of Biomedical Sciences, University at Albany, New York, NY, USA
e-mail: anil.ojha@health.ny.gov

C. J. Hurst (ed.), *The Connections Between Ecology and Infectious Disease*,
Advances in Environmental Microbiology 5,
https://doi.org/10.1007/978-3-319-92373-4_5

human body but cause disease only under conditions of compromised host immunity. Mycobacteria represent a large number of species that belong to this category, of which the etiological agent of tuberculosis (TB) *Mycobacterium tuberculosis* (Mtb) is the most prominent and pervasive species. One in every three individuals on this planet is estimated to be infected with Mtb, of which only a fraction (about 5–10%) present the disease symptoms of clinical TB (WHO 2012). The remainder population, clinically termed as possessing latent TB, carry the risk of reactivation into symptomatic disease upon compromised immunity from various origins such as HIV infection (WHO 2012). The complexity of the global TB epidemic is further compounded by the inefficient treatment regimens, which involve at least 6-month-long administration of multiple antibiotics (Jindani et al. 2003). This extraordinary level of Mtb persistence against both the host immunity as well as antibiotics raises pertinent questions about the underpinning causes, which seem to be shared by several other mycobacterial species that also share many genetic and phenotypic characteristics with the TB pathogen. In addition to Mtb and related species that form Mtb complex (MTC), several nontuberculous mycobacteria (NTMs) also colonize human hosts. Of major significance among the NTMs are *Mycobacterium avium*, *Mycobacterium abscessus*, and *Mycobacterium ulcerans*, all of which establish chronic infection in diverse host niches and are extremely difficult to eradicate (Gonzalez-Santiago and Drage 2015; Stout et al. 2016).

Identification of factors underlying the persistent characteristics of mycobacteria has been a hot pursuit for a long time. Among the most established contributing factors is the thick, waxy, and uniquely architectured bacterial cell wall, constituted of a myriad of atypical lipids and lipoproteins, which forms a barrier against physicomechanical stress as well as chemical diffusion (Brennan and Nikaido 1995). In addition to physical recalcitrance, physiological adaptation through acquisition of a nonreplicating but viable state is also argued to facilitate mycobacterial survival under external threats, presumably by a "ride-through" mechanism (Parrish et al. 1998). However, the molecular basis underlying the development of persistent bacilli remains unclear. An ever-mounting body of evidence in recent years suggests that the natural propensity of mycobacteria to grow into self-organized multicellular communities, called biofilms, significantly facilitates their survival (Ojha et al. 2005, 2015; Richards and Ojha 2014), presumably through both physical occlusion as well as physiological adaptation. This chapter reviews the biofilm paradigm and its implications in the understanding of mycobacterial persistence in the context of public health.

5.2 Biofilm Paradigm: The Whole Is More Than Sum of All

5.2.1 Early Discoveries

In discovering the etiological agents of infectious diseases including TB, Robert Koch laid down the key criteria for establishing the causal relationship between pathogen and the disease. These criteria, called Koch's postulates, later had tremendous influence on the field of microbiology. Among the most striking contribution of the Koch's postulates is the widely held view of microbes as free-living unicellular organisms. This was founded on the widespread practice of excessive manipulation of culturing conditions in the laboratory, presumably to fulfill the second tenet of the postulate that requires clonally pure culture of a presumptive pathogen. Several microbiologists in the early twentieth century however cautioned on the view of microbial unicellularity as a laboratory artifact by providing evidence of surface-associated sessile occurrence of microbes in communities (Henrici 1933; Zobell 1943), although little attention was paid to their arguments in the following years, perhaps because of the needs for pure cultures in the rapidly emerging field of molecular microbial genetics. In the early 1980s, Costerton and colleagues reestablished the view of microbial plurality through a series of microscopic observations of infected tissues, medical implant devices, and environmental specimens (Costerton et al. 1981; Marrie et al. 1982). The overriding conclusion of Costerton's studies was that microbes invariably grow and live in complex multicellular communities, which appeared to be encased in extracellular polymeric substance. These microbial communities were called biofilms. Later investigations rendered further insights into the inner architecture of biofilms, providing evidence of water channels and movement of solutes in extracellular space (de Beer et al. 1994). An implication of this finding was an ordered organization of resident microbes in biofilms, raising basic questions about the underlying mechanisms. Clinical significance of biofilms was rendered by studies that found unusually high levels of antibiotic tolerance associated with microbes in biofilms as compared to their planktonic counterparts (Nickel et al. 1985). Interestingly, the biofilm-associated antibiotic tolerance is a metastable phenotype that disappears when cells are re-dispersed into planktonic suspensions (Nickel et al. 1985; Ojha et al. 2015), implying that tolerance phenotype develops within the environmental context of biofilms.

5.2.2 Molecular Insights into Developmental Process of Biofilms

Investigations using genetically tractable species like *Pseudomonas aeruginosa*, *Escherichia coli*, *Bacillus subtilis*, *Vibrio cholera*, and *Staphylococcus aureus* have provided knowledge about the molecular process underlying the development of biofilms and differentiation of resident microbes within the architecture (Chai et al. 2008; Hall-Stoodley et al. 2004; Kolter and Greenberg 2006; Kolter and Losick 1998). Although each species appears to have evolved with specific environmental and genetic requirements for growth and development of biofilms, they also share multiple aspects of the developmental processes. Biofilms of a majority of the species proceed through distinct stages of development, initiated by attachment of single-cell planktonic bacilli on a substratum, followed by aggregated growth of bacilli into a multicellular community which matures through encapsulation of the community by extracellular matrix (Hall-Stoodley and Stoodley 2002). The processes of attachment, aggregation, and maturation (matrix synthesis) are regulated by distinct set of genetic network (Hall-Stoodley and Stoodley 2002). Upon attachment to a substratum, bacteria change their global gene expression pattern, and those changes include downregulation of their motility-associated flagellar biosynthesis genes with concomitant upregulation of adhesins, pillin, and other substratum-anchoring proteins (Richards and Ojha 2014). This transcriptional reprogramming facilitates transition from a motile existence as planktonic cells to their surface-associated sessile growth. Substratum attachment by *Pseudomonas aeruginosa* occurs in two stages. First the cells attach "reversibly" to a surface and then transition into an irreversibly attached state through expression of a surface-attachment protein, SadB (Caiazza and O'Toole 2004; Hinsa et al. 2003). This two-stage process is argued as a niche sampling process by the organism, although evidence of such stepwise attachment in other microbes remains undiscovered. Genetic distinctions between the process of substratum attachment and intercellular aggregation can be derived from studies in *V. cholerae* and *S. aureus*. *Vibrio cholerae* expresses Bap1, a matrix-associated protein, for substratum attachment and Rmb1 for cell-to-cell aggregation (Absalon et al. 2011). Similarly, *S. aureus* uses MSCRAMMs (microbial surface components recognizing adhesive matrix molecules) for substratum attachment but polysaccharide intercellular adhesion for intercellular aggregation (Conrady et al. 2008; Cramton et al. 1999; Cucarella et al. 2001; Heilmann et al. 1996). While substratum and intercellular aggregation are key early-stage events in the developmental process, the process of maturation is never completed without the synthesis of extracellular polymeric substances (EPS) that constitute the matrix in which the cells are "glued" together. Polysaccharides, proteins, DNA, and lipids generally constitute EPS, although relative abundance of these molecules in EPS varies across the species (Branda et al. 2005; Fux et al. 2005).

5.2.3 Spatial Differentiation of Microbes in Biofilms

Development of three-dimensional architecture is associated with occlusion of nutrient and oxygen to the resident cells within biofilms (Kolter and Losick 1998; Stoodley et al. 2002). Thus, the interior cells must adapt to the self-generated restrictions on oxygen and nutrient supplies in order to grow and survive. Indeed, gene expression analyses in *B. subtilis* and *P. aeruginosa* using transcriptionally fused reporter systems have shown that cells in spatially different regions of biofilms produce different sets of gene products (Chai et al. 2008; Rani et al. 2007; Werner et al. 2004).

Strikingly only a minority of cells in *B. subtilis* biofilms are engaged in matrix synthesis. Moreover, the signaling mechanisms for matrix synthesis in *B. subtilis* are intricately linked to a sporulation pathway through an upstream regulator, SpoA (Chai et al. 2008). The linkage between the two biological processes—with seemingly disparate energy commitment—could possibly be viewed as a part of a feedback loop mechanism that signals bacilli to switch from a high-energy commitment state (biofilm producer) to a conservation state (sporulation) upon experiencing starvation. This is suggested by the distinct subpopulations of matrix producers and spore producers within a single biofilm of *B. subtilis*. Similarly, active metabolism in biofilms of *P. aeruginosa*, measured by transcription and DNA replication, was seen only in the top 25% of the total depth of the biofilms (Rani et al. 2007; Werner et al. 2004). The sharp decrease in metabolism between interior and exterior locations within a biofilm correlates with the microavailability of oxygen, and most cells (~90%) residing under low oxygen conditions seem to remain viable (Werner et al. 2004), suggesting a programmed metabolic slowdown in the majority of the cells. Further support for metabolic differentiation was obtained by real-time microsensor measurements, which indicated an anoxic environment and associated metabolic inactivity, without loss of viability, within the interiors of the dental biofilms (von Ohle et al. 2010).

Besides physiological and metabolic heterogeneity, resident cells in biofilms also display structural and morphological heterogeneities (Wimpenny et al. 2000). In *E. coli* biofilms, the fibrous surface proteins known as curli form the outer regions of the architecture encasing ovoid bacterial cells, whereas cells at the bottom of the biofilms are elongated and entangled in mesh-like structures (Serra et al. 2013). Importantly, development of these structural patterns is orchestrated by stationary phase sigma factor (RpoS) as well as alarmone ppGGpp and the secondary messenger cyclic di-GMP (Serra et al. 2013).

5.2.4 Interspecies Interactions in Polymicrobial Biofilms

Although the studies described above were performed in monospecies biofilms, the findings have profound implications in the context of polymicrobial biofilms. The changes in microenvironment which occur within the biofilms of primary colonizers could create new niches for the growth of secondary colonizers (Elias and Banin 2012). The secondary colonizers could also induce wide-ranging defense mechanisms in the primary colonizer (Rendueles and Ghigo 2012), that together maintain species balance within the community. A continuum of interspecies dynamics is perhaps best exemplified by the studies of polymicrobial biofilms in oral ecosystem, in which the primary colonizers, *Streptococcus mutans* and *Actinomyces* species, are important for growth of *Lactobacillus* species at a later time (Benedict et al. 1992). Similarly, the EPS of another primary colonizer in oral biofilms, *Streptococcus gordonii*, facilitate recruitment of *Porphyromonas gingivalis* (Kuboniwa et al. 2006), which appears to support subsequent recruitment of *Treponema denticola* (Yamada et al. 2005). Lactic acid produced by *Streptococcus oralis* is utilized by *Veillonella* sp. for colonization in oral biofilms (Periasamy and Kolenbrander 2010).

Evidence of competition for niche control is also widespread in polymicrobial biofilms (Rendueles and Ghigo 2012). For example, a surfactin required for swarming motility in *B. subtilis* inhibits biofilm formation of multiple species including *E. coli* and *Salmonella enterica* (Mireles et al. 2001), while polysaccharides produced by uropathogenic *E. coli*, called capsule 2, inhibits surface attachment and aggregation of various gram-positive and gram-negative species (Valle et al. 2006). In addition to preventing colonization, many species also employ strategies to actively disrupt an existing biofilms of other species. For example, dispersin B from *Actinobacillus actinomycetemcomitans* is a hydrolase of a major EPS component, poly-*N*-acetylglucosamine, that can hydrolyze the matrix and disperse biofilms of several bacterial species (Kaplan et al. 2004). Many bacterial species target other matrix components like DNA and proteins to break down the preexisting biofilms of other bacilli (Iwase et al. 2010; Nijland et al. 2010). A cross-kingdom competition for niche control was demonstrated by a study on mixed biofilms of *Candida albicans* and *Pseudomonas aeruginosa* that often co-colonize in lungs of cystic fibrosis patients. (Hogan and Kolter 2002). *Pseudomonas aeruginosa* biofilms appear to impair the colonization of *Candida* species through secreted factors (Holcombe et al. 2010), while farnesol from *Candida albicans* can inhibit quinolone synthesis in *P. aeruginosa* and associated gene regulation mechanisms that control many attributes of that bacterial species including biofilm formation and virulence (Cugini et al. 2007).

5.2.5 Intercellular Communication in Biofilms

From the studies described above, it is evident that microbes are sensing not only their environment but also other microbial cells around them. The most well-studied type of intercellular sensing among microbes is quorum sensing, in which individual cells sense the density of their own species (intraspecies quorum sensing) as well as other species (interspecies quorum sensing) through accumulation of secreted signals that trigger a response only above a threshold concentration (Camilli and Bassler 2006; Waters et al. 2008). While quorum sensing controls a wide range of bacterial phenotypes in both planktonic and biofilm cultures, activities in biofilms directly impact the structural integrity and subsequent existential consequences on the involved population of cells. Quorum sensing is employed by *Vibrio* sp. to exit biofilms, perhaps to recolonize a new niche (Hammer and Bassler 2003), whereas the mechanism appears to induce aggregation and matrix synthesis in other species (Hogan et al. 2004). In the context of polymicrobial communities, interspecies quorum sensing can induce either cooperative or competitive behavior in the participating cells depending on the involved species. An interesting example for coexistence of agonistic as well as antagonistic effects of quorum sensing can be found in oral microflora. Interspecies quorum sensing between *S. gordonii* and *P. gingivalis* induce biofilms of both these species (McNab et al. 2003). However, a serine protease called challisin secreted by *S. gordonii* can interfere with activity of competence-stimulating peptide (CSP) of *S. mutans*, thereby impairing the ability of *S. mutans* to from biofilms (Senadheera and Cvitkovitch 2008). Whereas these dynamics likely play themselves out during development of dental biofilms and associated caries, additional such scenarios are likely possible.

5.2.6 *Persistence of Microbes in Biofilms*

As stated earlier in this chapter, one of the defining aspects of microbial existence in biofilms is their recalcitrance to external threats including but not limited to physicomechanical stress, chemical stress, antibiotics, and host immunity (Fux et al. 2005; Stoodley et al. 2002; Van Acker et al. 2014). We have at present only a limited understanding of the factors underlying such tolerance. Studies of monospecies biofilms have revealed that tolerance primarily arises from two factors: (a) diffusion barrier created by extracellular matrix and (b) physiological adaptation to unique microenvironments (Mah and O'Toole 2001). The property of matrix as a diffusion barrier however depends on both the bacterial species as well as the nature of chemical challenge (Mah and O'Toole 2001). While *P. aeruginosa* could restrict the diffusion of piperacillin, antibiotics like vancomycin and rifampicin are restricted by *S. epidermis* (Mah and O'Toole 2001). These differences could perhaps explain perceptions of greater drug tolerance by polymicrobial biofilms, relative to homospecies biofilms of the participating communities (Elias and Banin 2012). It is argued that cooperative behavior among various species is the primary basis of induced resistance in polymicrobial biofilms. This is evident from induced resistance in a slime-negative mutant of *S. epidermis* when cultured in biofilms with *C. albicans* (Adam et al. 2002). While the slime-negative mutant of *S. epidermis* was sensitive to vancomycin in a monospecies culture, it acquired resistance to that antibiotic when co-cultured with *C. albicans* (Adam et al. 2002).

Physiological tolerance to antibiotics in biofilms is linked to their slower growth rate and adaptation to self-generated stresses from limiting nutrients and oxygen, although the extent of resistance varies for different antibiotics (Fux et al. 2005; Mah and O'Toole 2001). By comparing chemostat planktonic cultures with their biofilm counterparts for many bacterial species, investigators have discovered that planktonic cells were comparable to biofilms in ciprofloxacin sensitivity under slow-growing condition but gained sensitivity when growth rate was increased (Desai et al. 1998; Duguid et al. 1992). Downregulation of metabolisms, and subsequent decrease in molecules targeted by antibiotics, is postulated to give rise to growth rate-related persistence (Lewis 2008). Toxin-antitoxin (TA) modules like *mazEF*, *relE*, and *hipA*, as well as *glpD* and *plsB*, have been implicated in maintaining persistent cells under antibiotic exposure by regulating growth and cell division (Lewis 2008). However, the role of these genes in controlling biofilm-associated persistence remains unknown. Adaptation to stresses in biofilms is supported both by upregulation of the stress-related sigma factor RpoS and by many nutrient assimilation pathways in the multicellular architectures (Mah and O'Toole 2001; Serra et al. 2013). The linkage between these pathways and stress tolerance is supported by the fact that mutation in RpoS sensitizes *P. aeruginosa* biofilms to antibiotics (Mah and O'Toole 2001).

5.2.7 Dispersal of Microbes from Biofilms

Dispersal of cells from established biofilms is the final developmental stage. This process is particularly significant from a clinical perspective for novel intervention strategies to disrupt infectious biofilms and thus facilitate rapid clearance. Dispersal could either be auto-induced or forced by a competing species as described in Sect. 5.2.4. Auto-induced dispersion is considered either a bet-hedging process to diversify the colonization capacity or an exit strategy from an existing condition (McDougald et al. 2012; Petrova and Sauer 2016). Generally, dispersal is triggered by environmental cues, such as nutrient availability, oxygen tension, nitric oxide, temperature, etc., and requires auto-induced signaling pathways that involve self-produced extracellular molecules. For example, carbon limitations induce detachment of many bacterial species from surfaces (Delaquis et al. 1989; Gjermansen et al. 2005; Sauer et al. 2004). Dispersal of *Pseudomonas putida* under conditions of carbon limitation is triggered by degradation of a major matrix component, LapA, by a protease LapG. Expression of LapG is negatively regulated by a secondary messenger cyclic di-GMP, such that downshift of the messenger under a condition of nutrient limitation induces LapG activity (Gjermansen et al. 2010) favoring microbial dispersal. In *V. cholerae*, high cell densities result in the induction of a master regulator HapR which represses cyclic di-GMP and EPS synthesis, triggering dispersal of cells from biofilms (Hammer and Bassler 2003). In addition to downregulation of EPS synthesis in the above instances, other species utilize degradation mechanisms to dissolve matrix material for dispersal purposes (Karatan and Watnick 2009). Localized degradation of matrix is associated with production of hydrolases, proteases, and nucleases to degrade EPS, environmental DNA (eDNA), and proteins of the matrix that are thought to facilitate programmed dispersal of cells from biofilms (Allison et al. 1998; Baty et al. 2000; Gjermansen et al. 2010; Mann et al. 2009). Interestingly, localized and programmed cellular death in biofilms of *P. aeruginosa* by rhamnolipids has also been linked to large-scale dispersal of cells (Boles et al. 2005). Another mechanism of dispersal in *P. aeruginosa* involves localized lysis of cells, presumably to provide nutrients to neighboring interior cells as an exit cue (Purevdorj-Gage et al. 2005).

5.3 Biofilm Paradigm in Mycobacteria

5.3.1 A Historical Perspective

Long before microbial multicellularity was hypothesized and conceptualized, microscopic observations of Mtb found "cords" or parallel filamentous bundles of bacilli (Jones 1896). However, despite a linkage between cording and virulence of Mtb, these studies went underappreciated due to the influence of Koch's postulates (Silva et al. 1985). In fact, the propensity of mycobacteria to cord was considered a major hurdle in mycobacteriology that involved clonal purity of strains, to the extent that deliberate inclusion of dispersing agents like Tween 80 in medium has been the mainstream technique of in vitro culturing of Mtb for over six decades (Dubos and Middlebrook 1948). This is despite the fact that cording has been attributed to the virulence of Mtb, and detergent has been shown to compromise cell wall integrity and bacterial resistance to antibiotics (Hunter et al. 2006).

5.3.2 Manifestation of Mycobacterial Multicellularity

Detergent-free liquid cultures of Mtb and other mycobacteria invariably form microscopic aggregates of varied morphologies, from unstructured clusters to cords (Richards and Ojha 2014). If left unperturbed, these aggregates organize on the air-medium interface into macroscopic textured pellicles (Richards and Ojha 2014). Similarly, on a solid substratum, the aggregates form colonies (Richards and Ojha 2014). These morphological developments are not restricted to Mtb, but are characteristics of all mycobacterial species including the environmental species like *Mycobacterium smegmatis*. In the past decade, these morphological developments of mycobacteria have been closely examined in the context of the broader paradigm of biofilms, and it is increasingly becoming clear that the aggregated growth of mycobacteria fits into the definitions of biofilms from multiple criteria. First, the formation of pellicles has specialized genetic requirements, which are dispensable for planktonic growth (Islam et al. 2012; Richards and Ojha 2014). Second, the pellicles harbor 100–1000-fold more drug-tolerant persisters than do planktonic cultures of similar cell density, and mutations that disrupt pellicle formation also reduce the frequency of persisters (Ojha et al. 2008). Third, mycobacteria in pellicles have distinct sets of gene-expression profiles as compared to their planktonic counterparts, suggesting that physiological adaptation of cells is associated with their multicellular architecture (Ojha and Hatfull 2007). For example, *M. smegmatis* cells in pellicles induce high levels of siderophore synthesis genes, required for iron acquisition, despite having enough iron in the medium to keep the genes repressed in planktonic cells (Ojha and Hatfull 2007). A reasonable interpretation from this observation is that cells in pellicles likely experience localized depletion of iron due to limited availability from diffusion restriction, leading to induction of the iron sequestration pathway. Consistent with this idea, multiple nutrient transporters are also induced in pellicle cultures of *M. smegmatis*, as compared to their planktonic counterparts. Induction of several candidate multidrug efflux pumps in pellicles suggests the necessity of efficient efflux pathways in pellicle cells for maintaining metabolic homeostasis (Ojha and Hatfull 2007). Taken together, these bodies of evidence indicate that the aggregated growth of mycobacteria represents an organized multicellular community of cells, similar to the biofilms of other microbial species.

5.3.3 *Biofilms of Environmental Mycobacteria and Their Impact on Human Health*

Our environment is replete with mycobacterial species, particularly NTMs and rapid-growing mycobacteria (RGMs), which inhabit the soil and water ecosystems as well as engineered water distribution systems (van Ingen et al. 2009; Whiley et al. 2012). Perhaps the most striking evidence of mycobacterial prevalence in our immediate surroundings emerges from a study that reports large abundance of NTMs in polymicrobial communities found in showerheads in American cities (Feazel et al. 2009). Coexistence of NTMs in polymicrobial communities not only highlights the shared habitat for mycobacteria and other bacterial species but also opens up questions about their interactions with the other participating species (Whiley et al. 2012). One example of such interactions is elucidated by a role of *Acinetobacter calcoaceticus* in nucleating co-aggregation of *Mycobacterium mucogenicum* in polymicrobial biofilms formed in water distribution systems (Simoes et al. 2008). The public health impact of polymicrobial biofilms with mycobacteria is accentuated by high incidence of contamination with NTMs and RGMs of the water distribution systems in nosocomial settings (van Ingen et al. 2009; Whiley et al. 2012; Williams et al. 2005). This is further aggravated by high levels of mycobacterial tolerance to chloramine, a common disinfectant used for potable water treatment in urban and hospital settings (Steed and Falkinham 2006; van Ingen et al. 2009; Whiley et al. 2012; Williams et al. 2005). Of the NTMs and RGMs, *Mycobacterium avium* complex and *Mycobacterium abscessus* stand out for their alarmingly high rate of incidence in clinical settings.

5.3.3.1 *Mycobacterium avium Complex*

Mycobacterium avium complex (MAC) refers to a select group of very closely related mycobacterial species—*Mycobacterium avium* ssp. *avium*, *Mycobacterium avium* ssp. *hominissuis*, *Mycobacterium avium* ssp. *paratuberculosis*, *Mycobacterium avium* ssp. *paratuberculosis*, and *Mycobacterium intracellulare* (Whiley et al. 2012). Members of the MAC group cause pulmonary and extrapulmonary infections in immunocompromised individuals. Prior to the development of AIDS therapeutics, about 40% of AIDS patients developed MAC infection and a vast majority of them ultimately succumbed to it (Karakousis et al. 2004). In addition, patients undergoing transplant surgery also carry high risk of MAC infection, thereby increasing the burden of MAC-related mortality in developed countries (Whiley et al. 2012). Additionally, it must be noted that MAC infections, particularly those caused by *M. intracellulare*, are also prevalent in individuals with potentially increased susceptibility even in the absence of overtly obvious compromised immunity (Field et al. 2004; Han et al. 2005; Sugita et al. 2000). Infections caused by MAC are also attributed to Crohn's disease (Pierce 2009), although causal relationship between MAC infection and those disease symptoms remains controversial (Chamberlin et al. 2001). We can recognize that MAC infections share characteristics of other mycobacterial infections including the fact of their being highly recalcitrant to antibiotics, and successful treatment often requires multiple drugs administered simultaneously over a period of 12 months.

Relationship between MAC biofilms and their persistent and virtually untreatable infections can be viewed from two perspectives. First, the ability of MAC to establish biofilms in environmental polymicrobial communities and their extraordinary recalcitrance to disinfectants increases our exposure to these species, thereby increasing the incidence of nosocomial infections associated with MAC. Second, biofilms of MAC can facilitate the persistence of these microbes against the challenges of host immunity and antibiotic treatment (Falkinham 2007; Rose and Bermudez 2014), thereby making the infections often chronic and untreatable. Interestingly, mutants of *M. avium* that fail to form biofilms in vitro also have impaired ability to invade and translocate bronchial epithelial cells (Yamazaki et al. 2006a), suggesting that colonization of host tissues could also be associated with biofilm formation of MAC. Thus, pharmacological or physical disruption of MAC biofilms could potentially be an effective strategy to not only prevent but also treat infections. Structural and genetic analyses of *M. avium* biofilms have revealed interesting insights that could possibly yield suitable targets for disruption. Bermudez and colleagues recently showed the presence of eDNA in biofilms of clinical isolates of *M. avium* and also demonstrated that treatment of *M. avium hominissuis* isolate 5 (MAH 5) biofilms with DNAse reduced the biomass by half (Rose et al. 2015), thus opening the possibility of using DNAse as one of the agents to disrupt biofilms. Gene expression analysis revealed induced expression of several key genes in biofilms of *M. avium*, among the most notable of which were genes involved in the tricarboxylic acid (TCA) cycle, GDP-mannose biosynthesis, and glycopeptidolipid (GPL) biosynthesis (Yamazaki et al. 2006b), suggesting that these genes might serve as candidates for targeting biofilms.

5.3.3.2 *Mycobacterium ulcerans*

Mycobacterium ulcerans is another NTM and it causes a highly persistent skin and soft tissue infection, called Buruli ulcer, which often remains refractory to treatment with most of the common antibiotics (Yotsu et al. 2015). *Mycobacterium ulcerans* is largely associated with freshwater ecosystems, and it is believed that contact with contaminated water leads to infection (Garchitorena et al. 2015). Although the precise mechanisms of transmission of *M. ulcerans* remain elusive, both biotic and abiotic factors have been attributed to its dynamics in aquatic ecosystem and subsequent infection (Garchitorena et al. 2015). However, the outcome of infection with *M. ulcerans* is highly dependent on the pathogen's ability to produce a toxin called mycolactone, which is considered to be a primary factor for tissue damage (George et al. 1999). Interestingly, the synthesis of mycolactone has been linked to the ability of *M. ulcerans* to form biofilms (Marsollier et al. 2005), thus raising the possibility of a direct relationship between biofilms and pathogenesis of *M. ulcerans*.

A molecular insight into this relationship however has been limited by genetic intractability of the pathogen that demands new tools of molecular genetics for further investigation.

5.3.3.3 *Mycobacterium abscessus*

Mycobacterium abscessus is an environmental RGM found both in natural aquatic ecosystems as well as in engineered water distribution systems. *M. abscessus* can cause chronic infections in lungs and post-traumatic wounds on skin upon contact with contaminated water (Benwill and Wallace 2014). Infections with *M. abscessus* have been increasingly common in recent years (Benwill and Wallace 2014). Individuals with cystic fibrosis are particularly at high risk of developing infection with *M. abscessus*. The often necessity of a 24-month-long multidrug treatment regimen for *M. abscessus* infections poses numerous challenges, particularly in the context of immunocompromised patients. The chronic and drug-tolerant characteristics of *M. abscessus* infections share similarities with the characteristics of other microbial species that establish biofilms in host tissues. Recent advancements in genetic techniques have offered insight into the molecular relationship between biofilms and pathogenesis of *M. abscessus* (Medjahed and Reyrat 2009; Medjahed and Singh 2010). Strains which demonstrate rough colony morphology when growing on agar plates generally are devoid of glycopeptidolipids (GPL) on the cell surface, these strains can readily form cords and biofilms, and they can colonize the host more efficiently as compared to smooth colony mutants (Bernut et al. 2014; Catherinot et al. 2007; Nessar et al. 2011). However, deficiency of GPL-negative mutants in substratum attachment also suggests that initial colonization of the pathogen is distinct from aggregated growth and biofilm formation, which is perhaps more critical for its pathogenesis and persistence (Nessar et al. 2011). This idea is clearly elucidated by a recent finding that biofilm defect in *M. smegmatis* due to deficiency in aggregation can be restored by suppressor mutation in GPL biosynthesis (Yang et al. 2017). In another interesting study, the switch from rough to smooth colonies was found to be a reversible process induced by exposure to subinhibitory concentrations of aminoglycosides, suggesting an active process of adaptation by the pathogen to a drug-tolerant state under challenged condition (Tsai et al. 2015).

5.3.4 *In Vitro Biofilms of Mtb Complex: From an Assay for Drug Discovery to the Riddle of Persistent TB*

Even though the evidence for Mtb biofilms in the host tissues remains to be established, the fact that these biofilms can exist as structured multicellular drug-tolerant communities in vitro (Richards and Ojha 2014) offers a new growth model to understand the mechanisms of physiological adaptation by the pathogen to environmental stresses, which in turn could be relevant for understanding its persistence inside the host. For example, mutation in *mma4*, which synthesizes oxygenated mycolic acids in Mtb, abrogates biofilm growth of Mtb (Sambandan et al. 2013). Given that oxygenated mycolic acids increase the envelope permeability (Takayama et al. 2005), it could be argued that a *mma4*-dependent envelope permeability is perhaps critical for enhanced nutrient uptake of Mtb in restrictive biofilm environments. A role of *mma4* in Mtb survival inside of its host further argues for overlap between the survival mechanisms employed by the pathogen during development of biofilms in vitro and growth inside the host. Similarly, identity of genetic loci for antibiotic tolerance in biofilms could potentially lead to novel targets for therapeutics against persistent TB.

5.4 Future Outlook

In summary, the large abundance of mycobacteria in our environment and their extraordinary ability for persistence against the common antibacterial agents poses a significant public health risk. Novel mitigation strategies would require better understanding of the molecular mechanisms underlying their persistence characteristics and process of biofilm development. Urgency in development of new molecular genetics tools for manipulation of mycobacterial pathogens like *M. abscessus* is necessitated by the rapidly increase in their clinical incidence. New and powerful high-throughput sequencing platforms and advanced microscopic techniques have the potential to lead to new methods for investigations into these areas of research.

Compliance with Ethical Standards

Funding This study was funded by the National Institute of Health (Grant # AI132422).

Conflict of Interest Anil K. Ojha declares that he has no conflict of interest.

Ethical Approval This article does not contain any studies with human participants or animals performed by the author.

References

Absalon C, Van Dellen K, Watnick PI (2011) A communal bacterial adhesin anchors biofilm and bystander cells to surfaces. PLoS Pathog 7:e1002210. https://doi.org/10.1371/journal.ppat.1002210

Adam B, Baillie GS, Douglas LJ (2002) Mixed species biofilms of *Candida albicans* and *Staphylococcus epidermidis*. J Med Microbiol 51:344–349. https://doi.org/10.1099/0022-1317-51-4-344

Allison DG, Ruiz B, SanJose C, Jaspe A, Gilbert P (1998) Extracellular products as mediators of the formation and detachment of *Pseudomonas fluorescens* biofilms. FEMS Microbiol Lett 167:179–184

Baty AM 3rd, Eastburn CC, Techkarnjanaruk S, Goodman AE, Geesey GG (2000) Spatial and temporal variations in chitinolytic gene expression and bacterial biomass production during chitin degradation. Appl Environ Microbiol 66:3574–3585

Benedict MI, Wulff LM, White RB (1992) Current parental stress in maltreating and nonmaltreating families of children with multiple disabilities. Child Abuse Negl 16:155–163

Benwill JL, Wallace RJ Jr (2014) *Mycobacterium abscessus*: challenges in diagnosis and treatment. Curr Opin Infect Dis 27:506–510. https://doi.org/10.1097/QCO.0000000000000104

Bernut A, Herrmann JL, Kissa K, Dubremetz JF, Gaillard JL, Lutfalla G, Kremer L (2014) *Mycobacterium abscessus* cording prevents phagocytosis and promotes abscess formation. Proc Natl Acad Sci USA 111:E943–E952. https://doi.org/10.1073/pnas.1321390111

Boles BR, Thoendel M, Singh PK (2005) Rhamnolipids mediate detachment of *Pseudomonas aeruginosa* from biofilms. Mol Microbiol 57:1210–1223. https://doi.org/10.1111/j.1365-2958.2005.04743.x

Branda SS, Vik S, Friedman L, Kolter R (2005) Biofilms: the matrix revisited. Trends Microbiol 13:20–26

Brennan PJ, Nikaido H (1995) The envelope of mycobacteria. Annu Rev Biochem 64:29–63. https://doi.org/10.1146/annurev.bi.64.070195.000333

Caiazza NC, O'Toole GA (2004) SadB is required for the transition from reversible to irreversible attachment during biofilm formation by *Pseudomonas aeruginosa* PA14. J Bacteriol 186:4476–4485. https://doi.org/10.1128/JB.186.14.4476-4485.2004

Camilli A, Bassler BL (2006) Bacterial small-molecule signaling pathways. Science 311:1113–1116. https://doi.org/10.1126/science.1121357

Catherinot E et al (2007) Hypervirulence of a rough variant of the *Mycobacterium abscessus* type strain. Infect Immun 75:1055–1058. https://doi.org/10.1128/IAI.00835-06

Chai Y, Chu F, Kolter R, Losick R (2008) Bistability and biofilm formation in *Bacillus subtilis*. Mol Microbiol 67:254–263. https://doi.org/10.1111/j.1365-2958.2007.06040.x

Chamberlin W et al (2001) Review article: *Mycobacterium avium* subsp. paratuberculosis as one cause of Crohn's disease. Aliment Pharmacol Ther 15:337–346

Conrady DG, Brescia CC, Horii K, Weiss AA, Hassett DJ, Herr AB (2008) A zinc-dependent adhesion module is responsible for intercellular adhesion in staphylococcal biofilms. Proc Natl Acad Sci USA 105:19456–19461. https://doi.org/10.1073/pnas.0807717105

Costerton JW, Irvin RT, Cheng KJ (1981) The role of bacterial surface structures in pathogenesis. Crit Rev Microbiol 8:303–338

Cramton SE, Gerke C, Schnell NF, Nichols WW, Gotz F (1999) The intercellular adhesion (ica) locus is present in *Staphylococcus aureus* and is required for biofilm formation. Infect Immun 67:5427–5433

Cucarella C, Solano C, Valle J, Amorena B, Lasa I, Penades JR (2001) Bap, a *Staphylococcus aureus* surface protein involved in biofilm formation. J Bacteriol 183:2888–2896. https://doi.org/10.1128/JB.183.9.2888-2896.2001

Cugini C, Calfee MW, Farrow JM 3rd, Morales DK, Pesci EC, Hogan DA (2007) Farnesol, a common sesquiterpene, inhibits PQS production in *Pseudomonas aeruginosa*. Mol Microbiol 65:896–906. https://doi.org/10.1111/j.1365-2958.2007.05840.x

de Beer D, Stoodley P, Roe F, Lewandowski Z (1994) Effects of biofilm structures on oxygen distribution and mass transport. Biotechnol Bioeng 43:1131–1138. https://doi.org/10.1002/bit.260431118

Delaquis PJ, Caldwell DE, Lawrence JR, McCurdy AR (1989) Detachment of *Pseudomonas fluorescens* from biofilms on glass surfaces in response to nutrient stress. Microb Ecol 18:199–210. https://doi.org/10.1007/BF02075808

Desai M, Buhler T, Weller PH, Brown MR (1998) Increasing resistance of planktonic and biofilm cultures of *Burkholderia cepacia* to ciprofloxacin and ceftazidime during exponential growth. J Antimicrob Chemother 42:153–160

Dubos RJ, Middlebrook G (1948) The effect of wetting agents on the growth of tubercle bacilli. J Exp Med 88:81–88

Duguid IG, Evans E, Brown MR, Gilbert P (1992) Growth-rate-independent killing by ciprofloxacin of biofilm-derived *Staphylococcus epidermidis*; evidence for cell-cycle dependency. J Antimicrob Chemother 30:791–802

Elias S, Banin E (2012) Multi-species biofilms: living with friendly neighbors. FEMS Microbiol Rev 36:990–1004. https://doi.org/10.1111/j.1574-6976.2012.00325.x

Falkinham JO 3rd (2007) Growth in catheter biofilms and antibiotic resistance of *Mycobacterium avium*. J Med Microbiol 56:250–254. https://doi.org/10.1099/jmm.0.46935-0

Feazel LM, Baumgartner LK, Peterson KL, Frank DN, Harris JK, Pace NR (2009) Opportunistic pathogens enriched in showerhead biofilms. Proc Natl Acad Sci USA 106:16393–16399. https://doi.org/10.1073/pnas.0908446106

Field SK, Fisher D, Cowie RL (2004) *Mycobacterium avium* complex pulmonary disease in patients without HIV infection. Chest 126:566–581. https://doi.org/10.1378/chest.126.2.566

Fux CA, Costerton JW, Stewart PS, Stoodley P (2005) Survival strategies of infectious biofilms. Trends Microbiol 13:34–40. https://doi.org/10.1016/j.tim.2004.11.010

Garchitorena A, Guegan JF, Leger L, Eyangoh S, Marsollier L, Roche B (2015) *Mycobacterium ulcerans* dynamics in aquatic ecosystems are driven by a complex interplay of abiotic and biotic factors. elife 4:e07616. https://doi.org/10.7554/eLife.07616

George KM, Chatterjee D, Gunawardana G, Welty D, Hayman J, Lee R, Small PL (1999) Mycolactone: a polyketide toxin from *Mycobacterium ulcerans* required for virulence. Science 283:854–857

Gjermansen M, Ragas P, Sternberg C, Molin S, Tolker-Nielsen T (2005) Characterization of starvation-induced dispersion in *Pseudomonas putida* biofilms. Environ Microbiol 7:894–906. https://doi.org/10.1111/j.1462-2920.2005.00775.x

Gjermansen M, Nilsson M, Yang L, Tolker-Nielsen T (2010) Characterization of starvation-induced dispersion in *Pseudomonas putida* biofilms: genetic elements and molecular mechanisms. Mol Microbiol 75:815–826. https://doi.org/10.1111/j.1365-2958.2009.06793.x

Gonzalez-Santiago TM, Drage LA (2015) Nontuberculous mycobacteria: skin and soft tissue infections. Dermatol Clin 33:563–577. https://doi.org/10.1016/j.det.2015.03.017

Hall-Stoodley L, Stoodley P (2002) Developmental regulation of microbial biofilms. Curr Opin Biotechnol 13:228–233

Hall-Stoodley L, Costerton JW, Stoodley P (2004) Bacterial biofilms: from the natural environment to infectious diseases. Nat Rev Microbiol 2:95–108

Hammer BK, Bassler BL (2003) Quorum sensing controls biofilm formation in *Vibrio cholerae*. Mol Microbiol 50:101–104

Han XY, Tarrand JJ, Infante R, Jacobson KL, Truong M (2005) Clinical significance and epidemiologic analyses of *Mycobacterium avium* and *Mycobacterium intracellulare* among patients without AIDS. J Clin Microbiol 43:4407–4412. https://doi.org/10.1128/JCM.43.9.4407-4412.2005

Heilmann C, Schweitzer O, Gerke C, Vanittanakom N, Mack D, Gotz F (1996) Molecular basis of intercellular adhesion in the biofilm-forming *Staphylococcus epidermidis*. Mol Microbiol 20:1083–1091

Henrici AT (1933) Studies of freshwater bacteria: I. A direct microscopic technique. J Bacteriol 25:277–287

Hinsa SM, Espinosa-Urgel M, Ramos JL, O'Toole GA (2003) Transition from reversible to irreversible attachment during biofilm formation by *Pseudomonas fluorescens* WCS365 requires an ABC transporter and a large secreted protein. Mol Microbiol 49:905–918
Hogan DA, Kolter R (2002) Pseudomonas-Candida interactions: an ecological role for virulence factors. Science 296:2229–2232. https://doi.org/10.1126/science.1070784
Hogan DA, Vik A, Kolter R (2004) A *Pseudomonas aeruginosa* quorum-sensing molecule influences *Candida albicans* morphology. Mol Microbiol 54:1212–1223
Holcombe LJ et al (2010) *Pseudomonas aeruginosa* secreted factors impair biofilm development in *Candida albicans*. Microbiology 156:1476–1486. https://doi.org/10.1099/mic.0.037549-0
Hunter RL, Olsen MR, Jagannath C, Actor JK (2006) Multiple roles of cord factor in the pathogenesis of primary, secondary, and cavitary tuberculosis, including a revised description of the pathology of secondary disease. Ann Clin Lab Sci 36:371–386
Islam MS, Richards JP, Ojha AK (2012) Targeting drug tolerance in mycobacteria: a perspective from mycobacterial biofilms. Expert Rev Anti-Infect Ther 10:1055–1066. https://doi.org/10.1586/eri.12.88
Iwase T et al (2010) *Staphylococcus epidermidis* Esp inhibits *Staphylococcus aureus* biofilm formation and nasal colonization. Nature 465:346–349. https://doi.org/10.1038/nature09074
Jindani A, Dore CJ, Mitchison DA (2003) Bactericidal and sterilizing activities of antituberculosis drugs during the first 14 days. Am J Respir Crit Care Med 167:1348–1354
Jones AC (1896) On the so-called Tubercle Bacilli Report of the sixty-sixth meeting of the British Association for the Advancements of Science 1015–1016
Kaplan JB, Ragunath C, Velliyagounder K, Fine DH, Ramasubbu N (2004) Enzymatic detachment of *Staphylococcus epidermidis* biofilms. Antimicrob Agents Chemother 48:2633–2636. https://doi.org/10.1128/AAC.48.7.2633-2636.2004
Karakousis PC, Moore RD, Chaisson RE (2004) *Mycobacterium avium* complex in patients with HIV infection in the era of highly active antiretroviral therapy. Lancet Infect Dis 4:557–565. https://doi.org/10.1016/S1473-3099(04)01130-2
Karatan E, Watnick P (2009) Signals, regulatory networks, and materials that build and break bacterial biofilms. Microbiol Mol Biol Rev 73:310–347. https://doi.org/10.1128/MMBR.00041-08
Kolter R, Greenberg EP (2006) Microbial sciences: the superficial life of microbes. Nature 441:300–302
Kolter R, Losick R (1998) One for all and all for one. Science 280:226–227
Kuboniwa M et al (2006) *Streptococcus gordonii* utilizes several distinct gene functions to recruit *Porphyromonas gingivalis* into a mixed community. Mol Microbiol 60:121–139. https://doi.org/10.1111/j.1365-2958.2006.05099.x
Lewis K (2008) Multidrug tolerance of biofilms and persister cells. Curr Top Microbiol Immunol 322:107–131
Mah TF, O'Toole GA (2001) Mechanisms of biofilm resistance to antimicrobial agents. Trends Microbiol 9:34–39
Mann EE et al (2009) Modulation of eDNA release and degradation affects *Staphylococcus aureus* biofilm maturation. PLoS One 4:e5822. https://doi.org/10.1371/journal.pone.0005822
Marrie TJ, Nelligan J, Costerton JW (1982) A scanning and transmission electron microscopic study of an infected endocardial pacemaker lead. Circulation 66:1339–1341
Marsollier L et al (2005) Colonization of the salivary glands of *Naucoris cimicoides* by *Mycobacterium ulcerans* requires host plasmatocytes and a macrolide toxin, mycolactone. Cell Microbiol 7:935–943
McDougald D, Rice SA, Barraud N, Steinberg PD, Kjelleberg S (2012) Should we stay or should we go: mechanisms and ecological consequences for biofilm dispersal. Nat Rev Microbiol 10:39–50. https://doi.org/10.1038/nrmicro2695
McNab R, Ford SK, El-Sabaeny A, Barbieri B, Cook GS, Lamont RJ (2003) LuxS-based signaling in *Streptococcus gordonii*: autoinducer 2 controls carbohydrate metabolism and biofilm formation with *Porphyromonas gingivalis*. J Bacteriol 185:274–284

Medjahed H, Reyrat JM (2009) Construction of *Mycobacterium abscessus* defined glycopeptidolipid mutants: comparison of genetic tools. Appl Environ Microbiol 75:1331–1338. https://doi.org/10.1128/AEM.01914-08

Medjahed H, Singh AK (2010) Genetic manipulation of *Mycobacterium abscessus*. Curr Protoc Microbiol Chapter 10:Unit 10D 12. https://doi.org/10.1002/9780471729259.mc10d02s18

Mireles JR 2nd, Toguchi A, Harshey RM (2001) *Salmonella enterica* serovar typhimurium swarming mutants with altered biofilm-forming abilities: surfactin inhibits biofilm formation. J Bacteriol 183:5848–5854. https://doi.org/10.1128/JB.183.20.5848-5854.2001

Nessar R, Reyrat JM, Davidson LB, Byrd TF (2011) Deletion of the mmpL4b gene in the *Mycobacterium abscessus* glycopeptidolipid biosynthetic pathway results in loss of surface colonization capability, but enhanced ability to replicate in human macrophages and stimulate their innate immune response. Microbiology 157:1187–1195. https://doi.org/10.1099/mic.0.046557-0

Nickel JC, Ruseska I, Wright JB, Costerton JW (1985) Tobramycin resistance of *Pseudomonas aeruginosa* cells growing as a biofilm on urinary catheter material. Antimicrob Agents Chemother 27:619–624

Nijland R, Hall MJ, Burgess JG (2010) Dispersal of biofilms by secreted, matrix degrading, bacterial DNase. PLoS One 5:e15668. https://doi.org/10.1371/journal.pone.0015668

Ojha A, Hatfull GF (2007) The role of iron in *Mycobacterium smegmatis* biofilm formation: the exochelin siderophore is essential in limiting iron conditions for biofilm formation but not for planktonic growth. Mol Microbiol 66:468–483

Ojha A, Anand M, Bhatt A, Kremer L, Jacobs WR Jr, Hatfull GF (2005) GroEL1: a dedicated chaperone involved in mycolic acid biosynthesis during biofilm formation in mycobacteria. Cell 123:861–873

Ojha AK et al (2008) Growth of *Mycobacterium tuberculosis* biofilms containing free mycolic acids and harbouring drug-tolerant bacteria. Mol Microbiol 69:164–174

Ojha AK, Hatfull GF, Jacobs WR Jr (2015) Genetic dissection of mycobacterial biofilms. In: Parish T, Roberts D (eds) Mycobacteria protocols. Springer, New York

Parrish NM, Dick JD, Bishai WR (1998) Mechanisms of latency in *Mycobacterium tuberculosis*. Trends Microbiol 6:107–112

Periasamy S, Kolenbrander PE (2010) Central role of the early colonizer *Veillonella* sp. in establishing multispecies biofilm communities with initial, middle, and late colonizers of enamel. J Bacteriol 192:2965–2972. https://doi.org/10.1128/JB.01631-09

Petrova OE, Sauer K (2016) Escaping the biofilm in more than one way: desorption, detachment or dispersion. Curr Opin Microbiol 30:67–78. https://doi.org/10.1016/j.mib.2016.01.004

Pierce ES (2009) Possible transmission of *Mycobacterium avium* subspecies paratuberculosis through potable water: lessons from an urban cluster of Crohn's disease. Gut Pathog 1:17. https://doi.org/10.1186/1757-4749-1-17

Purevdorj-Gage B, Costerton WJ, Stoodley P (2005) Phenotypic differentiation and seeding dispersal in non-mucoid and mucoid *Pseudomonas aeruginosa* biofilms. Microbiology 151:1569–1576. https://doi.org/10.1099/mic.0.27536-0

Rani SA et al (2007) Spatial patterns of DNA replication, protein synthesis, and oxygen concentration within bacterial biofilms reveal diverse physiological states. J Bacteriol 189:4223–4233. https://doi.org/10.1128/JB.00107-07

Rendueles O, Ghigo JM (2012) Multi-species biofilms: how to avoid unfriendly neighbors. FEMS Microbiol Rev 36:972–989. https://doi.org/10.1111/j.1574-6976.2012.00328.x

Richards JP, Ojha AK (2014) Mycobacterial biofilms. Microbiol Spectr 2(5). https://doi.org/10.1128/microbiolspec.MGM2-0004-2013

Rose SJ, Bermudez LE (2014) *Mycobacterium avium* biofilm attenuates mononuclear phagocyte function by triggering hyperstimulation and apoptosis during early infection. Infect Immun 82:405–412. https://doi.org/10.1128/IAI.00820-13

Rose SJ, Babrak LM, Bermudez LE (2015) *Mycobacterium avium* possesses extracellular DNA that contributes to biofilm formation, structural integrity, and tolerance to antibiotics. PLoS One 10: e0128772. https://doi.org/10.1371/journal.pone.0128772

Sambandan D et al (2013) Keto-mycolic acid-dependent pellicle formation confers tolerance to drug-sensitive *Mycobacterium tuberculosis*. MBio 4:e00222-00213. https://doi.org/10.1128/mBio.00222-13

Sauer K, Cullen MC, Rickard AH, Zeef LA, Davies DG, Gilbert P (2004) Characterization of nutrient-induced dispersion in *Pseudomonas aeruginosa* PAO1 biofilm. J Bacteriol 186:7312–7326. https://doi.org/10.1128/JB.186.21.7312-7326.2004

Senadheera D, Cvitkovitch DG (2008) Quorum sensing and biofilm formation by *Streptococcus mutans*. Adv Exp Med Biol 631:178–188. https://doi.org/10.1007/978-0-387-78885-2_12

Serra DO, Richter AM, Klauck G, Mika F, Hengge R (2013) Microanatomy at cellular resolution and spatial order of physiological differentiation in a bacterial biofilm. MBio 4:e00103-00113. https://doi.org/10.1128/mBio.00103-13

Silva CL, Ekizlerian SM, Fazioli RA (1985) Role of cord factor in the modulation of infection caused by mycobacteria. Am J Pathol 118:238–247

Simoes LC, Simoes M, Vieira MJ (2008) Intergeneric coaggregation among drinking water bacteria: evidence of a role for *Acinetobacter calcoaceticus* as a bridging bacterium. Appl Environ Microbiol 74:1259–1263. https://doi.org/10.1128/AEM.01747-07

Steed KA, Falkinham JO 3rd (2006) Effect of growth in biofilms on chlorine susceptibility of *Mycobacterium avium* and *Mycobacterium intracellulare*. Appl Environ Microbiol 72:4007–4011. https://doi.org/10.1128/AEM.02573-05

Stoodley P, Sauer K, Davies DG, Costerton JW (2002) Biofilms as complex differentiated communities. Annu Rev Microbiol 56:187–209

Stout JE, Koh WJ, Yew WW (2016) Update on pulmonary disease due to non-tuberculous mycobacteria. Int J Infect Dis 45:123–134. https://doi.org/10.1016/j.ijid.2016.03.006

Sugita Y, Ishii N, Katsuno M, Yamada R, Nakajima H (2000) Familial cluster of cutaneous *Mycobacterium avium* infection resulting from use of a circulating, constantly heated bath water system. Br J Dermatol 142:789–793

Takayama K, Wang C, Besra GS (2005) Pathway to synthesis and processing of mycolic acids in *Mycobacterium tuberculosis*. Clin Microbiol Rev 18:81–101

Tsai SH, Lai HC, Hu ST (2015) Subinhibitory doses of aminoglycoside antibiotics induce changes in the phenotype of *Mycobacterium abscessus*. Antimicrob Agents Chemother 59:6161–6169. https://doi.org/10.1128/AAC.01132-15

Valle J, Da Re S, Henry N, Fontaine T, Balestrino D, Latour-Lambert P, Ghigo JM (2006) Broad-spectrum biofilm inhibition by a secreted bacterial polysaccharide. Proc Natl Acad Sci USA 103:12558–12563. https://doi.org/10.1073/pnas.0605399103

Van Acker H, Van Dijck P, Coenye T (2014) Molecular mechanisms of antimicrobial tolerance and resistance in bacterial and fungal biofilms. Trends Microbiol 22(6):326–333. https://doi.org/10.1016/j.tim.2014.02.001

van Ingen J, Boeree MJ, Dekhuijzen PN, van Soolingen D (2009) Environmental sources of rapid growing nontuberculous mycobacteria causing disease in humans. Clin Microbiol Infect 15:888–893. https://doi.org/10.1111/j.1469-0691.2009.03013.x

von Ohle C, Gieseke A, Nistico L, Decker EM, DeBeer D, Stoodley P (2010) Real-time microsensor measurement of local metabolic activities in ex vivo dental biofilms exposed to sucrose and treated with chlorhexidine. Appl Environ Microbiol 76:2326–2334. https://doi.org/10.1128/AEM.02090-09

Waters CM, Lu W, Rabinowitz JD, Bassler BL (2008) Quorum sensing controls biofilm formation in *Vibrio cholerae* through modulation of cyclic di-GMP levels and repression of vpsT. J Bacteriol 190:2527–2536. https://doi.org/10.1128/JB.01756-07

Werner E et al (2004) Stratified growth in *Pseudomonas aeruginosa* biofilms. Appl Environ Microbiol 70:6188–6196. https://doi.org/10.1128/AEM.70.10.6188-6196.2004

Whiley H, Keegan A, Giglio S, Bentham R (2012) *Mycobacterium avium* complex – the role of potable water in disease transmission. J Appl Microbiol 113:223–232. https://doi.org/10.1111/j.1365-2672.2012.05298.x

WHO (2012) The burden of diseases caused by TB Global Tuberculosis Report:8–28

Williams MM, Santo Domingo JW, Meckes MC (2005) Population diversity in model potable water biofilms receiving chlorine or chloramine residual. Biofouling 21:279–288. https://doi.org/10.1080/08927010500452695

Wimpenny J, Manz W, Szewzyk U (2000) Heterogeneity in biofilms. FEMS Microbiol Rev 24:661–671

Yamada M, Ikegami A, Kuramitsu HK (2005) Synergistic biofilm formation by *Treponema denticola* and *Porphyromonas gingivalis*. FEMS Microbiol Lett 250:271–277. https://doi.org/10.1016/j.femsle.2005.07.019

Yamazaki Y et al (2006a) The ability to form biofilm influences *Mycobacterium avium* invasion and translocation of bronchial epithelial cells. Cell Microbiol 8:806–814. https://doi.org/10.1111/j.1462-5822.2005.00667.x

Yamazaki Y, Danelishvili L, Wu M, Macnab M, Bermudez LE (2006b) *Mycobacterium avium* genes associated with the ability to form a biofilm. Appl Environ Microbiol 72:819–825. https://doi.org/10.1128/AEM.72.1.819-825.2006

Yang Y, Thomas J, Li Y, Vilcheze C, Derbyshire KM, Jacobs WR Jr, Ojha AK (2017) Defining a temporal order of genetic requirements for development of mycobacterial biofilms. Mol Microbiol 105:794–809. https://doi.org/10.1111/mmi.13734

Yotsu RR, Murase C, Sugawara M, Suzuki K, Nakanaga K, Ishii N, Asiedu K (2015) Revisiting Buruli ulcer. J Dermatol 42:1033–1041. https://doi.org/10.1111/1346-8138.13049

Zobell CE (1943) The effect of solid surfaces upon bacterial activity. J Bacteriol 46:39–56

Part III
The Ecology of Infectious Diseases Affecting Livestock and Wildlife

Chapter 6
Opportunistic Bacteria Associated with Mammalian Livestock Disease

Christon J. Hurst

Abstract This chapter presents a current compendium of knowledge regarding the opportunistically bacteria associated with those livestock mammals used as sources of meat and milk for consumption by humans. All of these mammalian species are terrestrial and the majority are bovids. However, as important as the bovids are to us, they do indeed represent only part of a diverse collection which includes rodents as small as the Edible dormouse and cervids as large as the moose or Eurasian elk. The opportunistic microorganisms mentioned in this chapter represent a critical concern with respect to human food supplies. This chapter lists the pathogens by genus along with their higher taxonomy, provides information regarding the ecology of these genera, and summarizes their associated disease information with respect to livestock mammals. The genus listings include the individual pathogen species by name and also identify those host mammals which are known to be affected by each pathogen species.

6.1 Introduction

Humans are by definition omnivores, which means that we can and indeed generally do eat as food almost anything whose ingestion is not either physically or chemically lethal. Animals certainly are an important source of dietary protein for most humans and as livestock those animals are most easily and inexpensively fed if either they are herbivores or can otherwise forage for themselves. The biggest factors in our choices of food animals are the affordability and availability of the animals and, of course, our level of hunger. A major challenge for us when keeping the animals as livestock is that we maintain their health, which includes a need to understand the pathogens which affect these animals.

C. J. Hurst (✉)
Universidad del Valle, Cali, Colombia

Cincinnati, OH, USA
e-mail: christonjhurst@fuse.net

C. J. Hurst (ed.), *The Connections Between Ecology and Infectious Disease*,
Advances in Environmental Microbiology 5,
https://doi.org/10.1007/978-3-319-92373-4_6

This is a companion article for Hurst (2016), which presented the subject of opportunistic algae, fungi and ichthyosporea associated with mammalian livestock disease.

6.2 Livestock and Their Use

Livestock, in its broader definition, can include any type of animal that either is domesticated, semi-domesticated, or captured from the wild and then maintained under agricultural supervision. That supervision is termed animal husbandry and it has made a big change in the cultural evolution of our species by allowing a longer term control of meat resources and enabling dairy industry. Maintaining livestock freed us from continually hunting wild animals in order to meet our food requirements and subsequently enabled the development of larger human communities. The list of animals considered to be livestock includes not only mammals, but also amphibians, most notably frogs; those numerous avian groups which are categorized as poultry; crustaceans, most notably decapods such as crab, crayfish, lobster, prawn and shrimp; fish including eels; insects, most notably bees, crickets and silk worms; gastropod molluscs including abalone and edible terrestrial snails; bivalve molluscs including clams, mussels and oysters; and among the reptiles we keep as livestock not only snakes and turtles, but also various members of the order Crocodilia including alligators, caimans, and crocodiles.

While the definition of livestock is not limited either to mammals or the other groups of animals that are intended as food for humans, this chapter must have a more limited focus and therefore considers only those mammals maintained for the goal of consumption by humans. Some of the mammals listed in this chapter as livestock serve us for more than a single function. In addition to eating them as a source of meat, many of the larger mammals are of course considered to produce sufficient milk and be adequately passive in terms of their temperament that they are used as dairy animals. Those animals utilized in dairy industries generally are represented by the bovids, but the category of dairy animals also includes such cervids as moose and reindeer, camelids including camel and to a far lesser extent llama, and among the equines we use both the horse and donkey as sources of milk. Body fat from many of these animals is considered a food in its own right and also perceived as an important cooking ingredient. The larger mammals often have been used as a source of labor and transportation, either as beasts of burden for animal portage or harnessed for use as drayage animals. Indeed the very largest often are ridden.

There are, of course, many additional mammalian species kept as livestock and primarily of note are three species perceived as fur animals with their meat generally used to feed other animals. These three fur animals are the American mink (*Neovison vison*) and the two remaining species of chinchilla, Short-tailed chinchilla (*Chinchilla chinchilla*) and Long-tailed chinchilla (*Chinchilla lanigera*). Still other terrestrial mammals such as rats are hunted and eaten by humans but they generally are not kept as livestock.

Component parts of livestock that neither are eaten by humans nor fed as meat to other animals are utilized for a variety purposes with a few of those examples mentioned here. Body fat and butter fat from mammals has been burned as a source of lighting. We make not only health and beauty products, but also household furnishings and decorations from animal materials. Body fat serves as an ingredient for making soap. Hydrolysed keratin is a common cosmetic ingredient. Animal body fat, butter fat, and waxes such as lanolin are used for skin emollients. The hides of larger animals often are used as a source of leather, collagen, and derivatives of collagen including gelatin. Bones and connective tissue similarly are used as a source of collagen for making gelatin. Bones also are used as an ingredient in ceramic furnishings. Mammals which have long hair that easily can be spun and woven often become a source of fiber for clothing and other domestic items made of cloth. Mammals are, of course, not the only livestock animals that we view as raw materials. As an example, the shells of crustaceans and molluscs can be used as ingredients in dietary supplements for other animals. Birds often produce desirable feathers used for decorative purposes including clothing. Chemical industries have developed commercial applications for a broad range of additional products manufactured from shells and feathers.

6.2.1 The Mammals Kept as Livestock for Human Consumption

The mammalian species addressed in this chapter are those used as sources of either meat, milk, or both, for human consumption. These mammals are listed in Table 6.1 and taxonomically described in this section. All of the listed mammals are used as sources of meat. Those mammals which also are used as sources of milk for human ingestion are members of the genera *Alces*, *Bos*, *Bubalus*, *Camelus*, *Capra*, *Cervus*, *Ovis*, and *Rangifer*.

Examining the list of mammals maintained as livestock intended for human consumption may reveal some interesting surprises. What we either now raise to eat or have raised in the past varies tremendously from one part of the world to another and the dietary choices include some cultural preferences whose origins often were distinctly local in nature. Our preferential decisions include aspects that are religious, perhaps most notably being the considerations regarding consumption of cattle and pig. There also is the truth that different societies often view as pets those same animals which other people would consider to be traditional foods, and this difference frequently has become a point of contention when cultural groups meet and intermix. Local taste preferences include choices of such mammals as cat, dog, dormouse, horse and guinea pig. It is unfortunately often considered easier to negatively criticise someone else's cultural practices when we should instead present

Table 6.1 Mammals kept as livestock for human consumption

Infraclass	Order	Family	Subfamily	Genus	Common name
Eutheria	Artiodactyla	Bovidae	Bovinae	*Bison*	American bison
				Bos	Cattle Yak Gayal
				Bubalus	Carabao Water buffalo
			Caprinae	*Capra* *Ovis*	Goat Sheep
		Camelidae	Camelinae	*Camelus* *Lama* *Vicugna*	Camel Llama Alpaca
		Cervidae	Odocoileinae	*Alces* *Rangife*	Moose Reindeer
			Cervinae	*Cervus*	Elk Red deer
				Dama	Fallow deer
		Suidae	Suinae	*Sus*	Pig
	Carnivora	Canidae Felidae	Caninae Felinae	*Canis* *Felis*	Dog Cat
	Lagomorpha	Leporidae	Leporinae	*Oryctolagus*	Rabbit
	Perissodactyla	Equidae	Equinae	*Equus*	Donkey (Ass) Horse
	Rodentia	Caviidae Hydrochaeridae Gliridae Myocastoridae	Caviinae Hydrochoerinae Glirinae Not assigned	*Cavia* *Hydrochoerus* *Glis* *Myocastor*	Guinea pig Capybara Dormouse Coypu (Nutria)
Metatheria	Diprotodontia	Macropodidae	Not assigned	*Macropus*	Kangaroo

All of these listed mammals are raised as sources of meat for human consumption. Some members of the following genera also are used as a sources of milk for human consumption: *Alces*, *Bos*, *Bubalus*, *Camelus*, *Capra*, *Cervus*, *Equus*, *Ovis*, and *Rangifer*. The Water buffalo also is called the Asian buffalo and sometimes those two names are combined as Asian water buffalo. The African buffalo or Cape buffalo does not seem to ever have been domesticated owing to its unpredictable nature and for which reason it is considered dangerous to humans. The terms yak, goat, sheep and pig respectively refer to the Domestic yak, Domestic goat, Domestic sheep, and Domestic pig. My usage of Camel refers to both the Bactrian camel and the Dromedary camel. The species indicated in this table as dormouse commonly is known as the Edible dormouse or Fat dormouse, it is live-trapped and then raised domestically. My usage of the term kangaroo refers as a group to three species, the Red kangaroo, Eastern grey kangaroo, and Western grey kangaroo. Of those three kangaroo species it is the Red kangaroo which seems to be the most suggested for domestication as livestock. Kangaroo indeed are bred in captivity and raised on farms but to date their production for human consumption generally is considered unprofitable because of cost competition against hunted animals. I have kept use of the order name Artiodactyla although Cetartiodactyla seems now to be on the road to replacing that order name acceptance and offer understanding of the differences between the many personal choices which are made by societal groups. We must not criticize what is eaten in other, and in particularly less wealthy cultures, simply because those who are fortunate enough to have wealthier lives can afford to have multiple options and be finicky eaters. We also should not allow our personal preferences to influence control over what human groups either can or cannot choose to eat.

6.2.2 Terminology for the Mammalian Livestock Groups Included in This Chapter

The following briefly summarizes some of the terminology that I have used in this chapter when referring to mammalian livestock.

The Carabao often is called a water buffalo, but currently these two names are not considered synonymous. For this chapter the common name Carabao corresponds to the species (*Bubalus carabanensis*). The common name buffalo is used in this chapter as a reference to the Water buffalo (*Bubalus bubalis*) which also is called the Asian buffalo, and sometimes those names are combined in common practice to form the name Asian water buffalo. The African buffalo or Cape buffalo (*Syncerus caffer*) seems not to have been domesticated owing to its unpredictable nature, for which reason it is considered dangerous to humans. The term bison, when used in this chapter, indicates the American bison (*Bison bison*) unless specifically identified as a referral of information pertaining to the European bison (*Bison bonasus*). The terms yak, goat, sheep and pig respectively refer to the Domestic yak (*Bos grunniens*), Domestic goat (*Capra hircus*), Domestic sheep (*Ovis aries*), and Domestic pig (*Sus scrofa*). My usage of camel refers to both the Bactrian camel (*Camelus bactrianus*) and the Dromedary camel (*Camelus dromedarius*).

Not all of the mammals listed in Table 6.1 are considered domesticated, a few such as the dormouse and kangaroo are captured wild and then raised in captivity. The species indicated in Table 6.1 as dormouse is that commonly known as the Edible dormouse or Fat dormouse (*Glis glis*), it is live-trapped and then raised domestically. My usage of the term kangaroo refers as a group to three species, the Red Kangaroo (*Macropus rufus*), Eastern grey kangaroo (*Macropus giganteus*), and Western grey kangaroo (*Macropus fuliginosus*). Of those three kangaroo species it is the Red kangaroo which seems to be the most suggested for domestication as livestock. Kangaroo indeed are bred in captivity and raised on farms but to date their production for human consumption generally is considered unprofitable because of cost competition against hunted animals.

Bovid is a category that refers to members of the family Bovidae, and is represented in this chapter by both bovines and the caprids, with members of the latter group also identified by the term caprines.

Bovine is a term which refers to members of the subfamily Bovinae, is represented in this chapter by the following species: the American bison (*Bison bison*); Cattle (*Bos taurus*);

Gayal (*Bos frontalis*); Yak (referring in this chapter to the Domesticated yak, *Bos grunniens*); Water Buffalo (*Bubalus bubalis*); and Carabao (*Bubalus carabanensis*). Neither of these bovines biological relatives the European bison (*Bison bonasus*) nor the African buffalo or Cape buffalo (*Syncerus caffer*), seem to have been either domesticated or otherwise considered as livestock and thus those two relatives are not included in this chapter. Interestingly, the American bison currently is raised agriculturally in Europe.

Canine is a term which represents those members of the subfamily Caninae that are grouped within the family Canidae. The category of canines is represented in this chapter by only one subspecies recognized as Dog or Domestic dog (*Canis lupus familiaris*).

Caprid and caprine are terms which refer to members of the subfamily Caprinae, and that group is represented in this chapter by the following species: goat, which for the purposes of this chapter represents only the Domestic goat (*Capra hircus*), and sheep, which herein refers to only the Domestic sheep (*Ovis aries*).

Camelid is a term which refers to members of the family Camelidae, and they are represented in this chapter by the following animals: Alpaca (*Vicugna pacos*); camel, referring in this chapter to both the Bactrian camel (*Camelus bactrianus*) and the Dromedary camel (*Camelus dromedarius*); and Llama (*Lama glama*).

Caviid is a term which refers to members of the family Caviidae, a group that is represented in this chapter by the Guinea pig (*Cavia porcellus*).

Cervid is a term which refers to members of the family Cervidae, and they are represented in this chapter by the following animals: Elk (*Cervus canadensis*); Fallow deer (*Dama dama*); Moose or Eurasian Elk (*Alces alces*); Red deer (*Cervus elaphus*); and Reindeer, also called Caribou (*Rangifer tarandus*). The information provided for pathogens of moose does not distinguish between species of *Alces*, all of which are termed to be moose.

Equine is a term that refers to members of the subfamily Equinae, which includes the genus Equus. That group is represented in this chapter by two species, those being the Donkey or Ass (*Equus asinus*) and the Horse or Domestic Horse (*Equus ferus caballus*).

Feline is a term which refers to members of the subfamily Felinae within the family Felidae, as represented in this chapter by only that species know as Cat or Domestic cat (*Felis catus*).

Glirid is a term which refers to members of the family Gliridae, a group that is represented in this chapter by the name dormouse, herein referring to only the Edible dormouse or Fat dormouse (*Glis glis*).

Leporid is a term which refers to members of the family Leporidae, and that group is represented in this chapter by the rabbit, also known as the European rabbit or Common rabbit (*Oryctolagus cuniculus*).

Macropod is a term which refers to members of the family Macropodidae, and three of its member species are represented in this chapter under the general name kangaroo; those are the Eastern grey kangaroo (*Macropus giganteus*), Western grey kangaroo (*Macropus fuliginosus*), and Red Kangaroo (*Macropus rufus*).

Rodent is a term that refers to members of the order Rodentia, which includes both the caviids and glirids as well as the hydrochaerids. That order also includes the Coypu or Nutria (*Myocastor coypus*).

Ruminant is a term which in this chapter refers only to those mammals belonging to the suborder Ruminantia, as represented by members of the families Bovidae and Cervidae. Camelids also are foregut fermenters but not classed as ruminants because of differences in stomach structure.

Suid is a term which refers to members of the family Suidae, and that group is represented in this chapter by only one species, the Pig or Domestic pig (*Sus scrofa*).

6.3 Opportunistic Pathogenicity

Opportunistic pathogens are those which typically cause disease subsequent to an opportunity. Some opportunistic pathogens are environmental residents that may live naturally in soil and water. But, if one of these opportunistic environmental organisms arrives at a location either on or within a potential host animal or plant which suits the metabolic capabilities and requirements of that environmental organism, then an infection can ensue and the host may become diseased. Many other opportunistic pathogens reside naturally either at various times or exclusively as symbionts or commensals of host plants and animals, and yet should the host become weakened in its capacity to mount a protective defense, that otherwise harmless symbiotic or commensal organism may reveal its pathogenic potential. The ability of a host to defend against the pathogenic potential inherent in a symbiotic or commensal organism often involves a component of site specificity. For example, a microorganism which normally is commensal in either the mouth or colon of an animal may cause disease if it reaches a cutaneous wound on that same animal. The term 'primary pathogen' often is used as a defining reference to those pathogenic organisms which can cause disease in otherwise healthy hosts. Correspondingly, opportunistic pathogens are frequently, although often confusingly, termed 'secondary pathogens' perhaps suggesting that opportunistic pathogens cannot be the sole attacker of a host. Opportunistic pathogens very often can act alone as the sole cause of pathogenicity related to an infection. Perhaps a clearer usage of the term secondary pathogen would involve a more restricted reference to uniquely those organisms whose pathogenicity can be expressed only as a compounding factor to an underlying infection that has been caused by some other pathogen.

There are numerous factors which affect opportunistic pathogenicity and although they strongly overlap, it is possible to group these as: host factors, microbial factors, ecological factors, and medical factors. Some of them are included in Table 6.2. A general introduction to disease transmission was published by Hurst and Murphy (1996).

6.3.1 Host Factors

There are several component host factors including immunosuppression and immune dysfunction which can increase the susceptibility of mammals to pathogenic attack.

Table 6.2 Factors Affecting Opportunistic Pathogenicity

Host factors
Immunosuppression
Concurrent infection
Diabetes
Dietary deficiency
Injury
Medically induced effects
Stress
Underlying immune dysfunction
Secretory systems
Adaptive immunity, also termed specific defense
B cells
T cells
Innate immunity, also termed nonspecific defense
Defensins
Extrachromosomal histone H2B
Lysozymes
Macrophage
Magainins
Pattern recognition receptors
Transferrins
Microbial factors
Zoonotic potential and host range
Ecological factors
Demography of an infectious disease
Presence of a susceptible animal host
Source of pathogenic organism
Environmental presence
Presence of another infected host serving as reservoir
Presence of suitable vector
Medical factors
Severity and duration of disease and symptoms
Acute versus chronic
Overt versus subacute
Appropriate diagnosis
Prevention and treatment
Antimicrobial compounds
Microbial treatments
Probiotics
Prebiotics
Synbiotics
Vaccines

6.3.1.1 Immunosuppression

Immunosuppression can be present because of such conditions as concurrent infections, diabetes, dietary deficiencies, injury, medically induced suppression which generally is a side effect of treatment for some other illness, and stress whose components include not only climatic factors like seasonal weather but also the effects of transport and relocating animals including capture and release programs.

6.3.1.2 Immune Dysfunction

Underlying immune dysfunction generally is due to genetic deficiencies and often the impaired immune functions can be improved by selective breeding programs. Table 6.2 mentions the category of secretory system dysfunctions, representing either the absence, overproduction, or functional failure of metabolic components including those normally associated with mucosal surfaces of the respiratory and genitourinary tracts. An example of secretory failure is the chloride channel dysfunction which causes cystic fibrosis and increases susceptibility to respiratory infections. Dysfunction also can occur in the adaptive and innate immune systems, which in turn will compromise those systems and often results in the host more easily being targeted by potentially pathogenic organisms.

6.3.1.2.1 Adaptive Immunity

Adaptive immunity also is termed specific defense. In mammals, its main components are B cells and T cells. The B cells produce secretory proteins including antibodies as well as signaling proteins termed cytokines. The central role of T cells is termed cell mediated immunity, and its component categories include Helper T cells which assist other lymphocytes, Cytotoxic T cells which destroy tumor cells and virus-infected cells, Memory T cells which allow the immune system to respond more quickly to subsequent infections by similar pathogens, Suppressor T cells which help maintain immunological tolerance, and Natural killer T cells which are perceived as connecting the adaptive and innate immune systems. Some of the T cells also release cytokines.

6.3.1.2.2 Innate Immunity

Innate immunity also is termed nonspecific defense. There are numerous components which comprise the innate immune system and only some examples are mentioned here. Defensins are cationic proteins that typically bind to and assist in the destruction of microbes by forming pore-like defects. Defensins are present in some types of immune cells and many are secretory in nature. Extrachromosomal histone H2B represents a category of proteins that typically seem to remain in the cytoplasm where they aid in antiviral defense by attacking intracellular double stranded DNA. Lysozymes are proteins that damage bacterial cell walls by catalyzing hydrolysis, they are present in some types of immune cells and also are secretory. Macrophage are a category of phagocytes which act in conjunction with both adaptive and innate immune responses. The macrophage can engulf and then destroy microbes, and yet some pathogens have evolved a capability to survive inside the host animal by establishing residence within macrophage. Magainins are secretory peptides which act at least in part by permeabilizing cell membranes. Pattern recognition receptors also are proteins, of which some types are secretory, that act by recognizing molecules termed pathogen-associated molecular patterns. Transferrins are secretory plasma proteins that bind and control the level of available iron.

6.3.2 Microbial factors

Among the microbial factors are zoonotic potential and host range. These are evolutionarily related to the success of past host-pathogen interactions and determined by metabolic requirements of the pathogen in conjunction with the pathogens ability to resist the hosts defenses. Zoonotic potential of environmental microbes can be influenced by ambient environmental conditions that affect genetic expression of the pathogens metabolism. An example of environmental preconditioning occurs when ambient environmental conditions such as elevated temperature activate the genetic expression of microbial metabolic traits that act as pathogenic traits, priming the capability of a microbe to initiate infection when it enters the warm body of a mammalian host.

6.3.3 Ecological Factors

Ecological factors can influence the demography of an infectious disease, often acting in conjunction to determine the likelihood of pathogen acquisition. From the perspective of a pathogen, demographic success requires a source of the pathogenic organism, simultaneous presence of a susceptible animal host, and some way in which the pathogen can encounter the host. The source of many pathogenic organisms arises from their possessing a natural environmental presence, such presence may range from ubiquitous to geographically limited and also can be seasonal. Other pathogenic species are acquired from infected hosts and importantly for these pathogens the likely simultaneous presence of both infected reservoir hosts and susceptible hosts can be geographically limited and also may be seasonal. Some pathogens require transferral by a suitable vector, but the presence of their potential vector species likewise can be geographically limited and also include a component of seasonality.

6.3.4 Medical Factors

There are a large number of medically related factors which can influence the incidence of opportunistic pathogenicity and outcome. Prevention of disease is the best goal. Among the disease prevention measures, those with proven success are good general animal husbandry practices, use of vaccines, and administration of prophylactic antimicrobials. Vaccines often protect the majority of animals but also will kill some animals and so their usage involves considering the net balance of risk. Prophylactic use of antimicrobials clearly can benefit animals but potentially also may have a negative affect by increasing the antibiotic resistance capabilities of microbes.

In addition to understanding that environmental conditions and host-microbe interactions can affect the occurrence of opportunistic infections, we must recognize the role of microbe-microbe interactions. It is for this reason that preventative treatments based upon the concept of assisting an animal's protective commensal microbial populations have seen an increased level of interest in recent years. These treatments often are categorized as probiotics, prebiotics, and synbiotics. Probiotics are microbes administered to commensally compete against pathogens, but some of the microbes used as probiotics are themselves opportunistic pathogens and their administration can cause disease. Prebiotics are compounds administered with the intention of benefiting commensal microbes, with an example being nutritional supplements. Synbiotics is a concept that represents simultaneous administration of probiotic microbes and prebiotic compounds.

When disease does occur, obtaining an appropriate diagnosis in a timely fashion better allows for successful treatment of the affected animal. Disease may be acute, meaning that it has either a rapid onset or short duration, versus chronic, meaning that the disease has long term persistence. Overt disease is more likely to be recognized and adequately treated as compared with subacute disease. Following successful diagnosis, the issues of key importance in supporting the affected animals become reducing the duration of disease and severity of symptoms. Sometimes supportive care is sufficient, but when that accompanied by antimicrobial compounds becomes inadequate then further measures must be taken including the possibility of destroying the affected animals to protect healthy members of the host population.

6.4 Listing of Opportunistic Bacteria Affecting Livestock Mammals

Many of the listed bacteria naturally are environmental organisms and most typically they reside either in soil or water. Others are naturally associated with plants. Often these environmental organisms are saprophytically associated with decaying organic material and their association with animals typically arises from either a passive presence on the skin, accidental introduction into wounds, ingestion or inhalation. Some of the listed fungi either are opportunistic dermatophytes or commensal residents of the mucosal tissues, and many of these microbes may have either no natural environmental presence or only a brief presence in the natural environment as contaminants associated with carcasses and naturally shed materials including animal excretions.

Some pathogen species are very specific with respect to their host range and in particular that tendency exists with viruses, while other pathogens are generalists capable of infecting many different host animals. The bacterial pathogens tend to be generalists in terms of their infective capability and you could presume that any association listed between a member of these pathogen groups with one member of a mammalian genus could suggest that the pathogen similarly has an ability to infect other members of that same mammalian genus. Some disease associations found to exist for the European bison but not for the American bison are listed in this chapter under the presumption that a pathogen which affects the European bison likely would also be capable of affecting the American bison. I have listed all of the potentially opportunistic associations that I could find by internet researching. The fact that a disease association between some particular pathogen species and host species cannot be found listed in this chapter should not result in the readers presumption that such an association does not exist.

Some mammalian groups seem to be represented in more detail among these listings. That fact represents differences in the amount of modern veterinary research which is conducted regarding the different pathogen and host relationships. The ability to conduct those studies depends upon availability of research funding, the relative value of the animal in question plus acceptable cost of treating opportunistic infections where the animal is raised, and indeed financial capacity for treating the animal. More research is done on larger animals and on those animals considered as pets because these categories tend to have more financial value per individual animal. Historically traditional veterinary studies would always have been a part of animal husbandry. Those historical studies identified diseases but generally not the affecting pathogens. I suggest the following references as good sources of additional information on these opportunistic pathogens: Kahn and Line (2010), Pathogen Safety Data Sheets and Risk Assessment (2015), Sumbali (2011), The Center for Food Security and Public Health (2015), and Vetbook (2015).

Note: use of brackets around an assigned species name indicates that either part or all of that assigned name has not officially been approved.

6.4.1 Acholeplasma

Phylum: Tenericutes; Class: Mollicutes; Order: Acholeplasmatales; Family: Acholeplasmataceae; Genus: *Acholeplasma*. *Acholeplasma* species rely upon saprophytic nutrition which is an extracellular chemoheterotrophic digestion process that digests and decays organic material. The members of this genus vary with regard to the expansiveness of their host range. Their associated symptomatology varies from conjunctivitis or keratoconjunctivitis to systemic infections including those of the lung leading to pneumonia, and of either the fetus or placenta leading to abortion. *Acholeplasma laidlawii* is the only mollicute that is capable of existing free of any host and it can do so provided that the environment provides its necessary nutritional needs. Those species belonging to this microbial genus which have been indicated to have a potentially opportunistic disease association with mammalian livestock are: *Acholeplasma axanthum* (wide host range in mammals), *Acholeplasma cavigenitalium* (guinea pig), *Acholeplasma equifetale* (horse). *Acholeplasma hippikon* (horse), *Acholeplasma laidlawii* (wide host range in mammals including horse and camel), and *Acholeplasma oculi* (wide mammalian host range including sheep, goat, horse, camel).

6.4.2 Achromobacter

Phylum: Proteobacteria; Class: Betaproteobacteria; Order: Burkholderiales; Family: Alcaligenaceae; Genus: *Achromobacter*. The species *Achromobacter xylosoxidans* is an aquatic organism that causes lung infections including pulmonary nodules (pig, rabbit, coypu).

6.4.3 Acinetobacter

Phylum: Proteobacteria; Class: Gammaproteobacteria; Order: Pseudomonadales; Family: Moraxellaceae; Genus: *Acinetobacter*. Members of the genus *Acinetobacter* commonly are found in soil and water including fresh water and sewage. They notably can grow in drinking water distribution systems. The *Acinetobacter* are associated with necrotising fasciitis, urinary tract infections, and pneumonia. Those species belonging to this microbial genus which have been indicated to have a potentially opportunistic disease association with mammalian livestock are: *Acinetobacter baumannii* (cattle, pig), *Acinetobacter beijerinckii* (livestock), *Acinetobacter calcoaceticus* (cattle, coypu), *Acinetobacter gyllenbergii* (potentially in livestock), *Acinetobacter haemolyticus* (cattle, sheep, pig, coypu), *Acinetobacter lwoffii* (cattle, goat, sheep), *Acinetobacter radioresistens* (livestock), *Acinetobacter rudis* (cattle), *Acinetobacter schindleri* (ruminants), and *Acinetobacter ursingii* (potentially in livestock).

6.4.4 Actinobacillus

Phylum: Proteobacteria; Class: Gammaproteobacteria; Order: Pasteurellales; Family: Pasteurellaceae; Genus: *Actinobacillus*. The members of this genus are naturally part of the subgingival ecosystem. Members of the genus *Actinobacillus* can affect lymph nodes of the head, notably the face, mouth, esophagus and neck, and their symptoms include the formation of granulomas; granulomas also can form in the lungs associated with pleuropneumonia, and in the udder. Those species belonging to this microbial genus which have been indicated to have a potentially opportunistic disease association with mammalian livestock are: *Actinobacillus capsulatus* (alpaca, llama), *Actinobacillus equuli* (horse), *Actinobacillus lignieresii* (camel, cattle, horse, pig, sheep, dog), *Actinobacillus pleuropneumoniae* (cattle, sheep, pig), *Actinobacillus suis* (pig, horse, cattle, goat, alpaca, dog, cat), and *Actinobacillus ureae* (found in coypu although a disease association remains unclear).

6.4.5 Actinobaculum

Phylum: Actinobacteria; Class: Actinobacteria; Order: Actinomycetales; Family: Actinomycetaceae; Genus: *Actinobaculum*. The members of this genus typically are found as saprophytes in soil and water. The species *Actinobaculum suis* is importantly known as the cause of ascending urinary tract infections in swine and that disease is known as Porcine Cystitis-Pyelonephritis Complex (pig).

6.4.6 Actinomyces

Phylum: Actinobacteria; Class: Actinobacteria; Order: Actinomycetales; Family: Actinomycetaceae; Genus: *Actinomyces*. The members of this genus typically are found in soil and compost. The disease symptoms associated with members of the genus *Actinomyces* are varied. They include secondary invasion of decayed teeth and mandible leading to oral abscesses, osteomyelitis characterized by dislodgement of teeth, mandibular fractures (called lumpy jaw) and cervical abscesses. This bacterial genus can cause soft tissue infections including bursitis, abscesses in the gastrointestinal tract and lungs, meningitis, meningoencephalitis, and endophthalmitis. Those species belonging to this microbial genus which have been indicated to have a potentially opportunistic disease association with mammalian livestock are: *Actinomyces bovis* (cattle, sheep, pig, moose, caribou, presumably the cause of lumpy jaw in kangaroo), *Actinomyces canis* (dog), *Actinomyces hordeovulneris* (dog), *Actinomyces hyovaginalis* (goat, sheep, pig), *Actinomyces israelii* (pig, cattle), *Actinomyces naeslundii* (pig), *Actinomyces ruminicola* (cattle), *Actinomyces suimastitidis* (pig), and *Actinomyces viscosus* (dog).

6.4.7 Aerococcus

Phylum: Firmicutes; Class: Bacilli; Order: Lactobacillales; Family: Aerococcaceae; Genus: *Aerococcus*. The members of this bacterial genus are saprophytes commonly isolated from air and dust, plus they also have been isolated from vegetation and soil. *Aerococcus suis* has been noted to cause meningitis (pig). *Aerococcus urinae* is found in urine and associated with urogenital infections including bovine reproductive disease (cattle, horse). *Aerococcus urinaeequi* similarly is found in urine of cattle and horse, this bacterial species previously was named *Pediococcus urinaeequi*, and although it causes disease in humans it seems to not yet have been found causal of livestock infections. *Aerococcus vaginalis* may be associated with urogenital inflammation (cattle). *Aerococcus viridans* is broadly infective and its host range includes fish, lobsters, and sea turtles, with respect to mammalian livestock this bacterial species is noted for causing clinical and also subclinical mastitis particularly in cattle and it also causes Bovine Severe Respiratory Syndrome (cattle, pig).

6.4.8 Aeromonas

Phylum: Proteobacteria; Class: Gammaproteobacteria; Order: Aeromonadales; Family: Aeromonadaceae; Genus: *Aeromonas*. The members of this genus typically are found in aquatic environments. Members of the genus *Aeromonas* typically are associated with diarrhea, hemorrhages, and septicemia. Those species belonging to this microbial genus which have been indicated to have a potentially opportunistic disease association with mammalian livestock are: *Aeromonas caviae* (guinea pig), *Aeromonas hydrophila* (dog, cattle, horse, pig, sheep), and *Aeromonas veronii* (found in coypu although a disease association remains unclear).

6.4.9 Afipia

Phylum: Proteobacteria; Class: Alphaproteobacteria; Order: Rhizobiales; Family: Bradyrhizobiaceae; Genus: *Afipia*. The species *Afipia felis* typically is found residing in hospital water supplies. It causes wound infections and also has been associated with aseptic facial granulomas and pyoderma (cat).

6.4.10 Alcaligenes

Phylum: Proteobacteria; Class: Betaproteobacteria; Order: Burkholderiales; Family: Alcaligenaceae; Genus: *Alcaligenes*. Members of this genus typically are found in soil and aquatic environments. The species *Alcaligenes faecalis* correspondingly is found in moist environmental areas such as soil and water, and additionally is found in the alimentary tract of vertebrates. The species *Alcaligenes faecalis* is an opportunistic pathogen typically associated with postoperative infections in humans and thus its presence in livestock animals might also prove to have a disease association (found in coypu although a disease association remains unclear).

6.4.11 Anaplasma

Phylum: Proteobacteria; Class: Alphaproteobacteria; Order: Rickettsiales; Family: Anaplasmataceae; Genus: *Anaplasma*. The members of this genus typically reside in host blood cells as obligately intracellular microorganisms. The clinical signs associated with *Anaplasma* infections are highly variable, ranging from subclinical infection in animals under a year of age to severe peracute disease in naive adults. Symptoms include infertility, anemia, jaundice, dark yellow urine, weight loss, meningitis, encephalitis, encephalomyelitis, abortion and death. Recovered animals typically become persistent carriers and reservoirs of infection for life. Those species belonging to this microbial genus which have been indicated to have a potentially opportunistic disease association with mammalian livestock are: *Anaplasma bovis* (cattle and possibly others as well), *Anaplasma marginale* (American bison), *Anaplasma ovis* (goat, sheep), *Anaplasma phagocytophilum* (sheep, cattle, dog, horse, cat, llama), and *Anaplasma platys* (dog).

6.4.12 Arcanobacterium

Phylum: Actinobacteria; Class: Actinobacteria; Order: Actinomycetales; Family: Actinomycetaceae; Genus: *Arcanobacterium*. The members of this genus typically are soil dwelling and they infect through either commensal activities or via trauma. Members of the species *Arcanobacterium* are considered normal flora of the pharynx but can cause internal ear infections, upper respiratory infections including pharyngitis and sinusitis, endocarditis, lower respiratory infections including pneumonia, and necrotising fasciitis. Notably, the bacterial species previously classified as *Arcanobacterium pyogenes* now is classified as *Trueperella pyogenes*. Those species belonging to this microbial genus which have been indicated to have a potentially opportunistic disease association with mammalian livestock are: *Arcanobacterium canis* (dog), *Arcanobacterium haemolyticum* (dog, cattle), *Arcanobacterium hippocoleae* (dog), *Arcanobacterium phocae* (dog), and *Arcanobacterium pluranimalium* (dog).

6.4.13 Arcobacter

Phylum: Proteobacteria; Class: Epsilonproteobacteria; Order: Campylobacterales; Family: Campylobacteraceae; Genus: *Arcobacter*. Arcobacter species are found in feces including those of animals that may not be displaying symptomatology and

Arcobacter are highly abundant in sewage, which may represent their natural ecology. *Arcobacter butzleri* has been associated with enteritis and diarrhea (pig, cattle, horse). *Arcobacter skirrowii* is associated with diarrhea (sheep) and aborted fetuses (pig, sheep, cattle). *Arcobacter thereius* has been isolated from tissue of an aborted fetus (pig).

6.4.14 Arthrobacter

Phylum: Actinobacteria; Class: Actinobacteria; Order: Micrococcales; Family: Micrococcaceae; Genus: *Arthrobacter*. The members of this genus commonly are found in soil. Members of the genus *Arthrobacter* typically are associated with genital infections including vaginitis. The species belonging to this microbial genus which has been indicated to have a potentially opportunistic disease association with mammalian livestock is: *Arthrobacter gandavensis* (cattle).

6.4.15 Atlantibacter

Phylum: Proteobacteria; Class: Gammaproteobacteria; Order: Enterobacteriales; Family: Enterobacteriaceae; Genus: *Atlantibacter*. The species *Atlantibacter hermannii* has been isolated from soil and presumably resides naturally in soil, but *Atlantibacter hermannii* also is broadly infective of mammals notably including cattle. This species generally is considered nonpathogenic but has been isolated from wounds, eye infections, periodontitis, sputum and stool. *Atlantibacter hermannii* previously was named *Escherichia hermannii*.

6.4.16 Bacillus

Phylum: Firmicutes; Class: Bacilli; Order: Bacillales; Family: Bacillaceae; Genus: *Bacillus*. The members of this genus typically are found in soil. Many of the species belonging to the genus *Bacillus*, including some sold as probiotics, are opportunistic and there have been instances in which *Bacillus* administered as probiotics resulted in overt infectious disease. The range of opportunistic disease caused by *Bacillus* species includes actinomycotic-like lesions as well as chronic infections of the eye, kidney, skin and udder. Their ability to cross the placenta can result in abortion. This opportunistic nature contrasts with that of some other member species, notably *Bacillus anthracis*, which are considered to be a obligate pathogens. Toxins produced by many of the *Bacillus* species can complicate the subject of infection. Curiously notable among these is *Bacillus cereus*, whose growth in improperly stored steamed rice has resulted in a foodborne association for humans that is termed fried rice syndrome. Those species belonging to this microbial genus which have been indicated to have a potentially opportunistic disease association with mammalian livestock are: *Bacillus cereus* (broadly infective of mammals including camel and water buffalo), *Bacillus licheniformis* (bovines), *Bacillus mycoides* (cattle), *Bacillus pumilus* (broadly infective of ruminants including cattle), and *Bacillus subtilis* (cattle, sheep).

6.4.17 Bacteroides

Phylum: Bacteroidetes; Class: Bacteroidia; Order: Bacteroidales; Family: Bacteroidaceae; Genus: *Bacteroides*. The members of this genus typically are found naturally existing only in a symbiotic relationship with the mammal within which they inhabit the mucins covering colonic mucosa. Various members of the genus *Bacteroides* contribute to inflammatory bowel disease, meningitis, meningoencephalitis, periodontal disease, and pleuropneumonia. Those species belonging to this microbial genus which have been indicated to have a potentially opportunistic disease association with mammalian livestock are: *Bacteroides fragilis* (sheep, cattle, horse, pig), *Bacteroides helcogenes* (pig), *Bacteroides ovatus* (cow), and *Bacteroides pyogenes* (pig).

6.4.18 Bartonella

Phylum: Proteobacteria; Class: Alphaproteobacteria; Order: Rhizobiales; Family: Bartonellaceae; Genus: *Bartonella*. The members of this genus typically are obligate intracellular parasites and known to be carried by mites, fleas, body lice and ticks. Members of the genus *Bartonella* have been associated with wound infections and oral inflammation. They also may complicate other diseases such as respiratory infections. Those species belonging to this microbial genus which have been indicated to have a potentially opportunistic disease association with mammalian livestock are: *Bartonella alsatica* (rabbit), *Bartonella australis* (kangaroo), *Bartonella bovis* (cattle), *Bartonella capreoli* (deer, elk), *Bartonella chomelii* (cattle), *Bartonella clarridgeiae* (cat is the presumed host), *Bartonella henselae* (cat), *Bartonella melophagi* (sheep), *Bartonella rochalimae* (dog), *Bartonella schoenbuchensis* (cattle, deer), and *Bartonella vinsonii* (dog).

6.4.19 Bergeriella

Phylum: Proteobacteria; Class: Betaproteobacteria; Order: Neisseriales; Family: Neisseriaceae; Genus: *Bergeriella*. The species *Bergeriella denitrificans* has been isolated from biofilm present on oral, throat, and nasopharynx mucosa (found in guinea pig but the disease association remains unclear.

6.4.20 Bergeyella

Phylum: Bacteroidetes; Class: Flavobacteriia; Order: Flavobacteriales; Family: Flavobacteriaceae; Genus: *Bergeyella*. The species *Bergeyella porcorum* has been isolated from the lungs and tonsils of pigs and may be normal flora in those locations. The species *Bergeyella zoohelcum* is considered normal oral flora and that seems to be its natural environment. The species *Bergeyella zoohelcum* has, however, been associated with bite wounds and potentially with bacteremia after ingestion of goat blood (cat, dog, goat).

6.4.21 Bordetella

Phylum: Proteobacteria; Class: Betaproteobacteria; Order: Burkholderiales; Family: Alcaligenaceae; Genus: *Bordetella*. The species *Bordetella bronchiseptica* can persist in the environment but presumably is not natural to the environment, instead it naturally resides in the respiratory tract. In particular, the association of this bacterial species with rabbits is nearly asymptomatic although it has been associated with both acute pneumonia and also can affect the upper respiratory tract. Infection outbreaks may have a high morbidity but mortality usually is low (broadly infective for terrestrial mammals including dog, cat, pig, rabbit, sheep, goat, kangaroo, guinea pig, donkey, horse). It is important to note that *Bordetella bronchiseptica* often acts synergistically with *Pasteurella multocida* to cause disease with paroxysmal cough and sneezing being among the most notable signs and those can be accompanied by oculonasal discharge and rhinitis with the possible consequence of sudden death.

6.4.22 Borrelia

Phylum: Spirochaetes; Class: Spirochaetia; Order: Spirochaetales; Family: Borreliaceae; Genus: *Borrelia*. The normal ecology of this genus is to be maintained as an infection of ticks and it might not have an ecology that is independent of host vertebrates and invertebrates. The species belonging to *Borrelia* are associated with systemic disease which can affect the joints, as well as causing renal, neurologic, and cardiac abnormalities. The most common syndrome includes recurrent lameness accompanied by anorexia, lethargy, and lymphadenopathy. Renal failure generally is fatal and this bacterial genus also can cause abortion. Those species belonging to this microbial genus which have been indicated to have a potentially opportunistic disease association with mammalian livestock are: *Borrelia coriaceae* (cattle), and *Borrelia theileri* (cattle).

6.4.23 Borreliella

Phylum: Spirochaetes; Class: Spirochaetia; Order: Spirochaetales; Family: Borreliaceae; Genus: *Borreliella*. Possibly, as with *Borrelia*, the normal ecology of this genus is to be maintained as an infection of ticks and it might not have an ecology that is independent of host vertebrates and invertebrates. Those species belonging to this microbial genus which have been indicated to have a potentially opportunistic disease association with mammalian livestock are: *Borreliella afzelii* (Edible dormouse), and *Borreliella burgdorferi* (dog, cat, deer, horse, cattle, sheep, Edible dormouse). *Borreliella afzelii* formerly was named *Borrelia afzelii* and *Borreliella burgdorferi* formerly was named *Borrelia burgdorferi*.

6.4.24 Brachyspira

Phylum: Spirochaetes; Class: Spirochaetia; Order: Brachyspirales; Family: Brachyspiraceae; Genus: *Brachyspira*. The normal environment for members of this genus is within the intestine and they typically also are transiently identifiable in moist feces excreted by infected animals. The *Brachyspira* cause intestinal edema with haemorrhage. Those species belonging to this microbial genus which have been indicated to have a potentially opportunistic disease association with mammalian livestock are: *Brachyspira canis* (dog), *Brachyspira hyodysenteriae* (pig, guinea pig), and *Brachyspira pilosicoli* (pig).

6.4.25 Brucella

Phylum: Proteobacteria; Class: Alphaproteobacteria; Order: Rhizobiales; Family: Brucellaceae; Genus: *Brucella*. The *Brucella* can survive for periods of years in soil and be found in aerosols, but their natural ecology seems to be residing in association with a host mammal. Infections caused by the member species of *Brucella* often demonstrate no outward clinical signs. Chronically infected animals may present with evidence of only decreased milk production. Other symptoms include epididymitis, seminal vesiculitis, orchitis, testicular abscess, spondylitis and arthritis. They can cause placentitis and endometritis. The most severe consequences include unfavorable reproductive outcomes such as mid to late term abortion, stillborn offspring, and neonatal weakness that results in death. The disease name associated with this bacterial genus bears specific mention, and it is brucellosis. Those species belonging to this microbial genus which have been indicated to have a potentially opportunistic disease association with mammalian livestock are: *Brucella abortus* (capybara, cervids including elk, yak, camel, cattle, bison, water buffalo, llama, goat, sheep), *Brucella canis* (dog), *Brucella melitensis* (camel, goat, sheep), *Brucella neotomae* (rodents), *Brucella ovis* (sheep goat), and *Brucella suis* (dog, rabbit, reindeer, pig).

6.4.26 Burkholderia

Phylum: Proteobacteria; Class: Betaproteobacteria; Order: Burkholderiales; Family: Burkholderiaceae; Genus: *Burkholderia*. The members of this genus typically are found in soil and water. The severity of disease outcome associated with infections caused by members of the genus *Burkholderia* varies tremendously depending upon the combination of bacterial and host species that are involved. In general, the members of this genus are associated with ulcerative nodules of the skin, internal organs including lung, and the upper respiratory tract. One member of this genus, *Burkholderia mallei*, causes an acute form of cervical lymphadenopathy in horse and donkey which is called glanders. Glanders can be extremely deleterious and the causative infection can be transmitted to cats that subsequently are fed meat from diseased animals. Those cats infected by consuming infected horse meat consequently develop localized nodules on nasal mucosa and bloody nasal discharge. *Burkholderia mallei* also causes a chronic form represented by generalized lymphadenopathy and ulcerated skin nodules. Thus, it should be understood that the severity of illness associated with *Burkholderia mallei* varies greatly and is dependent upon the site of infection. Those species belonging to this microbial genus which have been indicated to have a potentially opportunistic disease association with mammalian livestock are: *Burkholderia cepacia* (cattle, rabbit, sheep), *Burkholderia mallei* (cat, camel, goat, kangaroo, coypu), and *Burkholderia pseudomallei* (camel, alpaca, kangaroo, cattle, pig, goat, horse, cat, dog, sheep, pig, guinea pig).

6.4.27 Campylobacter

Phylum: Proteobacteria; Class: Epsilonproteobacteria; Order: Campylobacterales; Family: Campylobacteraceae; Genus: *Campylobacter*. The members of this genus typically colonize the intestinal mucosa and often seem to be commensal. They correspondingly are found in manure and they do survive to be found in environmental water and soil. The disease symptomatology associated with members of the genus *Campylobacter* varies from none to severe gastroenteritis. They can also cause periodontal disease, infertility, abortion and stillbirth. Those species belonging to this microbial genus which have been indicated to have a potentially opportunistic disease association with mammalian livestock are: *Campylobacter coli* (cattle, pig, seldom in camel), *Campylobacter fetus* (sheep, goat, cattle, camel, alpaca, llama, buffalo), and *Campylobacter jejuni* (horse, cattle, sheep, goat, rabbit, kangaroo, camel).

6.4.28 Chlamydia

Phylum: Chlamydiae; Class: Chlamydiia; Order: Chlamydiales; Family: Chlamydiaceae; and Genus: *Chlamydia*. The members of this genus seem unable to survive outside a eukaryotic host cell and for many their lifecycle in fact requires growth within a eukaryotic host cell, they can be found within free living amoeba. The members of this genus are associated with conjunctivitis, keratoconjunctivitis, rhinitis, abortion, and pneumonitis. Those species belonging to this microbial genus which have been indicated to have a potentially opportunistic disease association with mammalian livestock are: *Chlamydia abortus* (cattle, sheep, goat, fallow deer, red deer), *Chlamydia caviae* (guinea pig) *Chlamydia felis* (cat), *Chlamydia pecorum* (sheep, goat, cattle, pig), *Chlamydia pneumoniae* (camel, cattle), *Chlamydia psittaci* (sheep, cattle, goat, cat, pig, horse, dog), *Chlamydia suis* (pig), and *Chlamydia trachomatis* (cattle, pig, guinea pig). *Chlamydia abortus* previously was named *Chlamydophila abortus*, *Chlamydia caviae* previously was named *Chlamydophila caviae*, and *Chlamydia felis* previously was named *Chlamydophila felis*.

6.4.29 Chromobacterium

Phylum: Proteobacteria; Class: Betaproteobacteria; Order: Neisseriales; Family: Chromobacteriaceae; Genus: *Chromobacterium*. The species *Chromobacterium violaceum* is considered to reside naturally in water and soil of tropical and sub-tropical regions. Infections cause by the species *Chromobacterium violaceum* begin as either localized skin infections or localized lymphadenitis. They can progress to septicemia and abscesses of internal organs including the liver, lung, spleen, skin, lymph nodes, and brain resulting in fatal multiorgan failure (broadly infective as an opportunistic pathogen of mammals including macropods).

6.4.30 Chryseobacterium

Phylum: Bacteroidetes; Class: Flavobacteriia; Order: Flavobacteriales; Family: Flavobacteriaceae; Genus: *Chryseobacterium*. The members of this genus typically are found soil, plants, water, and milk. The *Chryseobacterium* often are found to be present without an apparent disease association, and yet they can cause a variety of disease consequences including mastitis and meningoencephalitis. Those species belonging to this microbial genus which have been indicated to have a potentially opportunistic disease association with mammalian livestock are: *Chryseobacterium bovis* (cattle, also found in coypu although a disease association remains unclear for coypu), *Chryseobacterium gleum* (coypu—disease association remains unclear), *Chryseobacterium haifense* (cattle), *Chryseobacterium indologenes* (coypu—disease association remains unclear), and *Chryseobacterium oranimense* (cattle).

6.4.31 Citrobacter

Phylum: Proteobacteria; Class: Gammaproteobacteria; Order: Enterobacteriales; Family: Enterobacteriaceae; Genus: *Citrobacter*. The members of this genus mainly are found in the open environment including soil, water, and sewage. The species *Citrobacter freundii* is opportunistic and can cause a number of different disease syndromes including urinary tract infections, joint infections, pneumonia, sepsis, meningitis, brain abscesses, and neonatal sepsis (camel, guinea pig, coypu).

6.4.32 Clostridioides

Phylum: Firmicutes; Class: Clostridia; Order: Clostridiales; Family: Peptostreptococcaceae; Genus: *Clostridioides*. The species *Clostridioides difficile* typically is found in soil, aquatic sediments, sewage, river water, lake water, and swimming pool water. *Clostridioides difficile* can cause symptoms ranging from diarrhea to life-threatening inflammation of the colon (pig, guinea pig, rabbit, bovines, equines). *Clostridioides difficile* previously was classified as *Clostridium difficile*.

6.4.33 Clostridium

Phylum: Firmicutes; Class: Clostridia; Order: Clostridiales; Family: Clostridiaceae; Genus: *Clostridium*. The members of this genus typically are found in soil, aquatic sediments, and sewage. The members of this genus are infamous for the toxins which they produce and some species such as *Clostridium tetani* and *Clostridium botulinum* are associated with muscular spasms and paralytic toxicoses. Respiratory or cardiac paralysis can occur and usually causes death, typically a mortality of 80% is expected with those infections. *Clostridium botulinum* also causes a chronic toxicosis. But, given that understanding, *Clostridium tetani* causes illness only when associated with deep narrow wounds that typically result from punctures. *Clostridium moniliforme* causes ascending urinary tract infections, meningitis and meningoencephalitis. *Clostridium perfringens* can cause gas gangrene, a myonecrosis with gas production. The bacterial species listed here typically reside naturally in the intestine although they can cause ulcerative colitis, chronic intestinal toxicosis, myonecrosis (gangrene), haemoglobinuria, infectious necrotic hepatitis, malignant edema, and abomasitis. Those species belonging to this microbial genus which have been indicated to have a potentially opportunistic disease association with mammalian livestock are: *Clostridium botulinum* (cattle, chicken, deer, donkey), *Clostridium cadaveris* (horse, sheep, guinea pig, rabbit), *Clostridium chauvoei* (bison, cattle, deer, sheep), *Clostridium moniliforme* (dog), *Clostridium novyi* (cattle, sheep, bison), *Clostridium perfringens* (elk, moose, alpaca, llama, cattle, guinea pig, camel, dog, goat, sheep, red deer, fallow deer, pig, yak, kangaroo), *Clostridium septicum* (cattle, sheep, goat, pig), and *Clostridium tetani* (cattle, donkey). *Clostridium moniliforme* previously was named *Eubacterium moniliforme*.

6.4.34 Corynebacterium

Phylum: Actinobacteria; Class: Actinobacteria; Order: Corynebacteriales; Family: Corynebacteriaceae; Genus: *Corynebacterium*. The members of this genus typically are found in soil and water, they also are commensal on the skin and mucous membranes. The *Corynebacterium* species are associated with a wide range of symptoms including ascending urinary tract infections causing cystitis and pyelonephritis. They also cause septic peritonitis, lymphadenitis, respiratory infections, mastitis, posthitis (infection of the penus forskin) and vulvitis. Those species belonging to this microbial genus which have been indicated to have a potentially opportunistic disease association with mammalian livestock are: *Corynebacterium amycolatum* (cattle), *Corynebacterium auris* (cattle), *Corynebacterium auriscanis* (dog), *Corynebacterium bovis* (cattle, camel), *Corynebacterium camporealensis* (sheep), *Corynebacterium canis* (dog), *Corynebacterium cystitidis* (cattle, sheep, goats), *Corynebacterium glucuronolyticum* (pig), *Corynebacterium jeikeium* (cat), *Corynebacterium lactis* (cattle), *Corynebacterium mastitidis* (sheep), *Corynebacterium pilosum* (cattle), *Corynebacterium pseudotuberculosis* (goat, sheep, horse, cattle, camel, water buffalo, llama, alpaca, cervids), *Corynebacterium renale* (cattle), *Corynebacterium striatum* (cattle), *Corynebacterium suicordis* (pigs), *Corynebacterium ulcerans* (camel, cat, cattle), *Corynebacterium ulceribovis* (cattle), *Corynebacterium urealyticum* (cat, dog), and *Corynebacterium uterequi* (horse).

6.4.35 Coxiella

Phylum: Proteobacteria; Class: Gammaproteobacteria; Order: Legionellales; Family: Coxiellaceae; Genus: *Coxiella*. The species *Coxiella burnetii* is an opportunistic commensal that survives in the environment. It is shed in urine (particularly in association with pregnancy), milk, and feces. During birth, this species can be found in high levels within the amniotic fluids and placenta. *Coxiella burnetii* can be environmentally acquired by inhalation of dust from contaminated materials. The species *Coxiella burnetii* causes Q fever with both acute and chronic presentations. It often presents as subclinical disease with symptoms including anorexia, infertility, metritis, late term abortion, or retained placenta. Abortions often are followed by recovery of the dam without complications (cattle, sheep, goat, camel).

6.4.36 Cutibacterium

Phylum: Actinobacteria; Class: Actinobacteria; Order: Propionibacteriales; Family: Propionibacteriaceae; Genus: *Cutibacterium*. The species *Cutibacterium acnes* is virtually ubiquitous as a commensal in the oral cavity and on the skin of animals, including residence both in and around sweat glands and sebaceous glands, and surprisingly it can become a lethal opportunist. In particular, *Cutibacterium acnes* which causes the benign skin condition termed acne also causes chronic inflammatory disease of the eyes, and additionally is a cause of placentitis and infection of the fetus leading to abortion (cattle, dog). *Cutibacterium acnes* previously was classified as *Propionibacterium acnes*.

6.4.37 Dermatophilus

Phylum: Actinobacteria; Class: Actinobacteria; Order: Micrococcales; Family: Dermatophilaceae; Genus: *Dermatophilus*. The species *Dermatophilus congolensis* may be a saprophyte in soil but its presumed normal habitat and ecology is as a commensal on the skin. The species *Dermatophilus congolensis* causes exudative dermatitis (broadly infective of mammals although affecting more frequently cattle, horse, sheep, goat).

6.4.38 Dichelobacter

Phylum: Proteobacteria; Class: Gammaproteobacteria; Order: Cardiobacteriales; Family: Cardiobacteriaceae; Genus: *Dichelobacter*. The species *Dichelobacter nodosus* survives in soil but its normal ecology may be in association with the paw or hoof where it can cause interdigital dermatitis (moose, sheep, kangaroo, cattle).

6.4.39 Edwardsiella

Phylum: Proteobacteria; Class: Gammaproteobacteria; Order: Enterobacteriales; Family: Enterobacteriaceae; Genus: *Edwardsiella*. The *Edwardsiella* typically are found in natural water, including water in which aquarium fish are raised. The members of this species typically are associated with the gastrointestinal tract of healthy fish. They can cause necrotic skin wounds and gastroenteritis, from the gastrointestinal tract they can spread extraintestinally to become septicemia. Those species belonging to this microbial genus which have been indicated to have a potentially opportunistic disease association with mammalian livestock are: *Edwardsiella ictaluri* (dog), and *Edwardsiella tarda* (dog, pig, cattle).

6.4.40 Ehrlichia

Phylum: Proteobacteria; Class: Alphaproteobacteria; Order: Rickettsiales; Family: Anaplasmataceae; Genus: *Ehrlichia*. The members of this genus typically are obligately intracellular pathogens and their ecology is similar to that of the bartonella which involves invertebrate (mites, lice, fleas, ticks) and vertebrate host cycles. Illnesses caused by members of the genus *Ehrlichia* typically are non-specific and range from either subclinical or mild to severe including abrupt death. The associated symptoms include lethargy, thrombocytopenia, anorexia, diarrhea, lymphadenopathy, weight loss, muscle or joint pain, hepatocellular enzyme abnormalities, lesions of the visceral membranes, and respiratory or neurologic signs. *Ehrlichia canis* has been identified as causing fatal central nervous system disease in a dog for which the fungal species *Cladophialophora bantiana* (that fungal species previously was named *Xylohypha bantiana*) caused a contributing secondary infection in the form of a systemic mycosis affecting the liver, spleen, kidney, and adrenal glands. Those species belonging to this microbial genus which have been indicated to have a potentially opportunistic disease association with mammalian livestock are: *Ehrlichia canis* (dog), *Ehrlichia chaffeensis* (cat, dog), *Ehrlichia ewingii* (dog), *Ehrlichia ovina* (sheep), and *Ehrlichia ruminantium* (cattle, sheep, goat, cervids, buffalo).

6.4.41 Enterobacter

Phylum: Proteobacteria; Class: Gammaproteobacteria; Order: Enterobacteriales; Family: Enterobacteriaceae; Genus: *Enterobacter*. The members of this genus typically are found in soil and water, and they are as well found on seeds and plants. Many *Enterobacter* species frequently are found in the gastrointestinal tract, feces, and dairy products. Members of this genus produce a variety of illnesses among which are respiratory infections including pleuropneumonia and septic peritonitis. Those species belonging to this microbial genus which have been indicated to have a potentially opportunistic disease association with mammalian livestock are: *Enterobacter cancerogenus* (coypu—disease association remains unclear), *Enterobacter cloacae* (guinea pig, rabbit, coypu), and *Enterobacter hormaechei* (sheep).

6.4.42 Enterococcus

Phylum: Firmicutes; Class: Bacilli; Order: Lactobacillales; Family: Enterococcaceae; Genus: *Enterococcus*. The members of this genus are normal intestinal flora and characteristically are found in sewage contaminated water. Members of the genus *Enterococcus* typically are associated with urinary system infections. Those species belonging to this microbial genus which have been indicated to have a potentially opportunistic disease association with mammalian livestock are: *Enterococcus avium* (dog), *Enterococcus canintestini* (dog), *Enterococcus canis* (dog), *Enterococcus casseliflavus* (capybara), *Enterococcus durans* (camel), *Enterococcus faecalis* (capybara, moose, cattle, coypu, dog), *Enterococcus faecium* (llama, moose, coypu, dog), *Enterococcus gallinarum* (coypu—disease association remains unclear), *Enterococcus hirae* (llama), and *Enterococcus villorum* (pig).

6.4.43 Erysipelothrix

Phylum: Firmicutes; Class: Erysipelotrichia; Order: Erysipelotrichales; Family: Erysipelotrichaceae; Genus: *Erysipelothrix*. The members of this genus typically reside on the skin, but they also can be found in sewage and fecally contaminated soil. The disease symptoms associated with members of the genus include anorexia, depression, stilted gait, exercise intolerance, lameness, enlarged joints and death. The species *Erysipelothrix rhusiopathiae* can cause an acute rash of the upper dermis (erysipelas) often appearing patterned. *Erysipelothrix tonsillarum* often causes tonsillitis. Those species belonging to this microbial genus which have been indicated to have a potentially opportunistic disease association with mammalian livestock are: *Erysipelothrix rhusiopathiae* (very broadly infective of vertebrate animals perhaps most notably pig, kangaroo, sheep, cattle, moose, and rabbit, but the host range also encompasses fish, birds, reptiles and invertebrates including lobster) and *Erysipelothrix tonsillarum* (pig).

6.4.44 Escherichia

Phylum: Proteobacteria; Class: Gammaproteobacteria; Order: Enterobacteriales; Family: Enterobacteriaceae; Genus: *Escherichia*. The members of this genus typically are intestinal commensals, but they also can be found naturally in subtropical and tropical soil and water. Infections caused by the bacterial genus *Escherichia* typically can be asymptomatic but also include enteritis, mastitis, endometritis, sepsis, septic peritonitis, ascending urinary tract infections, peritonitis, pleuropneumonia, otitis externa, meningitis and meningoencephalitis. The most commonly noted member of this genus is *Escherichia coli*, which normally is an intestinal commensal whose various genetic strains are infamous for causing potentially fatal disease when ingested by an inappropriate, non-host adapted animal species. Those species belonging to this microbial genus which have been indicated to have a potentially opportunistic disease association with mammalian livestock are: *Escherichia albertii* (cattle, pig, rabbit), *Escherichia coli* (broadly infective of terrestrial mammals including pig, sheep, goat, cattle, capybara, water buffalo, yak, horse, dog, rabbit, cat, camel, alpaca, llama, kangaroo, elk, moose, red deer, guinea pig, reindeer, coypu), *Escherichia fergusonii* (cattle, goat, horse), and *Escherichia vulneris* (widely indicated as affecting animals but specific information was not found on associations with mammalian livestock).

6.4.45 Eubacterium

Phylum: Firmicutes; Class: Clostridia; Order: Clostridiales; Family: Eubacteriaceae; Genus: *Eubacterium*. The members of this genus typically are found as saprophytes in soil and water. Some also are normal inhabitants of human skin and cavities, occasionally causing infection of soft tissue. The species *Eubacterium rangiferina* is associated with reindeer although the disease association remains unclear.

6.4.46 Ewingella

Phylum: Proteobacteria; Class: Gammaproteobacteria; Order: Enterobacteriales; Family: Yersiniaceae; Genus: *Ewingella*. The species *Ewingella americana* can be found in domestic sources of water including air conditioning units, ice baths and wound irrigation systems. In human, this bacterial species typically is associated with wound infections, pneumonia and peritonitis (coypu—disease association remains unclear).

6.4.47 Finegoldia

Phylum: Firmicutes; Class: Tissierellia; Order: Tissierellales; Family: Peptoniphilaceae; Genus: *Finegoldia*. It is not known if the members of this species naturally have an environmental presence. Their understood ecology is as commensals representing part of the normal skin biota although they are as well found in the oral cavity, genitourinary tract, and intestine. The *Finegoldia* are associated with infections of the skin and soft tissues, additionally causing infections of internal organs following surgery. Those species belonging to this microbial genus which have been indicated to have a potentially opportunistic disease association with mammalian livestock are: *Finegoldia magna* (coypu, dog), *Finegoldia sp. feline oral taxon 341* (cat), and *Finegoldia sp. feline oral taxon 342* (cat).

6.4.48 Francisella

Phylum: Proteobacteria; Class: Gammaproteobacteria; Order: Thiotrichales; Family: Francisellaceae; Genus: *Francisella*. The species *Francisella tularensis* seems to be obligately intracellular and broadly associated with vertebrates as well as invertebrate host animals. It primarily infects macrophages in vertebrate host organisms and hemocytes of invertebrates. This bacterial species is capable of lengthy survival in the environment where it can be found in water, grassland and haystacks. The fact that *Francisella tularensis* can survive in *Acanthamoeba castellanii*, which is a free living opportunistically pathogenic soil amoeba, suggests that amoeba might represent an environmental reservoir for *Francisella tularensis* and hints at some of the complex ecology that can be involved with opportunistic infections. The species *Francisella tularensis* causes the disease tularemia which bears notation for the severity of its associated illness. This bacterial species generally is considered to affect all endothermic animals. Despite being quite lethal in some host species such as human, *Francisella tularensis* may qualify as an opportunistic pathogen in some rodents and other mammalian groups (rabbit, Edible dormouse, coypu).

6.4.49 Fusobacterium

Phylum: Fusobacteria; Class: Fusobacteria; Order: Fusobacteriales; Family: Fusobacteriaceae; Genus: *Fusobacterium*. The natural ecology of this genus seems to be residence in the oral mucosa where its member species typically cause subgingival dental plaque. No information could be found that would suggest *Fusobacterium* species exist openly in nature. Members of the genus *Fusobacterium* cause calf diphtheria and necrotic abscesses particularly of the tongue, throat, jaw and liver. They can cause pleuropneumonia, meningitis, and meningoencephalitis. *Fusobacterium necrophorum* can contribute to the interdigital dermatitis associated with *Dichelobacter nodosus*. Those species belonging to this microbial genus which have been indicated to have a potentially opportunistic disease association with mammalian livestock are: *Fusobacterium canifelinum* (cat, dog), *Fusobacterium equinum* (horse), *Fusobacterium necrophorum* (cattle, kangaroo, bison, water buffalo, alpaca, llama, cervids including elk and reindeer), *Fusobacterium nucleatum* (dog), and *Fusobacterium russii* (dog).

6.4.50 Gemella

Phylum: Firmicutes; Class: Bacilli; Order: Bacillales; Family: (not assigned); Genus: *Gemella*. The members of this genus primarily are found in the mucous membranes of the oral cavity and upper digestive tract, although they can grow in drinking water distribution systems. Members of the genus *Gemella* typically cause infection of previously damaged tissue and have been associated with abscesses including those of the brain. Those species belonging to this microbial genus which have been indicated to have a potentially opportunistic disease association with mammalian livestock are: *Gemella cuniculi* (rabbit) and *Gemella morbillorum* (cattle, coypu, rabbit, pig, goat, sheep, horse).

6.4.51 Haemophilus

Phylum: Proteobacteria; Class: Gammaproteobacteria; Order: Pasteurellales; Family: Pasteurellaceae; Genus: *Haemophilus*. The species [*Haemophilus*] *parasuis* is found in the upper respiratory tract of healthy pigs and no information was found suggesting that it naturally exists in the open environment. As an opportunistic pathogen, this bacterial species is associated with meningoencephalitis, peritonitis including inflammation of the internal viscera, and polyarthritis. It can additionally contribute to bacterial pneumonia (pig).

6.4.52 Hafnia

Phylum: Proteobacteria; Class: Gammaproteobacteria; Order: Enterobacteriales; Family: Hafniaceae; Genus: *Hafnia*. The members of this genus typically are found in soil, water and sewage. They also are commensals of the gastrointestinal tract. Members of the genus *Hafnia* cause systemic infections including septicemia and pneumonia. They also are suspected of causing gastroenteritis. Those species belonging to this microbial genus which have been indicated to have a potentially opportunistic disease association with mammalian livestock are: *Hafnia alvei* (rabbit, guinea pig, coypu, sheep, cattle, goat) and *Hafnia paralvei* (cattle).

6.4.53 Hathewaya

Phylum: Firmicutes; Class: Clostridia; Order: Clostridiales; Family: Clostridiaceae; Genus: *Hathewaya*. The species *Hathewaya histolytica* can be found in feces and soil. *Hathewaya histolytica* causes gas gangrene, a myonecrosis with gas production (guinea pig, rabbit, cattle). *Hathewaya histolytica* previously was classified as *Clostridium histolyticum*.

6.4.54 Helicobacter

Phylum: Proteobacteria; Class: Epsilonproteobacteria; Order: Campylobacterales; Family: Helicobacteraceae; Genus: *Helicobacter*. The members of this genus typically are found in surface water, soil, dung, and flies. When associated with a host animal, the *Helicobacter* typically live in the lining of the upper gastrointestinal tract. Members of the genus *Helicobacter* characteristically are associated with gastrointestinal symptoms, notably including ulcerative gastritis. They also can cause hepatic disease in some host species. Alpaca and llama seem particularly susceptible to stress-related stomach ulcers caused by *Helicobacter pylori*. Those species belonging to this microbial genus which have been indicated to have a potentially opportunistic disease association with mammalian livestock are: *Candidatus Helicobacter bovis* (cattle), *Helicobacter canadensis* (rodents, pig), *Helicobacter canis* (cat, dog), *Helicobacter equorum* (horse), *Helicobacter felis* (cat), and *Helicobacter pylori* (broadly infective of mammals including alpaca, llama, camel, goat, sheep, cattle, cat, horse, donkey).

6.4.55 Histophilus

Phylum: Proteobacteria; Class: Gammaproteobacteria; Order: Pasteurellales; Family: Pasteurellaceae; Genus: *Histophilus*. The species *Histophilus somni* is considered to be an obligate parasite, typically found commensally on mucosal surfaces of the reproductive and respiratory tracts of healthy animals. It also can be acquired environmentally from inhalation of dust and aerosolized water containing the bacteria. This bacterial species is associated with otitis media and otitis interna, pneumonia, and abortion (alpaca, cattle, llama, donkey).

6.4.56 Janthinobacterium

Phylum: Proteobacteria; Class: Betaproteobacteria; Order: Burkholderiales; Family: Oxalobacteraceae; Genus: *Janthinobacterium*. The species *Janthinobacterium lividum* naturally lives in soil and is capable of causing opportunistic infections including fatal septicemia in human. Its associated with rabbits is of potential interest (rabbit—disease association remains unclear).

6.4.57 Klebsiella

Phylum: Proteobacteria; Class: Gammaproteobacteria; Order: Enterobacteriales; Family: Enterobacteriaceae; Genus: *Klebsiella*. The members of this genus typically are found in soil, water, plants, and insects.

Members of the genus *Klebsiella* also are normal fecal and oral cavity inhabitants. They can, however, cause septicemia, septic peritonitis, multiple organ inflammation, abscesses including those of the liver, endometritis, pleuropneumonia and meningitis. Those species belonging to this microbial genus which have been indicated to have a potentially opportunistic disease association with mammalian livestock are: *Klebsiella aerogenes* (capybara, camel, coypu, goat), *Klebsiella oxytoca* (guinea pig) and *Klebsiella pneumoniae* (capybara, coypu, llama, camel, cattle, equines, goat).

6.4.58 Kocuria

Phylum: Actinobacteria; Class: Actinobacteria; Order: Micrococcales; Family: Micrococcaceae; Genus: *Kocuria*. The members of this genus typically are associated with soil and also are commensals of the skin. The disease symptoms associated with members of this genus represent an association with urinary tract infections and they also can cause peritonitis in the form of cholecystitis (inflammation of the gallbladder). Those species belonging to this microbial genus which have been indicated to have a potentially opportunistic disease association with mammalian livestock are: *Kocuria kristinae* (sheep) and *Kocuria varians* (sheep).

6.4.59 Kytococcus

Phylum: Actinobacteria; Class: Actinobacteria; Order: Micrococcales; Family: Dermacoccaceae; Genus: *Kytococcus*. The species *Kytococcus sedentarius* is a marine dwelling microbe which causes internal infections following damage to mucosal tissues (sheep).

6.4.60 Lawsonia

Phylum: Proteobacteria; Class: Deltaproteobacteria; Order: Desulfovibrionales; Family: Desulfovibrionaceae; Genus: *Lawsonia*. The species *Lawsonia intracellularis* is obligately intracellular and although it can be found in feces from infected animals the duration of its environmental survival is uncertain. This species typically causes enteritis in the form of a proliferative enteropathy (dog, guinea pig, rabbit, pig, horse). There has to have been some fun found in all of this serious microbiology research and so I will relate to you that when I was a very young child my father managed a Lawson's Dairy store in Dayton, Ohio. My role in the family business was washing the store windows. The name of that chain of stores is unrelated to this bacterial genus and I never got diseased by their products.

6.4.61 Leclercia

Phylum: Proteobacteria; Class: Gammaproteobacteria; Order: Enterobacteriales; Family: Enterobacteriaceae; Genus: *Leclercia*. The species *Leclercia adecarboxylata* most likely is ubiquitous in the natural environment. It has been found in food, water, and terrestrial molluscs. As an opportunistic pathogen *Leclercia adecarboxylata* causes infection of wounds and mastitis (cattle, coypu).

6.4.62 Legionella

Phylum: Proteobacteria; Class: Gammaproteobacteria; Order: Legionellales; Family: Legionellaceae; Genus: *Legionella*. The species *Legionella pneumophila* is naturally associated with environmental freshwater. The typical source is warm water, and that association includes such domestic reservoirs as hot water storage tanks and recirculating heated water systems. It can be acquired from free living amoeba such as *Acanthamoeba* and also by inhalation of vesicles produced by *Acanthamoeba*. *Legionella pneumophila* causes pneumonia (dog, goat, horse, cattle, sheep, pig, guinea pig, rabbit).

6.4.63 Lelliottia

Phylum: Proteobacteria; Class: Gammaproteobacteria; Order: Enterobacteriales; Family: Enterobacteriaceae; Genus: *Lelliottia*. The species *Lelliottia amnigena* is found in drinking water, surface water, and soil. This species is known to cause endophthalmitis in human and thus its association with other mammals is of potential interest (coypu—disease association remains unclear). I would suggest researching under the previous name of this microbe which was *Enterobacter amnigenus*.

6.4.64 Leptospira

Phylum: Spirochaetes; Class: Spirochaetia; Order: Leptospirales; Family: Leptospiraceae; Genus: *Leptospira*. The natural habitat for members of this genus is shallow stagnant water including puddles, ponds, lakes, and bogs. Members of the genus *Leptospira* typically cause either none or modest symptomatology in carrier animals, but in non-adapted animals they can cause persistent infections that vary in severity from inapparent to fatal. Kidney infections are a typical consequence of the disease, and can include renal failure that often is accompanied by concurrent hepatitis. Persistent infections associated with members of this bacterial genus typically involve the kidney and genital tract. Abortion commonly can result within the first 5 weeks after infection. Those species belonging to this microbial genus which have been indicated to have a potentially opportunistic disease association with mammalian livestock are: *Leptospira fainei* (pig), *Leptospira inadai* (rodents), *Leptospira interrogans* (alpaca, llama, camel, dog, elk, cattle, pig, American bison), *Leptospira weilii* (kangaroo), and *Leptospira wolffii* (water buffalo, coypu, dog).

6.4.65 Listeria

Phylum: Firmicutes; Class: Bacilli; Order: Bacillales; Family: Listeriaceae; Genus: *Listeria*. The members of this genus typically are distributed widely throughout the environment, particularly in soil and fecal matter. Members of the genus *Listeria* characteristically are associated with septicemia and meningoencephalitis, but there also is an ophthalmic form. Those species belonging to this microbial genus which have been indicated to have a potentially opportunistic disease association with mammalian livestock are: *Listeria ivanovii* (ruminants, mainly sheep) and *Listeria monocytogenes* (alpaca, llama, reindeer, horse, cattle, camel, sheep, goat, pig, donkey, coypu, additionally found in several species of wallaby which belong to the genus *Macropus* and therefore *Listeria monocytogenes* also likely occurs in kangaroo).

6.4.66 Macrococcus

Phylum: Firmicutes; Class: Bacilli; Order: Bacillales; Family: Staphylococcaceae; Genus: *Macrococcus*. The members of this genus tend to live naturally on the skin and are found in milk suggesting a potential involvement with mastitis although a definitive disease association remains unclear. Those species belonging to this microbial genus which may have a potentially opportunistic disease association with mammalian livestock are: *Macrococcus bovicus* (cattle), *Macrococcus brunensis* (llama), *Macrococcus carouselicus* (horse), *Macrococcus caseolyticus* (horse, sheep, pig, cattle, goat), *Macrococcus equipercicus* (horse), *Macrococcus hajekii* (llama), and *Macrococcus lamae* (llama).

6.4.67 Mannheimia

Phylum: Proteobacteria; Class: Gammaproteobacteria; Order: Pasteurellales; Family: Pasteurellaceae; Genus: *Mannheimia*. The members of this genus typically are found in pasture grass, drinking water, soil, and straw bedding. They normally are commensals that variously inhabit the nasopharynx, tonsils and rumen. The *Mannheimia* often cause secondary respiratory infections, they also have been associated with otitis media and otitis interna, focal proliferative granulomatous inflammation of subcutaneous adipose tissue, mastitis which can become gangrenous, pneumonia, and meningoencephalitis. They also are a major cause of stress related respiratory illness in animals that recently have been transported, termed shipping fever or transit fever. Additional information on *Mannheimia haemolytica* can be found by researching its previous name which was *Pasteurella haemolytica*. Those species belonging to this microbial genus which have been indicated to have a potentially opportunistic disease association with mammalian livestock are: *Mannheimia caviae* (guinea pig), *Mannheimia glucosida* (sheep), *Mannheimia granulomatis* (cattle), *Mannheimia haemolytica* (camel, alpaca, llama, cattle, sheep, goat, horse, donkey), and *Mannheimia ruminalis* (sheep).

6.4.68 Micrococcus

Phylum: Actinobacteria; Class: Actinobacteria; Order: Micrococcales; Family: Micrococcaceae; Genus: *Micrococcus*. The members of this genus typically are found in water, dust, and soil. Members of the genus *Micrococcus* typically are associated with mastitis. There have been general associations between members of this bacterial genus with camelids and equines, but in those studies the discovered *Micrococcus* organisms were not identified by species. Those species belonging to this microbial genus which have been indicated to have a potentially opportunistic disease association with mammalian livestock are: *Micrococcus cohnii* (rabbit) and *Micrococcus luteus* (cattle, water buffalo).

6.4.69 Morganella

Phylum: Proteobacteria; Class: Gammaproteobacteria; Order: Enterobacteriales; Family: Morganellaceae; Genus: *Morganella.*

The species *Morganella morganii* is found in soil, water, and fecal flora. *Morganella morganii* has a commensal relationship in the intestinal tract. As a pathogen it is associated with urinary tract, wound and respiratory infections (kangaroo).

6.4.70 Moraxella

Phylum: Proteobacteria; Class: Gammaproteobacteria; Order: Pseudomonadales; Family: Moraxellaceae; Genus: *Moraxella*. The members of this genus generally are only isolated from animal tissues and fluids. They typically are commensals of the upper respiratory tract although they can cause infections of the lower respiratory tract. The members of this genus also are notable for their ability to colonize the ocular surface and cause both conjunctivitis and keratoconjunctivitis. They do as well cause infections of bite wounds. Those species belonging to this microbial genus which have been indicated to have a potentially opportunistic disease association with mammalian livestock are: *Moraxella boevrei* (sheep, goat), *Moraxella bovis* (cattle), *Moraxella bovoculi* (cattle, reindeer), *Moraxella canis* (cat, dog), *Moraxella caprae* (sheep), *Moraxella caviae* (guinea pig), *Moraxella cuniculi* (pig), *Moraxella equi* (horse), *Moraxella ovis* (sheep, goat, cattle), *Moraxella lacunata* (alpaca, llama), and *Moraxella porci* (pig).

6.4.71 Mycobacterium

Phylum: Actinobacteria; Class: Actinobacteria; Order: Corynebacteriales; Family: Mycobacteriaceae; Genus: *Mycobacterium*. Some species of *Mycobacterium* are free living, typically in soil and water including chlorinated tap water, and also in food sources. Other *Mycobacterium*, however, such as *Mycobacterium tuberculosis* seem to have evolved from being soil microorganisms and reached dependency upon animal hosts with concomitant loss of their environmental capability. Animals typically are infected by members of this genus either in utero or during the first few months of life and these early infections may be assymptomatic. Some members of this genus cause vulvitis and vaginitis. The various member species also can cause mastitis, lymphadenitis, chronic enteritis, systemic infections including peritonitis, and pneumonia. *Mycobacterium* species can contribute to urinary system infections associated with other microorganisms. Those species belonging to this microbial genus which have been indicated to have a potentially opportunistic disease association with mammalian livestock are: *Mycobacterium asiaticum* (red kangaroo), *Mycobacterium avium* (thus far identified in all of the mammalian livestock animals listed in this chapter except Edible dormouse, and capybara), *Mycobacterium bovis* (camel, llama, elk, fallow deer, bison, cattle, water buffalo, cat, goat, sheep, horse, pig, rabbit), *Mycobacterium caprae* (sheep, goat, cattle, pig, red deer), *Mycobacterium chelonae* (pig, cattle), *Mycobacterium farcinogenes* (cattle), *Mycobacterium fortuitum* (pig), *Mycobacterium genavense* (rabbit, kangaroo), *Mycobacterium intracellulare* (capybara, pig, cattle, horse), *Mycobacterium kansasii* (alpaca), *Mycobacterium lepraemurium* (rodents, cat), *Mycobacterium microti* (rodents, llama), *Mycobacterium porcinum* (cattle, pig), *Mycobacterium scrofulaceum* (cattle, buffalo, pig), *Mycobacterium smegmatis* (cat, dog, sheep), *Mycobacterium tuberculosis* (camel, horse, pig, cattle, rabbit), *Mycobacterium vaccae* (cattle), and *Mycobacterium xenopi* (pig).

6.4.72 Mycoplasma

Phylum: Tenericutes; Class: Mollicutes; Order: Mycoplasmatales; Family: Mycoplasmataceae; Genus: *Mycoplasma*. The members of this genus are ecologically diverse. Typically the *Mycoplasma* are not free-living albeit some are saprotrophic and members of this genus can be found as aquatic residents in natural water, water tanks, water baths, and on the cooling coils of refrigeration units. Many of these species can survive and replicate within protozoa which may in turn facilitate their existence in the natural environment. Members of the genus *Mycoplasma* have been associated with general loss of body condition, decreased milk production, and decreased reproductive rates. Specific disease associations include both upper and lower respiratory tract illnesses including sinusitis and pleuropneumonia (especially tuberculosis), otitis media and otitis interna, conjunctivitis, genital infections including posthitis, vulvitis and vaginitis, and chronic polyarthritis which can be severe and result in death. Carrier animals which demonstrate few or no signs of illness typically are found within mammalian populations. Those species belonging to this microbial genus which have been indicated to have a potentially opportunistic disease association with mammalian livestock are: *Candidatus Mycoplasma haemolamae* (alpaca, llama, donkey), *Mycoplasma arginini* (wide host range in mammals including sheep, goat, cattle, pig, cat, camel), *Mycoplasma arthritidis* (cattle, sheep, goat, rodents), *Mycoplasma bovigenitalium* (cattle), *Mycoplasma bovirhinis* (cattle), *Mycoplasma bovis* (bison, buffalo, camel, cat, cattle, deer, goat, pig, rabbit), *Mycoplasma bovoculi* (cattle), *Mycoplasma canis* (dog), *Mycoplasma capricolum* (goat, sheep, cattle, camel), *Mycoplasma caviae* (guinea pig), *Mycoplasma cavipharyngis* (guinea pig), *Mycoplasma conjunctivae* (sheep, goat), *Mycoplasma equigenitalium* (horse),

Mycoplasma equirhinis (horse), *Mycoplasma feliminutum* (cat), *Mycoplasma felis* (cat, horse), *Mycoplasma haemocanis* (dog), *Mycoplasma haemofelis* (cat), *Mycoplasma hyopharyngis* (pig), *Mycoplasma hyopneumoniae* (pig), *Mycoplasma hyorhinis* (pig), *Mycoplasma hyosynoviae* (pig), *Mycoplasma mucosicanis* (dog), *Mycoplasma mycoides* (bison, cattle, camel, goat, sheep, water buffalo, yak), *Mycoplasma ovipneumoniae* (sheep, goat), *Mycoplasma ovis* (sheep, goat), *Mycoplasma pulmonis* (rodents, horse), *Mycoplasma subdolum* (horse), *Mycoplasma suis* (pig), and *Mycoplasma wenyonii* (cattle).

6.4.73 Neisseria

Phylum: Proteobacteria; Class: Betaproteobacteria; Order: Neisseriales; Family: Neisseriaceae; Genus: *Neisseria*. The members of this genus naturally exist as colonizers of mucosal surfaces and although *Neisseria* can be found in fluids shed from the host, their environmental survival in those fluids is considered limited. Members of the genus *Neisseria* have a general association with dental plaque in many mammalian livestock animals including cattle, yak, cervids, kangaroo, sheep, cattle, llama and cat, and potentially those bacteria may contribute to oral cavity disease. More specific disease associations include otitis media, otitis interna, and meningitis. Those species belonging to this microbial genus which have been indicated to have a potentially opportunistic disease association with mammalian livestock are: *Neisseria animalis* (guinea pig), *Neisseria canis* (dog), *Neisseria dentiae* (cattle), *Neisseria meningitidis* (guinea pig), and *Neisseria weaveri* (dog).

6.4.74 Neorickettsia

Phylum: Proteobacteria; Class: Alphaproteobacteria; Order: Rickettsiales; Family: Anaplasmataceae; Genus: *Neorickettsia*. The bacterial species *Neorickettsia risticii* causes a syndrome whose symptoms include gastroenteritis, lymphadenopathy, abortion, and laminitis. This bacterial species may be transmitted by trematodes associated with snails, and in some cases the infection is acquired by the horse during ingestion of water that contains aquatic insects and helminths harboring the bacterially infected trematode metacercaria (horse). *Neorickettsia risticii* previously was named *Ehrlichia risticii*.

6.4.75 Nocardia

Phylum: Actinobacteria; Class: Actinobacteria; Order: Corynebacteriales; Family: Nocardiaceae; Genus: *Nocardia*. The members of this genus typically are found in soil that contains a high amount of organic material. Members of the genus *Nocardia* often are associated with abscesses. The symptoms associated with *Nocardia* infections include mastitis as well as other cutaneous and subcutaneous lesions, peritonitis, pneumonia, meningitis and meningoencephalitis. Those species belonging to this microbial genus which have been indicated to have a potentially opportunistic disease association with mammalian livestock are: *Nocardia asteroides* (cattle, water buffalo, dog, and camelids including llama), *Nocardia brasiliensis* (kangaroo), *Nocardia farcinica* (cattle), *Nocardia nova* (cattle), and *Nocardia otitidiscaviarum* (cattle).

6.4.76 Paeniclostridium

Phylum: Firmicutes; Class: Clostridia; Order: Clostridiales; Family: Peptostreptococcaceae; Genus: *Paeniclostridium*. The species *Paeniclostridium sordellii* is found in soil as well as animal feces and can be isolated from the intestine. There exist both pathogenic and non-pathogenic strains of this bacterial species and both groups are considered normal vaginal flora. Its pathogenic strains are known to cause toxic shock syndrome in humans and enteritis in sheep and cattle. The species *Paeniclostridium sordellii* additionally causes pneumonia, endocarditis, arthritis, peritonitis, myonecrosis, and infects the umbilical stump of newborns (alpaca, llama, guinea pig, cattle, sheep). The species *Paeniclostridium sordellii* previously was assigned the name *Clostridium sordellii.*

6.4.77 Pantoea

Phylum: Proteobacteria; Class: Gammaproteobacteria; Order: Enterobacteriales; Family: Erwiniaceae; Genus: *Pantoea*. The species *Pantoea agglomerans* can be isolated from plant surfaces including seeds and fruits, as well as human and animal feces. *Pantoea agglomerans* typically is a pathogen of fish in which it causes hemorrhaging ocular infections. In humans, this bacterial species is opportunistically associated with skin and eye infections. Following puncture wounds of humans caused by thorn and wood splinter injuries, this bacteria has been associated with septic arthritis, osteomyelitis and meningitis, and these illnesses particularly have been noted in immunosuppressed individuals and newborns. The discovery of this bacterial species in other mammals causes suspicion that it may similarly be an opportunistic pathogen in those other mammals (coypu, dog).

6.4.78 Pasteurella

Phylum: Proteobacteria; Class: Gammaproteobacteria; Order: Pasteurellales; Family: Pasteurellaceae; Genus: *Pasteurella*. The members of this genus naturally reside on mucosal membranes as a part of oral, respiratory, genital, and gastrointestinal floras; those which naturally reside in the mouth are shed by nasal secretions. They can survive for at least a period of weeks in water, fecal slurries, and on moist surfaces. Members of the genus *Pasteurella* are associated with general symptoms including depression. More specific illness associations for infections caused by *Pasteurella* include those typical of the upper and lower respiratory tracts among which pneumonia and pleuropneumonia are perhaps the most notable, and they additionally cause bite wound abscesses, otitis media and otitis interna, mastitis, arthritis, gastroenteritis, urinary system infections, meningitis and meningoencephalitis. The symptoms range in severity from subclinical to peracute and fatal. The *Pasteurella* often cause secondary respiratory infections. They also are a minor but still significant cause of stress related respiratory illness in animals that recently have been transported, termed shipping fever or transit fever. Those species belonging to this microbial genus which have been indicated to have a potentially opportunistic disease association with mammalian livestock are: [*Pasteurella*] *aerogenes* (this microorganism currently is considered Incertae sedis) (pig), *Pasteurella caballi* (pig, horse), *Pasteurella canis* (horse, dog, cattle, sheep), *Pasteurella dagmatis* (dog, cat), [*Pasteurella*] *mairii* (this microorganism currently is considered Incertae sedis) (pig), *Pasteurella multocida* (coypu, cervids including red deer and fallow deer, camel, alpaca, cattle, sheep, goat, pig, guinea pig, rabbit, cat, dog, kangaroo, yak), and *Pasteurella stomatis* (dog, cat).

6.4.79 Peptoniphilus

Phylum: Firmicutes; Class: Tissierellia; Order: Tissierellales; Family: Peptoniphilaceae; Genus: *Peptoniphilus*. The natural ecology of the species *Peptoniphilus indolicus* presumably includes soil and mud. It additionally is found in milk and other udder secretions. The typical disease associations for *Peptoniphilus indolicus* are wound infections and skin lesions including mastitis (cattle, pig). *Peptoniphilus indolicus* previously has been named *Peptococcus indolicus* and *Micrococcus indolicus*.

6.4.80 Peptostreptococcus

Phylum: Firmicutes; Class: Clostridia; Order: Clostridiales; Family: Peptostreptococcaceae; Genus: *Peptostreptococcus*. The members of this genus typically are found in soil. They also are commensal organisms living predominantly on the skin as well as in the mouth, vagina, gastrointestinal and urinary tracts. Members of the genus *Peptostreptococcus* characteristically are associated with periodontal disease but they also cause pleuropneumonia, meningitis and meningo-encephalitis. Those species belonging to this microbial genus which have been indicated to have a potentially opportunistic disease association with mammalian livestock are: *Peptostreptococcus anaerobius* (dog, cat, coypu), *Peptostreptococcus canis* (dog), and *Peptostreptococcus russellii* (found in pig although the disease association remains unclear).

6.4.81 Plesiomonas

Phylum: Proteobacteria; Class: Gammaproteobacteria; Order: Enterobacteriales; Family: Enterobacteriaceae; Genus: *Plesiomonas*. The species *Plesiomonas shigelloides* typically is found in freshwater. It also causes gastroenteritis which can be followed by septicemia if there is underlying immunodeficiency (cattle, goat, pig, cat, dog).

6.4.82 Porphyromonas

Phylum: Bacteroidetes; Class: Bacteroidia; Order: Bacteroidales; Family: Porphyromonadaceae; Genus: *Porphyromonas*. The members of this genus naturally reside in the mouth below the gingival surface. No information could be found that indicated their having an ecological presence in the natural environment. Members of the genus *Porphyromonas* have been associated with periodontal disease, vulvitis and vaginitis. Those species belonging to this microbial genus which have been indicated to have a potentially opportunistic disease association with mammalian livestock are: *Porphyromonas asaccharolytica* (dog), *Porphyromonas cangingivalis* (dog), *Porphyromonas canis* (dog), *Porphyromonas canoris* (dog), and *Porphyromonas crevioricanis* (dog). The previously named species *Porphyromonas cansulci*, which affects dog, has been reclassified and currently is considered part of the species *Porphyromonas crevioricanis*.

6.4.83 Prevotella

Phylum: Bacteroidetes; Class: Bacteroidia; Order: Bacteroidales; Family: Prevotellaceae; Genus: *Prevotella*. Although some members of this bacterial genus are found in flooded agricultural soils (rice fields), the members of this genus typically are considered to be oral, vaginal, rumenal and colonic flora. The most typical illness symptoms associated with members of the genus *Prevotella* are periodontal disease and foot abscesses, but the scope of illness caused by this bacterial genus also includes severe intestinal disease, bacteremia, systemic infections and osteomyelitis, infections of wounds such as those caused by bites, infections of the genital tract, and infections of both the upper and lower respiratory tracts. Those species belonging to this microbial genus which have been indicated to have a potentially opportunistic disease association with mammalian livestock are: *Prevotella buccae* (dog), *Prevotella melaninogenica* (cat, dog, cattle), and *Prevotella oris* (dog).

6.4.84 Proteus

Phylum: Proteobacteria; Class: Gammaproteobacteria; Order: Enterobacteriales; Family: Morganellaceae; Genus: *Proteus*. The members of this genus typically are saprophytes found in soil and water, decomposing animal matter, sewage, manure soil, human and animal feces. Members of the genus *Proteus* are associated with septic peritonitis, urinary tract infections, uterine infections, and otitis externa. Those species belonging to this microbial genus which have been indicated to have a potentially opportunistic disease association with mammalian livestock are: *Proteus mirabilis* (camel, cattle, cervids, kangaroo, goat), *Proteus penneri* (coypu—disease association remains unclear), and *Proteus vulgaris* (camel, equines, kangaroo, water buffalo, goat).

6.4.85 Providencia

Phylum: Proteobacteria; Class: Gammaproteobacteria; Order: Enterobacteriales; Family: Morganellaceae; Genus: *Providencia*. These listed members of the genus *Providencia* are associated with water, soil, and sewage. They also inhabit the skin and gastrointestinal tract. Members of the genus *Providencia* have been associated with bacteremia and also isolated from animals suffering from wounds including burns, urinary tract infections, pneumonia, meningitis, endocarditis, ocular infections, dermatitis, cellulitis, and cystitis associated with indwelling urinary catheters. Those species belonging to this microbial genus which have been indicated to have a potentially opportunistic disease association with mammalian livestock are: *Providencia alcalifaciens* (dog), *Providencia rettgeri* (kangaroo), and *Providencia rustigianii* (pig).

6.4.86 Pseudarthrobacter

Phylum: Actinobacteria; Class: Actinobacteria; Order: Micrococcales; Family: Micrococcaceae; Genus: *Pseudarthrobacter*. The members of this genus commonly are found in soil. Members of the genus *Pseudarthrobacter* also are associated with genital infections including vaginitis. The species belonging to this microbial genus which has been indicated to have a potentially opportunistic disease association with mammalian livestock is: *Pseudarthrobacter equi*, previously named *Arthrobacter equi* (horse).

6.4.87 Pseudomonas

Phylum: Proteobacteria; Class: Gammaproteobacteria; Order: Pseudomonadales; Family: Pseudomonadaceae; Genus: *Pseudomonas*. The members of this genus typically are found in water and soil, with an increased presence in manure amended soils. Members of the genus *Pseudomonas* have been associated with otitis media and otitis interna, wound infections notably including those associated with burns (I can attest to you from having studied microbiology in a hospital as a ten year old child, that *Pseudomonas* produces a characteristic and unforgettable sickeningly sweet odor in burn wounds), mastitis, lung infections, ascending urinary tract infections, and endometritis. Those species belonging to this microbial genus which have been indicated as having a potentially opportunistic disease association with mammalian livestock are: *Pseudomonas aeruginosa* (camel, cattle, cat, dog, donkey, guinea pig, horse, kangaroo, sheep, goat, water buffalo), *Pseudomonas alcaligenes* (cattle), *Pseudomonas fluorescens* (dog, coypu), *Pseudomonas luteola* (cat, dog, guinea pig, goat), *Pseudomonas mendocina* (cattle, coypu), and *Pseudomonas stutzeri* (sheep).

6.4.88 Pseudopropionibacterium

Phylum: Actinobacteria; Class: Actinobacteria; Order: Propionibacteriales; Family: Propionibacteriaceae; Genus: *Pseudopropionibacterium*. This species *Pseudopropionibacterium propionicum* is virtually ubiquitous as a commensal in the oral cavity and on the skin of animals, including residence both in and around sweat glands and sebaceous glands. It can infect the lacrimal apparatus and surprisingly it can become a lethal opportunist (cattle). In human this bacterial species has caused pelvic inflammatory disease following hysteroscopic surgery. *Pseudopropionibacterium propionicum* previously has been classified as *Propionibacterium propionicum*, *Propionibacterium propionicus*, *Actinomyces propionicus*, and *Arachnia propionica*.

6.4.89 Psychrobacter

Phylum: Proteobacteria; Class: Gammaproteobacteria; Order: Pseudomonadales; Family: Moraxellaceae; Genus: *Psychrobacter*. The species *Psychrobacter pulmonis* typically is found in soil, with an increased presence in manure amended soils. This microbial species has been associated with lung infections (sheep).

6.4.90 Rahnella

Phylum: Proteobacteria; Class: Gammaproteobacteria; Order: Enterobacteriales; Family: Yersiniaceae; Genus: *Rahnella*. The species *Rahnella aquatilis* typically is found in freshwater, soil, and in association with some invertebrates notably snails and beetles. The fact that *Rahnella aquatilis* interestingly also can be isolated from parasitic nematodes potentially suggests that those nematodes might be involved with transmission of this bacterial species. The species *Rahnella aquatilis* causes respiratory infections (cattle, goat).

6.4.91 Rhodococcus

Phylum: Actinobacteria; Class: Actinobacteria; Order: Corynebacteriales; Family: Nocardiaceae; Genus: *Rhodococcus*. The species *Rhodococcus equi* thrives in water and soil, particularly being found in areas that receive animal manure. This bacterial species causes peritonitis (goat, sheep, horse, cattle, camel, llama, alpaca, pig, cervids).

6.4.92 Rickettsia

Phylum: Proteobacteria; Class: Alphaproteobacteria; Order: Rickettsiales; Family: Rickettsiaceae; Genus: *Rickettsia*. The *Rickettsia* are obligate intracellular pathogens. Some members of the genus *Rickettsia* can survive in amoeba, and indeed some amoeba contain endosymbiotic *Rickettsia*. However, these listed species seem to have no presence in the natural environment and instead alternate in a cycle between arthropod and mammalian hosts. The arthropod hosts for these *Rickettsia* species tend to be ticks, mites, lice, or fleas. Members of the genus *Rickettsia* have been associated with a variety of non-specific illnesses and although these often are mild or non-clinical, the infections also can become severe to the point that they result in death of the animal. The infections can be either short term or long term. Sometimes the animals demonstrate no symptoms at all, albeit they meet the defining criterion of infection in response to the microorganism by producing serologically detectable levels of antibodies which react against the microorganism. The most common clinical signs associated with *Rickettsia* infections include lethargy, anorexia, ataxia, rash, swollen lymph nodes, and localized edema. Those species belonging to this microbial genus which have been indicated as having a potentially opportunistic disease association with mammalian livestock are: *Rickettsia aeschlimannii* (camel), *Rickettsia africae* (camel), *Rickettsia felis* (cat), *Rickettsia rickettsii* (capybara, coypu, cat, dog), *Rickettsia sibirica* (camel), and *Rickettsia typhi* (Edible dormouse—also called the Fat dormouse).

6.4.93 Rodentibacter

Phylum: Proteobacteria; Class: Gammaproteobacteria; Order: Pasteurellales; Family: Pasteurellaceae; Genus: *Rodentibacter*. Presumably, as with the *Pasteurella*, members of this genus naturally reside on mucosal membranes as a part of oral, respiratory, genital, and gastrointestinal floras and might be shed by nasal secretions with a potential for some environmental survival. The species belonging to this microbial genus which has been indicated to have a potentially opportunistic disease association with mammalian livestock is *Rodentibacter pneumotropicus*, previously named *Pasturella pneumotropica*, (guinea pig, dog, cat, rodents including coypu, camel).

6.4.94 Salmonella

Phylum: Proteobacteria; Class: Gammaproteobacteria; Order: Enterobacteriales; Family: Enterobacteriaceae; Genus: *Salmonella*. The species *Salmonella enterica* typically is found in compost, manure amended soil, and crops grown in manure amended soil. The association of mammals with *Salmonella enterica* varies from subclinical intestinal carriage to septicemia and death. This bacterial species is considered to be an opportunistic pathogen and most of the infections which it causes tend to produce relatively inconsequential disease including gastroenteritis and ocular lesions. *Salmonella enterica* also can be present in the tonsils. More severe disease outcomes can occur including septic peritonitis, bacteremia, multifocal petechial hemorrhage, inflammation of tissues lining the visceral organs (serositis), polyarthritis, bronchopneumonia, meningitis, and meningoencephalitis. Importantly, the septicemic infections caused by *Salmonella* often are fatal without prior observed clinical signs (camel, alpaca, llama, pig, cattle, coypu, cervids including elk and red deer, kangaroo, guinea pig, horse, goat, sheep, cat, dog).

6.4.95 Serratia

Phylum: Proteobacteria; Class: Gammaproteobacteria; Order: Enterobacteriales; Family: Yersiniaceae; Genus: *Serratia*. The *Serratia* are ubiquitous in the natural environment, being found in soil, water, and growing on plant material. Members of the genus *Serratia* colonize the respiratory and urinary tracts, and are associated with mastitis, respiratory illnesses, sepsis, and septic abortion. Those species belonging to this microbial genus which have been indicated as having a potentially opportunistic disease association with mammalian livestock are: *Serratia ficaria* (goat), *Serratia fonticola* (goat), *Serratia liquefaciens* (cattle, water buffalo, coypu), *Serratia marcescens* (capybara, coypu, cattle, sheep, goat), *Serratia odorifera* (goat), and *Serratia rubidaea* (cattle).

6.4.96 Shewanella

Phylum: Proteobacteria; Class: Gammaproteobacteria; Order: Alteromonadales; Family: Shewanellaceae; Genus: *Shewanella*. The species *Shewanella putrefaciens* typically is isolated from marine environments where it may have potential to remediate petroleum spills, and it is as well found anaerobic sandstones. This species of *Shewanella* is noted for causing open lesions in fish, and it opportunistically causes both soft tissue infections and bacteremia in human (coypu—disease association remains unclear).

6.4.97 Shigella

Phylum: Proteobacteria; Class: Gammaproteobacteria; Order: Enterobacteriales; Family: Enterobacteriaceae; Genus: *Shigella*. The *Shigella* typically are found in water, soil, and animal feces. Members of this genus can be carried asymptomatically. The disease that they typically cause is a form of enteritis which generally is self limiting but can become severe and fatal. They are the common cause of bacterial dysentery in human. Complications due to *Shigella* bacteremia are possible, mainly in immuno-compromised individuals, and those complications include arthritis, neuritis, vulvo-vaginitis, chronic colitis, conjunctivitis, and eventually death. Death can occur swiftly following the appearance of severe symptoms. Those species belonging to this microbial genus which have been indicated as having a potentially opportunistic disease association with mammalian livestock are: *Shigella dysenteriae* (camel, dog), *Shigella flexneri* (cattle), and *Shigella sonnei* (goat).

6.4.98 Sphingobacterium

Phylum: Bacteroidetes; Class: Sphingobacteriia; Order: Sphingobacteriales; Family: Sphingobacteriaceae; Genus: *Sphingobacterium.* The members of this genus typically are found in soil, plants, and water. The disease association with *Sphingobacterium multivorum* in livestock mammals remains unclear but this bacterial species is opportunistically pathogenic in humans where it causes respiratory tract infections and septicemia (found in cat, dog, and sheep, although the disease association with these livestock mammals remains unclear).

6.4.99 Sphingomonas

Phylum: Proteobacteria; Class: Alphaproteobacteria; Order: Sphingomonadales; Family: Sphingomonadaceae; Genus: *Sphingomonas*. The species *Sphingomonas paucimobilis* can be found in freshwater, seawater, and terrestrial habitats including plant root systems. This species can grow in drinking water distribution systems. *Sphingomonas paucimobilis* also can be found associated with equipment in hospitals including respiratory therapy items, humidifiers, water, air, bedside water bottles, sinks, and temperature probes. This species often is associated with bite wounds, the result of infection can include both ulcerations and abscess as well as bacteremia (cat, dog, rodents including coypu). The previous name of this species was *Pseudomonas paucimobilis*.

6.4.100 Staphylococcus

Phylum: Firmicutes; Class: Bacilli; Order: Bacillales; Family: Staphylococcaceae; Genus: *Staphylococcus*. The members of this genus typically are found as a minor component of the soil microbial community. In association with mammals they normally inhabitant the skin and mucous membranes including those of the upper respiratory tract. Members of the genus *Staphylococcus* commonly cause infections of the skin, eye, external ear and mammary gland, the severity of which can range from minor to severe including skin redness, pustules, and abscesses. The various *Staphylococcus* species can, however, affect every organ in the host's body. Encompassing the fullness of their disease consequences is a lengthy task and the description must include genito-urinary infections that may ascend the urinary tract, as well as infections which can begin in the lung to cause pneumonia and pleuropneumonia. *Staphylococcus* infections can progress to cause bacteremia, septic peritonitis, and endocarditis. These bacterial species also can affect the skeleton and joints causing osteomyelitis, and invade the central nervous system resulting in meningitis, encephalitis, meningoencephalitis and encephalomyelitis. These consequences of infection can prove fatal for immunocompromised animals. The toxins produced by *Staphylococcus* may generate symptoms that typically would be associated with food poisoning. *Staphylococcus* species also can cause food poisoning from enterotoxin contamination in meat. It is important to note that in older kangaroo *Staphylococcus aureus* sometimes is considered a primary pathogen rather than being opportunistic. Those species belonging to this microbial genus which have been indicated as having a potentially opportunistic disease association with mammalian livestock are: *Staphylococcus aureus* (coypu, dog, sheep, moose, camel, guinea pig, yak, water buffalo, pig, equines, cattle, kangaroo), *Staphylococcus auricularis* (cattle, coypu, sheep), *Staphylococcus capitis* (dog, kangaroo), *Staphylococcus caprae* (goat), *Staphylococcus chromogenes* (equines), *Staphylococcus cohnii* (dog, rabbit), *Staphylococcus delphini* (dog), *Staphylococcus epidermidis* (camel, cat, coypu, dog, kangaroo, horse, cattle, cervids), *Staphylococcus equorum* (cat, horse), *Staphylococcus felis* (cat), *Staphylococcus gallinarum* (sheep), *Staphylococcus haemolyticus* (horse, cat, dog), *Staphylococcus hominis* (coypu, dog), *Staphylococcus hyicus* (camel, coypu, cervids, pig, cattle), *Staphylococcus intermedius* (camel, cat, coypu, dog, equines, cattle), *Staphylococcus lentus* (dog, sheep), *Staphylococcus lugdunensis* (dog), *Staphylococcus pseudintermedius* (dog), *Staphylococcus saccharolyticus* (coypu—disease association remains unclear), *Staphylococcus saprophyticus* (cattle, sheep), *Staphylococcus schleiferi* (dog), *Staphylococcus sciuri* (pig, cattle, coypu, sheep, goat), *Staphylococcus warneri* (coypu, dog), and *Staphylococcus xylosus* (kangaroo, cattle, goat).

6.4.101 Stenotrophomonas

Phylum: Proteobacteria; Class: Gammaproteobacteria; Order: Xanthomonadales; Family: Xanthomonadaceae; Genus: *Stenotrophomonas*. The species *Stenotrophomonas maltophilia* ubiquitously is found in water, soil, and associated with plants. This bacterial species is notably resistant to antibiotics and produces severe infections in immunocompromised hosts. It is a growing source of latent pulmonary infections in human. *Stenotrophomonas maltophilia* typically causes infection following wounds, it also can cause pneumonia, urinary tract infections, and bacteremia (horse, dog, cat, goat, cattle, sheep, rabbit, coypu). The previous names assigned to *Stenotrophomonas maltophilia* include *Pseudomonas maltophilia* and a great deal of literature can be found by searching under that older bacterial species name.

6.4.102 Streptobacillus

Phylum: Fusobacteria; Class: Fusobacteria; Order: Fusobacteriales; Family: Leptotrichiaceae; Genus: *Streptobacillus*. Members of this genus are found in soil and contaminated water. *Streptobacillus felis* causes pneumonia (cat). *Streptobacillus moniliformis*, previously named *Actinomyces muris*, is perhaps the most notable member of this bacterial genus. The natural habitat of *Streptobacillus moniliformis* seems to be the oral cavity, nasopharynx, respiratory tract and middle ear of numerous rodent species, especially rats, with natural colonization rates in apparently healthy rats sometimes far greater than 50%, and *Streptobacillus moniliformis* seems to be an opportunistic pathogen in rats that occasionally causes otitis media, conjunctivitis, and pneumonia. *Streptobacillus moniliformis* may similarly affect those rodent species that are kept as livestock, and it does indeed cause throat infections (capybara) including cervical abscess (guinea pig), pneumonia (guinea pig), plus scrotal infections (capybara). *Streptobacillus moniliformis* causes rat bite fever in humans which is a systemic illness classically characterized by fever, rigor (fever with shivering and a feeling of chills), polyarthralgia, including more rarely pneumonitis, endocarditis and meningitis. The syndrome of rat bite fever caused by *Streptobacillus moniliformis* seems perhaps rare or unknown in livestock mammals. *Streptobacillus moniliformis* has, however, been carried and transmitted by cats, dogs, and pigs, thus potentially being an opportunistic pathogen in those livestock species. Cats, dogs, and pigs presumably acquire *Streptobacillus moniliformis* from eating rodents.

6.4.103 Streptococcus

Phylum: Firmicutes; Class: Bacilli; Order: Lactobacillales; Family: Streptococcaceae; Genus: *Streptococcus*. The members of this genus typically are found in water, soil, and vegetation. The *Streptococcus* are considered normal animal flora. However, members of the genus *Streptococcus* do cause infections whose severity ranges from asymptomatic to severe depending upon the age and immune status of the infected animal. These bacteria can cause mastitis and cellulitis, ocular discharge, otitis externa, otitis media and otitis interna, cervical lymphadenitis (glanders), and abscesses. *Streptococcus* infections can ascend the urinary tract and also cause endometritis. The bacteria belonging to this species can invade both the upper and lower respiratory tracts, cause septic peritonitis, pleuropneumonia, meningitis and meningoencephalitis. It is important to note that some subspecies of *Streptococcus equi* are considered primary pathogens rather than being opportunistic. Those species belonging to this microbial genus which have been indicated as having a potentially opportunistic disease association with mammalian livestock are: *Streptococcus agalactiae* (camel, cattle, dog, cat), *Streptococcus anginosus* (coypu, dog), *Streptococcus caballi* (horse), *Streptococcus canis* (dog, cat, cattle, rabbit), *Streptococcus caprae* (goat), *Streptococcus dentisani* (donkey), *Streptococcus dysgalactiae* (camel, equines, kangaroo, water buffalo, cattle), *Streptococcus entericus* (cattle), *Streptococcus equi* (alpaca, llama, camel, dog, horse, donkey, cattle, sheep, goat, pig, guinea pig), *Streptococcus equinus* (cattle, alpaca, sheep), *Streptococcus gallolyticus* (horse, dog, kangaroo, rabbit, guinea pig, goat), *Streptococcus orisasini* (donkey), *Streptococcus orisuis* (pig—disease association remains unclear), *Streptococcus ovis* (sheep), *Streptococcus pluranimalium* (cattle, sheep, goat, cat), *Streptococcus pneumoniae* (camel, guinea pig, equines, rabbit, rodents including coypu, dog), *Streptococcus porci* (pig), *Streptococcus porcinus* (pig), *Streptococcus porcorum* (pig), *Streptococcus pyogenes* (cattle, dog), *Streptococcus rupicaprae* (goat), *Streptococcus sanguinis* (llama), *Streptococcus suis* (pig, cattle, bison, sheep, goat), *Streptococcus uberis* (camel, water buffalo, cattle), *Streptococcus viridans* often generally termed 'Viridans streptococci' (alpaca, llama, cattle, sheep, goat, rabbit).

6.4.104 Treponema

Phylum: Spirochaetes; Class: Spirochaetia; Order: Spirochaetales; Family: Spirochaetaceae; Genus: *Treponema*. Members of the phylum Spirochaetes represent a diverse group of bacteria, many of whose members can be found in soil and marine sediments. Members of the genus *Treponema*, however, are strongly dependent upon their hosts and many seem to be obligate parasites meaning that they do not have a prolonged survival independent of a host. Various members of the genus *Treponema* normally dwell within the oral cavity and cause necrotic periodontal disease, others are associated with necrotic skin ulcers typically of the ear and foot, and some affect the genital organs. Many *Treponema* species also are found as natural residents in the rumen and colon of cattle and sheep where they contribute to the digestion process and where disease associated with their presence often remains unclear except for ulcerative colitis. All of these listed *Treponema* species are found in animal feces and presumably that presence plays a role in their transmission. There is a strong likelihood that fecal presence contributes to the risk of a common *Treponema* disease association for cattle which is called Bovine digital dermatitis. Those species belonging to this microbial genus which have been indicated as having a potentially opportunistic disease association with mammalian livestock are: *Candidatus Treponema suis* (pig), *Treponema brennaborense* (cattle), *Treponema bryantii* (cattle—disease association remains unclear), *Treponema denticola* (cattle—disease association remains unclear), *Treponema medium* (cattle—disease association remains unclear), *Treponema pedis* (cattle), *Treponema phagedenis* (cattle), *Treponema porcinum* (pig), *Treponema saccharophilum* (cattle—disease association remains unclear), and *Treponema zioleckii* (sheep—disease association remains unclear).

6.4.105 Trueperella

Phylum: Actinobacteria; Class: Actinobacteria; Order: Actinomycetales; Family: Actinomycetaceae; Genus: *Trueperella*. The members of this genus are found as soil dwellers and typically infect animals in association with commensal relationships or trauma related wounds. The *Trueperella* can be found in the urogenital tract where they cause both metritis and ascending urinary tract infections. They also can cause abscesses, otitis media and otitis interna, dermatitis of the footpads, mastitis, non-specific lesions involving the visceral organs, tenosynovitis (a type of infectious arthritis associated with the connective tissue surrounding tendons), lymphadenitis, gastroenteritis, pneumonia for which a severe outcome can include embolic pneumonia, and attack the upper respiratory tract. The *Trueperella* additionally have been isolated together with various other bacterial species under different disease conditions associated with opportunistic infections. The bacterial species *Trueperella bialowiezensis* and *Trueperella bonasi* are listed here because they affect the European bison and thus likely also would be infective of the American bison. The European bison has the taxonomic name *Bison bonasus* and currently seems not to be raised as a livestock animal. Notably, the previous names assigned to *Trueperella pyogenes* were *Corynebacterium pyogenes*, *Actinomyces pyogenes*, and *Arcanobacterium pyogenes*. Those species belonging to the genus *Trueperella* which have been indicated as having a potentially opportunistic disease association with mammalian livestock are: *Trueperella abortisuis* (cattle, pig), *Trueperella bialowiezensis* (European bison), *Trueperella bonasi* (European bison), and *Trueperella pyogenes* (broadly infective of mammals including camel, alpaca, llama, cattle, water buffalo, cervids including fallow deer and elk, pig, sheep, kangaroo, goat, horse).

6.4.106 Tsukamurella

Phylum: Actinobacteria; Class: Actinobacteria; Order: Corynebacteriales; Family: Tsukamurellaceae; Genus: *Tsukamurella*. Members of the genus *Tsukamurella* are saprophytes found in soil and water. The species *Tsukamurella inchonensis* is opportunistically associated with wound infections, as well as bacteremia, lung and gastric infections in human. For that reason, the discovery of this bacterial species in livestock mammals potentially is of agricultural interest (cattle and sheep, although the disease association for sheep remains unclear).

6.4.107 Tyzzerella

Phylum: Firmicutes; Class: Clostridia; Order: Clostridiales; Family: Lachnospiraceae; Genus: *Tyzzerella*. The species [*Clostridium*] *piliforme* (use of brackets around an assigned species name indicates that either part or all of that assigned name has not officially been approved) is found in soil and feces, and animals typically acquire it from contaminated food and water. This bacterial species also resides in organs of the digestive system and is associated with gastroenteritis. From residence in the ileum and cecum, [*Clostridium*] *piliforme* can pass through the hepatic portal into the liver, and from there progress to infect other tissues in the body. The taxonomic name [*Clostridium*] *piliforme* is not currently approved and this species has been assigned at least temporarily to the genus *Tyzzerella*. The previous name assigned to this species was *Bacillus piliformis*. This bacterial species opportunistically affects a number of mammalian livestock species and it is important to note that in rabbit this bacterial species may be a primary pathogen rather than opportunistic (cervids, cattle, kangaroo, alpaca, llama, rabbit).

6.4.108 Ureaplasma

Phylum: Tenericutes; Class: Mollicutes; Order: Mycoplasmatales; Family: Mycoplasmataceae; Genus: *Ureaplasma*. There is no indication that members of the genus *Ureaplasma* have a presence in the natural environment. The *Ureaplasma* are common inhabitants that colonize the oral cavity, oropharynx and genitourinary tract. Members of this genus are associated with urinary and possibly also respiratory tract infections, vulvitis, vaginitis, posthitis, salpingitis (infection and inflammation in the fallopian tubes) which can produce infertility, endometritis, abortion from early embryonic death, weak calves and decreased conception rates after pregnancy. Those species belonging to this microbial genus which have been indicated as having a potentially opportunistic disease association with mammalian livestock are: *Ureaplasma canigenitalium* (dog); *Ureaplasma cati* (cat); *Ureaplasma diversum* (cattle, pig); *Ureaplasma felinum* (cat), *Ureaplasma parvum* (sheep), and *Ureaplasma urealyticum* (rodents including guinea pig).

6.4.109 Veillonella

Phylum: Firmicutes; Class: Negativicutes; Order: Veillonellales; Family: Veillonellaceae; Genus: *Veillonella*. The members of this genus typically are found in saliva and also found in feces, which can play a significant role in their environmental transmission between hosts. The members of this bacterial species are considered to be normal residents of the intestine and oral mucosa of mammals, they also are present in the respiratory and genitourinary tracts. As infections, the *Veillonella* typically are associated with bite wounds. As opportunistic pathogens, notably in the case of *Veillonella parvula*, they have been implicated as a cause of endocarditis, meningitis, and osteomyelitis. Those species belonging to this microbial genus which have been indicated as having a potentially opportunistic disease association with mammalian livestock are: *Veillonella caviae* (guinea pig—disease association remains unclear) and *Veillonella parvula* (coypu, dog).

6.4.110 Vibrio

Phylum: Proteobacteria; Class: Gammaproteobacteria; Order: Vibrionales; Family: Vibrionaceae; Genus: *Vibrio*. The members of this genus are ubiquitous in the environment. They characteristically reside in and are acquired from water where they typically exist by growing on the chitin shell of aquatic macroinvertebrates and microinvertebrates including copepods. Correspondingly, the *Vibrio* are noted for their chitinase activity. Some members of this bacterial genus can be primary pathogens but most are opportunistic and cause secondary disease. Members of the genus *Vibrio* cause both acute as well as chronic forms of disease and often the symptoms are nonspecific. The *Vibrio* species often specifically attack the skin causing a darkening or erythema including petechial hemorrhages, wound infections and ulcerations. They also can cause otitis, produce sight-threatening ocular infections, attack the host's central nervous system including that of a developing fetus, cause gastroenteritis and coelomic distension, and cause respiratory problems. *Vibrio* also produce toxins that frequently play a role in pathogenesis. Some animal species that are noted for their toxicity can naturally harbor resident *Vibrio alginolyticus* as symbionts and this bacterial species has been suggested as one possible source of the potent neurotoxin tetrodotoxin associated with many vertebrates (puffer fish of the family Tetraodontidae, porcupine fish of the family Diodontidae, and the rough-skinned newt *Taricha granulosa*) and invertebrates (moon snails of the family Naticidae and blue-ringed octopus of the genus *Hapalochlaena*). Those species belonging to this microbial genus which have been indicated as having a potentially opportunistic disease association with mammalian livestock are: *Vibrio alginolyticus* (coypu) and *Vibrio cincinnatiensis* (cattle).

6.4.111 Yersinia

Phylum: Proteobacteria; Class: Gammaproteobacteria; Order: Enterobacteriales; Family: Yersiniaceae; Genus: *Yersinia*. The members of this genus can be environmentally present in urine and feces. It also is possible that they may be vectored through the environment in the free living amoeba *Acanthamoeba castellanii*, which naturally is found in soil and water. Members of the genus *Yersinia* cause mastitis, lymphadenopathy, abscess, hemorrhage, bacteremia, enteritis, abdominal pain, and pneumonia. The bacterial species *Yersinia pestis* is particularly notable because it causes the disease named plague. The severity of illness caused by *Yersinia pestis* ranges from subclinical to a chronic wasting, and peracute mortality can occur without other obvious signs of illness. In some host animals *Yersinia pestis* infections produce mortality rates as high as 80%–100%. However, infections caused by *Yersinia pestis* are subclinical in resistant host species and despite being quite lethal in human, *Yersinia pestis* may qualify as either a commensal or opportunistic pathogen in the Edible dormouse. Those species belonging to this microbial genus which have been indicated as having a potentially opportunistic disease association with mammalian livestock are: *Yersinia enterocolitica* (camel, horse, coypu, guinea pig, cervids, pig, cattle, cat, dog), *Yersinia pestis* (Edible dormouse), and *Yersinia pseudotuberculosis* (dog, cat, coypu, guinea pig, rabbit, cattle, pig, sheep, kangaroo, cervids, alpaca, llama, capybara).

Compliance with Ethical Standards

Conflict of Interest Christon J. Hurst declares that he has no conflict of interest.

Ethical Approval This article does not contain any studies with human participants or animals performed by the author.

References

Hurst CJ (2016) Opportunistic algae, fungi and ichthyosporea associated with mammalian livestock disease. In: Hurst CJ (ed) The rasputin effect: when commensals and symbionts become parasitic, advances in environmental microbiology, vol 3. Springer, Cham, pp 261–300. https://doi.org/10.1007/978-3-319-28170-4_10

Hurst CJ, Murphy PA (1996) The transmission and prevention of infectious disease. In: Hurst CJ (ed) Modeling disease transmission and its prevention by disinfection. Cambridge University Press, Cambridge, pp 3–54

Kahn CM, Line S (eds) (2010) The Merck veterinary manual, 10th edn. New York, Wiley

Pathogen Safety Data Sheets and Risk Assessment (2015) Public Health Agency of Canada, Ottawa. http://www.phac-aspc.gc.ca/lab-bio/res/psds-ftss/index-eng.php. Accessed 9 May 2015

Sumbali G (2011) The fungi, 2nd edn. Alpha Science International Limited, Oxford

The Center for Food Security and Public Health (2015) Iowa State University, Ames. http://www.cfsph.iastate.edu/. Accessed 9 May 2015

Vetbook (2015) Centre for veterinary education. The University of Sydney, Sydney. http://www.cve.edu.au/vetbook. Accessed 9 May 2015

Chapter 7
Infectious Disease in Wild Animal Populations: Examining Transmission and Control with Mathematical Models

Sergey S. Berg, James D. Forester, and Meggan E. Craft

Abstract The mathematical modeling of ecological interactions is an essential tool in predicting the behavior of complex systems across landscapes. The scientific literature is growing with examples of models used to explore predator-prey interactions, resource selection, population growth, and dynamics of disease transmission. These models provide managers with an efficient alternative means of testing new management and control strategies without resorting to empirical testing that is often costly, time-consuming, and impractical. This chapter presents a review of four types of mathematical models used to understand and predict the spread of infectious diseases in wild animals: compartmental, metapopulation, spatial, and contact network models. Descriptions of each model's uses and limitations are used to provide a look at the complexities involved in modeling the spread of diseases and the trade-offs that accompany selecting one modeling approach over another. Potential avenues for the improvement and use of these models in future studies are also discussed, as are specific examples of how each type of model has improved our understanding of infectious diseases in populations of wild animals.

S. S. Berg (✉)
Department of Computer and Information Sciences, University of St. Thomas, St. Paul, MN, USA
e-mail: berg@stthomas.edu

J. D. Forester
Department of Fisheries, Wildlife, and Conservation Biology, University of Minnesota, St. Paul, MN, USA

M. E. Craft
Department of Veterinary Population Medicine, University of Minnesota, St. Paul, MN, USA

C. J. Hurst (ed.), *The Connections Between Ecology and Infectious Disease*, Advances in Environmental Microbiology 5,
https://doi.org/10.1007/978-3-319-92373-4_7

7.1 Introduction

Infectious diseases of both humans and animals have played a large role in shaping the world as it is today, from the role of smallpox in the fall of the Aztec (McNeil 1998) to the 2001 foot-and-mouth disease epidemic that cost the United Kingdom over $6 billion (Thompson et al. 2002), and will continue to do so in the future. Although most research into this topic has focused on human and domesticated animals as hosts, the study of wildlife diseases has been growing rapidly since the 1980s (Anderson et al. 1981; Kawata 2010; McCallum 2016). This interest has been driven by desires to minimize risks to human health (e.g., zoonotic diseases such as leptospirosis; Munoz-Zanzi et al. 2014), decrease loss of livestock [e.g., transmission of brucellosis from elk (*Cervus canadensis*) to cattle (*Bos taurus*); Xie and Horan 2009], and preserve either native or endangered species [e.g., canine distemper in the black-footed ferret (*Mustela nigripes*); Thorne and William 1988].

Mathematical models have been a useful tool in the study of wildlife infectious diseases. These models have been used for tasks as varied as predicting the spatial spread of rabies in Eastern Europe (Källen et al. 1985), comparing the benefits of vaccination versus culling in eradicating bovine tuberculosis from England (Smith and Cheeseman 2002), investigating the role of bird feeders in promoting the spread of conjunctivitis in house finches (*Carpodacus mexicanus*) in North America (Hosseini et al. 2004), and even to show that flea-borne transmission is insufficient to maintain plague epizootics in Colorado prairie dogs (*Cynomys ludovicianus*; Webb et al. 2006). Here we present a review of four types of models used in studies of wildlife infectious diseases—compartmental, metapopulation, spatial, and contact network—and discuss their uses, limitations, and contributions to the understanding of infectious diseases. It is important to note that although we present these four models as separate for the purpose of this review, there is in fact substantial overlap between these categories (e.g., a metapopulation model can also exhibit properties of a spatial model).

7.2 Compartmental Models

The majority of disease models are based on categorizing hosts within a population based on their disease status. The most basic form of such a model describes the proportions of the population that are susceptible to (S), infected with (I), and recovered from (R) a particular disease (a complete list of model variables and parameters used in this chapter is provided in Table 7.1). For a pathogen that confers lifelong immunity from future infections, this compartmentalization results in the following set of differential equations that describe how individuals transition between different disease categories:

Table 7.1 Description of variables and parameters commonly used in mathematical models of infectious diseases

Symbol	Definition
S	Proportion of the population that is susceptible to a particular disease
I	Proportion of the population that is infected with a particular disease
R	Proportion of the population that is recovered from a particular disease
υ	Influx of new susceptible into the population through births
μ	Removal of individuals from the population through deaths
β	The product of contact rate and transmission probability for a particular disease
γ	Rate of recovery from a particular disease
X	Number of individuals that are susceptible to a particular disease
Y	Number of individuals that are infected with a particular disease
Z	Number of individuals that are recovered from a particular disease
N	Total number of individuals in the population
R_0	Basic reproductive ratio
λ	The force of infection for a particular disease
ρ_{ij}	The relative strength of transmission from one subpopulation, i, to a different subpopulation, j
P	The proportion of subpopulations infected with a particular disease
r	The reinfection (coupling) rate from an infected subpopulation to an uninfected subpopulation
e	The rate of local extinction for a subpopulation
D_X	Diffusion rate for individuals that are susceptible to a particular disease
D_Y	Diffusion rate for individuals that are infected with a particular disease
D_Z	Diffusion rate for individuals that are recovered from a particular disease
∇^2	Local diffusion rate of individuals through space
$\mathbf{G}$	Adjacency matrix that represents the presence, duration, or frequency of contact between individuals

$$\frac{dS}{dt} = \upsilon - \beta SI - \mu S, \tag{7.1}$$

$$\frac{dI}{dt} = \beta SI - \gamma I - \mu I, \tag{7.2}$$

$$\frac{dR}{dt} = \gamma I - \mu R, \tag{7.3}$$

where υ is the influx of new susceptible individuals through births, μ represents the removal of individuals through death, β is the product of contact rate and transmission probability, and γ is the recovery rate (Keeling and Rohani 2008). This formulation is known as the susceptible-infected-recovered (SIR) model, which was originally proposed by Kermack and McKendrick (1927). A slightly different version of these equations can be used to model the *number* of individuals in each disease category:

$$\frac{dX}{dt} = \upsilon - \beta XY/N - \mu X, \tag{7.4}$$

$$\frac{dY}{dt} = \beta XY/N - \gamma Y - \mu Y, \tag{7.5}$$

$$\frac{dZ}{dt} = \gamma Y - \mu Z, \tag{7.6}$$

where X, Y, and Z are the number of susceptible, infectious, and recovered individuals and N is the total number of individuals ($N = X + Y + Z$). Such compartmentalization ignores many details of the progression of the disease, as well as heterogeneity in individual responses and contact probabilities (i.e., all individuals have the same probability of contacting every other individual in the entire population), but nonetheless provides some very important insights into disease dynamics (Keeling and Eames 2005; Keeling and Rohani 2008; Kawata 2010).

Many modifications have been made to this basic framework, usually involving further subdivision of the disease classifications to reflect either more complex pathogen biology (e.g., waning immunity, fatal infection, or the presence of a latent period), multiple pathogens [e.g., canine distemper and sarcoptic mange in fishers (*Martes pennanti*); Keller et al. 2012], multiple hosts [e.g., louping ill virus in red grouse (*Lagopus lagopus scotica*), mountain hare (*Lepus timidus*), and red deer (*Cervus elaphus*); Gilbert et al. 2001], or greater structure within the host population (e.g., age, sex, or social status). These modifications include the susceptible-infectious (SI) and susceptible-exposed-infectious-recovered (SEIR) models, among others. For structured host populations, differences in contact rates between different subgroups often become important, replacing the single parameter β with a matrix that describes the transmission of infection between and within different groups. It is important to note that the assumption of random mixing still remains despite these modifications, although now it is limited to transmission within each subgroup (Keeling and Eames 2005; Keeling and Rohani 2008).

7.2.1 Uses and Limitations

Some of the first and most influential uses of compartmental models involved establishing the conditions necessary to ensure either disease persistence or eradication (Smith and Cheeseman 2002; Keeling and Rohani 2008). Two of the most important quantities in epidemiology emerged from such analyses. The first, R_0, defines the average number of secondary cases that would arise from an average primary case in an entirely susceptible population. The exact formulation of this variable depends on the type of model used (i.e., SIR versus SEIR); however, for the SIR model described above, it is given by $R_0 = \frac{\beta}{\gamma+\mu}$. The force of infection, λ, on the other hand, defines the per capita rate at which susceptible individuals contract the infection, and is equal to βI and $\beta Y/N$, for the proportions and number of individuals formulations, respectively. The values that these quantities take in different host-pathogen systems can determine the final outcome of the disease (e.g., R_0 must be greater than unity for a disease to spread and less than unity to ensure eventual disease eradication; Keeling and Eames 2005; Keeling and Rohani 2008; Kawata 2010).

Anderson et al. (1981), for example, used a compartmental model to demonstrate that vaccinating at random at least a proportion $p = 1 - 1/R_0$ of the population ensures the eventual eradication of rabies from foxes. Although numerous methods have been devised to improve the calculation of this value (e.g., incorporating age or transmission heterogeneity), the original formulation remains one of the most fundamental and often-used discoveries in epidemiology (Keeling et al. 2003; Garnett 2005; Yip et al. 2007). Multi-host models, on the other hand, have been used to investigate the effects of multiple hosts on maintaining infections in situations where a single host is insufficient to sustain the ongoing spread of the disease (e.g., Dobson and Meagher 1996; Gilbert et al. 2001).

The main limitation of these models is the mean-field assumption, wherein all individuals in the population contact each other with equal probability. Although certain modifications have been devised to address this concern (e.g., introducing host structure such that certain groups mix more preferentially with each other), it still remains the main drawback of compartmental models. Additionally, because such models are aspatial, they cannot be used to investigate the dynamics of host populations that exhibit significant spatial heterogeneity or to model the *spatial* spread of an invading disease (Keeling and Eames 2005; Keeling and Rohani 2008; Kawata 2010).

Another issue with compartmental models has centered on how best to model the transmission between susceptible and infectious individuals. For most of the field's history, density-dependent transmission was used because of its clear analogy to how the rate of interaction between two types of molecules is directly proportional to their densities. While this framework, which yields the familiar βSI transmission term, has been applied with some success, certain populations may not display such a characteristic. For a sexually transmitted disease in humans, for example, the number of transmissible contacts (i.e., sexual partners) is often independent of population density (Lloyd-Smith et al. 2004; Ryder et al. 2007). In these situations, a frequency-dependent rate of transmission is required, which alters the transmission term to $\beta SI/N$ (McCallum et al. 2001; Begon et al. 2002; Keeling and Rohani 2008).

Although these examples may give the impression that transmission operates only at two extremes (i.e., either independent or as a function of host density), in most situations, it is in fact a continuum of values and may even change throughout the course of an epidemic. Antonovics et al. (1995), for example, modeled this continuum based on the Type II Holling functional response curve, wherein contacts at low densities are proportional to the host density but eventually reach a maximum at higher densities. Because the exact formulation of the transmission term is crucial to a model's predictive and descriptive capabilities, as well as the qualitative behavior of the system itself, the relationship of contact rates to density is still an area of active research. In general, however, frequency-dependent transmission is used primarily in models of vector-borne pathogens and human populations, while density-dependent transmission is generally considered to be more applicable to plant and animal diseases (McCallum et al. 2001; Begon et al. 2002; Keeling and Rohani 2008).

Another common use of compartmental models has been to investigate the role of seasonality in producing either annual, semiannual, or even chaotic patterns of disease prevalence. Many wildlife populations, for example, experience significant changes in transmission rates throughout the year as a result of either flocking behavior (Hosseini et al. 2004), seasonal migration (Bradley and Altizer 2005), or congregation during the breeding and molting season (Swinton et al. 1998). An additional source of seasonality that is almost exclusive to wildlife populations arises from birth pulses which recruit a large number of susceptible individuals into the population at approximately the same time each year (Gremillion-Smith and Woolf 1988; Keeling and Rohani 2008). If a time-dependent component is incorporated into the transmission term, then increasing the amplitude of seasonality can increase the number of prevalence peaks throughout the year, eventually resulting in chaotic dynamics (Ireland et al. 2004).

7.2.2 Specific Applications

The earliest application of compartmental models to infectious diseases in wildlife was presented in the seminal work of Anderson et al. (1981), who developed a deterministic SEI (susceptible-exposed-infectious) model for the fatal rabies virus in European red foxes (*Vulpes vulpes*; Kawata 2010). This model demonstrated that the latent period of rabies could act as a time-delayed density-dependent regulator of fox population growth and result in the 4-year oscillations in population size and disease prevalence observed in fox populations throughout Europe. The authors also concluded that culling alone was unlikely to be effective once rabies becomes established within a fox population. Vaccination, on the other hand, significantly increased the odds of disease eradication, especially when combined with culling in regions with low fox density (Anderson et al. 1981).

Gilbert et al. (2001) used a multi-host approach to investigate how the persistence of a tick-borne virus (louping ill virus) is affected by the presence of grouse, hare, and deer in upland Britain. This model demonstrated that neither deer nor grouse alone could sustain such a virus, but that the combination of the two could and that the inclusion of hare into this system greatly increased the likelihood of virus persistence (Gilbert et al. 2001). These results were later used by Laurenson et al. (2003) to investigate whether decreasing hare density could be used as a strategy to decrease virus prevalence in grouse. They discovered that not only would this strategy decrease infections in grouse, but it would also increase their fecundity as a consequence of increased life expectancy. The authors concluded that mountain hares serve as a reservoir species that can maintain the virus as long as hare density remains above five individuals km^{-2}.

Packer et al. (2003) used a simple compartmental SI model to demonstrate a very interesting principle—that predator control programs aimed at increasing prey abundance could prove harmful to prey populations that are regulated by infectious diseases rather than by predation. The authors showed that because infected prey live longer in the absence of predators, they are able to infect many more susceptible animals than they would be able to otherwise. Although this effect is more pronounced when predators selectively remove infected prey (e.g., if the infection makes prey easier to capture or locate), nonselective predation is still beneficial to host populations by reducing the life span of infectious individuals. This principle was used to retroactively explain how the introduction of private gamekeepers (who removed predators) in red grouse habitats destabilized the grouse population in North Yorkshire, England (Packer et al. 2003).

7.2.3 *Future Directions*

As illustrated above, mean-field compartmental models have been used in a wide variety of settings and have provided the foundation for much of the epidemiological literature and theory. Unfortunately, the mean-field assumption and the aspatial nature of these models limit their ability to provide accurate predictions of the likely spread of a disease through a wildlife population. This limitation has led to the development of several more sophisticated methods to describe the spread of infectious diseases, which are detailed throughout the remainder of this chapter. Despite these advances in infectious disease modeling, there are still several aspects of disease ecology that can be best investigated with compartmental models.

For example, compartmental models can be used to explore the role of seasonal variation in mortality rates in the spread of infectious diseases. Previous research has investigated the role of seasonality in births (e.g., Ireland et al. 2004), demonstrating how increases and decreases to the magnitude of this seasonality can have profound consequences to the spread of the disease. Unfortunately, similar theory does not yet exist for the effects of altering the magnitude of seasonality in mortality rates. This avenue for future research would be particularly appealing because, unlike for births, newly developed theory could be tested by manipulating the timing, extent, and demographic characteristics of mortality through changes to harvest regulations. Because harvest constitutes a substantial portion of deaths in many wildlife populations (e.g., Erb et al. 2013; Ramsey et al. 2014), understanding how changes in harvest intensity will impact the spread of disease could prove to be a powerful tool for wildlife managers.

7.3 Metapopulation Models

Although the spread of infectious diseases is predominantly a localized process (i.e., between individuals in the same location), the movement of individuals between aggregate groups can help to facilitate the geographical spread of the disease. Metapopulation models, originally developed for application in ecology (Levins 1969; Hanski 1999), provide a simple but powerful means of capturing this spatial structure by subdividing the entire population into distinct groups known as subpopulations. These models typically assume that transmissible contacts will occur at a higher rate between individuals of the same group than between individuals of different groups, allowing each subpopulation to exhibit largely independent dynamics (Jesse et al. 2008; Ball et al. 2015).

In such a framework, the original SIR equations can be modified as

$$\frac{dX_i}{dt} = \upsilon_i N_i - \lambda_i X_i - \mu_i X_i, \tag{7.7}$$

$$\frac{dY_i}{dt} = \lambda_i X_i - \gamma_i Y_i - \mu_i Y_i, \tag{7.8}$$

$$\frac{dZ_i}{dt} = \gamma_i Y_i - \mu_i Z_i, \tag{7.9}$$

where i refers to parameters that are particular to subpopulation i and may vary substantially between subpopulations due to differences in the local environment (Broadfoot et al. 2001; Langlois et al. 2001; Keeling and Rohani 2008). Although most of the useful quantities derived from mean-field compartmental models still apply, they must now incorporate, ρ_{ij}, the relative strength of transmission from one subpopulation, i, to a different subpopulation, j (e.g., $\lambda_i = \beta_i \sum_j \rho_{ij} Y_j / N_i$ vs. $\lambda = \beta Y / N$; Keeling and Rohani 2008).

Although metapopulation structure is inherently based on the host population of interest, these models usually fall into one of three categories. Wind- and vector-borne transmission models, for example, assume that subpopulations are epidemiologically sessile, forcing coupling (i.e., the strength of transmission) to decrease with distance. These models have been used extensively to model the spatial dynamics of livestock diseases by treating each farm as a sessile subpopulation (e.g., Ferguson et al. 2001; Keeling 2001). Commuter models, which are most commonly used for human population, are instead based on the premise that permanent relocation between subpopulations is rare but temporary travel between them is sufficient for disease spread (Keeling and Rohani 2008). Finally, most models of animal diseases rely on either the migration or permanent movement of individuals to spread the pathogen (Keeling and Rohani 2008).

Once the appropriate relations are defined, the system of equations can be solved in either a deterministic or stochastic manner. Because this framework is able to capture the spatial clustering of the host population, it has been used extensively not only for the study of infectious diseases but also in the ecological literature to derive such concepts as the extinction-colonization balance, rescue effect, and isolation paradigm (Hanski 1998, 1999).

7.3.1 Uses and Limitation

One of the most common uses of metapopulation models is to investigate the influence of substructure on disease thresholds and persistence. Group size heterogeneity, for example, can increase disease persistence by allowing smaller subpopulations, where the pathogen would normally be unable to persist, to continually become reinfected from larger ones where the disease is maintained (Broadfoot et al. 2001). The level of substructure has also been shown to be an important consideration, with higher numbers of subdivision (e.g., dividing subpopulations into smaller groups) leading to longer disease outbreaks and increased persistence (Swinton et al. 1998). Similar approaches have also been used to understand the rate and heterogeneity of the spatial spread of a pathogen (e.g., Langlois et al. 2001; López et al. 2009).

A general trend that has emerged from this modeling framework is that given certain levels of aggregation and coupling, a pathogen can persist within the host population for much longer than would be predicted by a mean-field compartmental model (Swinton et al. 1998; Fulford et al. 2002; Keeling and Rohani 2008). Although many modeling studies have concluded that persistence increases with the intergroup connectivity, Jesse et al. (2008) demonstrated that the relationship is strongly nonlinear. More specifically, a peak in epidemic duration appears at small movement rates (due to asynchrony in the disease dynamics between subpopulations), followed by a global maximum at larger rates (when the pathogen "perceives" the metapopulation as a single well-mixed population). These results are particularly applicable to outbreak management because they demonstrated that while restricting animal movement (a common response to disease outbreaks in cattle farms) will reduce the total *number* of subpopulations that experience infection, it may also prolong epidemic duration instead of decreasing it. At these lower movement rates, any infectious animal that does move between subpopulations has a large probability of encountering a fully susceptible subpopulation, thereby preventing the disease from dying out for an increased length of time (similar to the rescue effect in ecology). Another important conclusion is that although increasing mixing between subpopulations via habitat corridors is an effective means of species conservation (Hanski 1999), their potential as conduits for the spread of infectious diseases should be carefully considered before implementation (Hess 1994, 1996).

Metapopulation models have also been used to evaluate the benefits of different control strategies that take into account the local nature of spatial transmission. Beyer et al. (2012), for example, demonstrated that rabies fade-out in Tanzania became increasingly more likely once vaccination reduced the number of susceptible dogs in each village subpopulation below 150. Fulford et al. (2002) used a similar approach to show that for subpopulations that occur in a chain configuration (e.g., along rivers or roads), limited culling applied to alternate patches was significantly more effective in reducing disease prevalence than extensive culling applied in large batches along the chain.

Apart from the fact that metapopulation models still assume homogeneous mixing (although at the intra-subpopulation level), their main limitation is that they require detailed information on the number of individuals in each subpopulation, which can be difficult to collect and computationally intensive to explicitly model (Keeling and Rohani 2008). An alternative approach, originally proposed by Levins (1969), models each subpopulation as being either pathologically empty (disease-free) or occupied (having infection). The intuitive way to conceptualize this is to assume that localized extinctions and successful recolonizations are extremely rare, so that each subpopulation spends the vast majority of its time either disease-free or close to the endemic equilibrium. For a large number of subpopulations with equal probabilities of infecting each other, the proportion of infected subpopulations, P, is governed by

$$\frac{dP}{dt} = r(1 - P)P - eP, \tag{7.10}$$

where r measures the reinfection (coupling) rate from an infected subpopulation to an uninfected one and e is the rate of local extinction (Keeling and Rohani 2008). Although this approach has been used much less extensively than the previously described standard approach, it has nonetheless been used to derive some interesting results [e.g., see Smith et al. (2002) below].

7.3.2 Specific Applications

Plowright et al. (2011) created a metapopulation model to provide a mechanistic understanding of the causal links between anthropogenic change and emergence of new zoonoses from wildlife in Australia. As urbanization destroyed most of the contiguous forest cover on the eastern coast, fruit bats (*Pteropus* spp.) began switching their diet from the more patchily distributed nectar and fruit sources to urban gardens that provide abundant food year-round. Unlike the old diet that required long-distance foraging and migration, the new one required minimal migration, which decreased the connectivity between subpopulations (identified by the location of their daytime roosts). By incorporating data on within- and between-camp transmission, the authors demonstrated that the decreased migration led to increased epidemic sizes and divergence in both amplitude and frequency from epidemics in rural camps. This was due to the fact that as fewer individuals migrated, the probability of infected hosts moving between camps decreased, which lowered the probability of camps becoming reinfected after local viral extinction and increased the time over which subpopulations could recruit susceptible individuals via births. This resulting decline in herd immunity across the metapopulation shifted disease dynamics toward sporadic, shorter, and more intense local epidemics with larger numbers of infected individuals. However, if this connectivity were to drop below a certain threshold level, then the resulting movement would become insufficient to reinfect distant camps. Thus, the same conditions that are currently responsible for large and sporadic epidemics could eventually lead to viral extinction, rendering the entire population vulnerable to a large synchronized outbreak (Plowright et al. 2011).

Another application of this modeling framework was recently employed by Pons-Salort et al. (2014) to explore the role of four different species of bats in the persistence of rabies across three cave-dwelling colonies on the Balearic Islands in Spain. By successively removing each of the four species from the transmission cycle, they were able to demonstrate that the common bent-wing bat (*Miniopterus schreibersii*) is the only species essential for the persistence of the virus, because it serves as a regional reservoir in the system. In fact, if this species were removed, then even increased contact rates and interisland exchanges between the remaining three species would be insufficient to maintain rabies infections. Additionally, Pons-Salort et al. (2014) demonstrated that contrary to expectation, the most abundant species of bat on the islands, the greater mouse-eared bat (*Myotis myotis*), plays a relatively minor role in the persistence of the virus.

A large wave of rabies infections began on the Virginia-West Virginia border in the mid-1970s and generated interest in how landscape structure (e.g., the location of roads, bodies of water, and human population centers) affects the spread of disease. Smith et al. (2002) used a Levins-type metapopulation model to study a small portion of this wave as it traveled through raccoons (*Procyon lotor*) in 169 townships across Connecticut from 1991 to 1996, concentrating on the underlying spatial heterogeneity of the habitat. A spatial SI model (used instead of an SIR because there is no recovery from infection at the township level) was compared to the observed data of reported rabies outbreaks and resulted in a best fit model where rivers reduced transmission by 87% compared to land boundaries; local transmission accounted for most but not all of the spatial spread; and human population density played a small but positive role in transmission (Smith et al. 2002). Although this formulation ignored within-subpopulation dynamics (i.e., a newly infected township can transmit infection as strongly as when endemic equilibrium is reached), it allowed for far richer parameterization without the need to estimate the number of raccoons in each township.

7.3.3 Future Directions

As illustrated above, metapopulation models have provided important insights into the complexities and spatial variation observed in disease prevalence and outbreaks. Despite the progress that has been made in the development and application of these models, however, there are a number of limitations that still require improvement.

For example, there is substantial interest in devising a practical framework for ascertaining the optimum timing and duration of vaccination pulses to promote the extinction of a pathogen at the population level. Although theoretical research has demonstrated that pulsed vaccination has the ability to synchronize epidemic behavior and thereby limit pathogen rescue effects, understanding how this benefit interacts with the resulting buildup of susceptibles between pulses can have enormous public and animal health consequences (Keeling and Rohani 2008).

Another exciting avenue for future research is to develop a method for using data from the first phase of an epidemic to accurately parameterize all aspects of a metapopulation model (Ball et al. 2015). Unfortunately, data collected during this phase are prone to a number of biases that are not yet fully understood (e.g., underestimation of fatality rates due to the time delay from disease onset to death; Nishiura et al. 2009). Overcoming these challenges could greatly improve the capability of metapopulation models to accurately predict the likely progression of an outbreak and the efficacy of different control strategies.

Finally, several studies have noted that assuming homogeneous mixing within subpopulations tends to overestimate the spatial spread, time to peak incidence, and the peak number of cases of epidemics when compared to more refined models (Ajelli et al. 2010; Keeling et al. 2010). Identifying situations where these differences are most prominent and devising methods to adjust model predictions accordingly still remains a major challenge in metapopulation modeling theory (Ball et al. 2015).

7.4 Spatial Models

Spatial models subdivide the area occupied by the host species into a grid and thus provide a way to model the spatial structure of a population without dividing it into distinct social groups. The most common approach is to model interactions as occurring only between individuals occupying either the same or neighboring grid cells while allowing individuals to move between them. These spatial lattice-based models can be thought of as a special case of metapopulation models (where each cell represents its own subpopulation) and as such are governed by a very similar set of equations:

$$\frac{dX_i}{dt} = \upsilon - \beta X_i \frac{\left(1 - \sum_j \rho_{ji}\right) Y_i + \sum_j \rho_{ij} Y_j}{\left(1 - \sum_j \rho_{ji}\right) N_i + \sum_j \rho_{ij} N_j} - \mu X_i, \tag{7.11}$$

$$\frac{dY_i}{dt} = \beta X_i \frac{\left(1 - \sum_j \rho_{ji}\right) Y_i + \sum_j \rho_{ij} Y_j}{\left(1 - \sum_j \rho_{ji}\right) N_i + \sum_j \rho_{ij} N_j} - \gamma Y_i - \mu Y_i, \tag{7.12}$$

$$\frac{dZ_i}{dt} = \gamma Y_i - \mu Z_i, \tag{7.13}$$

where ρ_{ij} is equal to one if i and j are neighbors and zero otherwise.

An alternative approach is to treat each grid cell as being large enough to only contain a single individual (i.e., a subpopulation of one). This results in a finite number of states for these subpopulations, wherein each cell is either empty or occupied by a susceptible, infectious, or recovered individual (Keeling and Rohani 2008; Miksch et al. 2013). This formulation follows a fire ecology framework originally developed by Bak et al. (1990), where each cell was either empty or occupied by a healthy or burning tree (reflecting a recovered, susceptible, and infectious host, respectively). Burned trees that died left empty spaces that could then be colonized (reflecting recovery and births), and fire (reflecting infection) could spread between neighboring trees. Although this framework makes a number of simplifying assumptions (e.g., trees do not always die from fire), it has nonetheless been used as the foundation for some interesting spatial models of infectious diseases [e.g., see Tischendorf et al. (1998) below].

7.4.1 Uses and Limitations

Like all mathematical models, lattice-based spatial models are a clear abstraction of reality because, except for agricultural settings such as orchards (e.g., Gibson 1997; Poggi et al. 2013), individuals do not exist on a regular grid. Thus, spatial models are mainly used to investigate how the effects of spatial separation and nonrandom mixing cause disease dynamics to deviate from mean-field approximations in a spatial framework. For example, these models have demonstrated that an invading infection is characterized by a clear wave-like spread because the disease must first spread to neighboring sites before reaching the rest of the population. Similar to metapopulation models, spatial models have also been used to evaluate the broad-scale consequences of different control efforts including vaccination (e.g., Tischendorf et al. 1998) and culling (e.g., White and Harris 1995; Deal et al. 2004).

Although spatial models are a clear abstraction of the spatial structure of most host populations, certain types of wildlife populations actually lend themselves very naturally to be modeled using such an approach. Populations that occur in either a one-dimensional chain (e.g., along a river) or loop (e.g., surrounding an impassable urban center) can be thought of as having only nearest-neighbor connectivity and as such can easily be modeled using a lattice-based approach.

Despite these uses, the separation of populations into a lattice framework has several drawbacks. Imposing a square grid onto a habitat, for example, forces interactions in particular directions (e.g., north, south, east, and west) to play a disproportionately larger role in disease transmission than interactions in other directions. The use of a hexagonal grid, however, helps to alleviate this problem and often better recreates observed spatial patterns (Anneville et al. 1998; Van Baalen and Rand 1998; Kao 2003; Birch et al. 2007). A more general issue is that the discretization of space forces intracell interactions to be far stronger than intercell ones, regardless of the actual distance between individuals in each cell. Two individuals that occupy adjacent corners of two separate cells, for example, would have weaker interactions than two who occupy the same cell but are at opposite ends, despite the fact that the former are closer to each other than the latter (Keeling and Rohani 2008).

Another disadvantage of this lattice-based approach is that dividing the space occupied by a population into discrete cells limits how precisely we can model the location of each individual (i.e., we only know which cell an individual is in but not where inside the cell they are). An alternative approach is to treat this space as continuous, which allows us to model the *exact* location of each individual in the population. This approach is essentially an extension of the lattice-based framework, where the size of each cell has become infinitely small, and is the foundation of reaction-diffusion models (Keeling and Rohani 2008). The standard reaction-diffusion model assumes that individuals move randomly throughout the landscape and that infections are only transmitted between individuals sharing the same location (Källen et al. 1985; Murray et al. 1986; Yachi et al. 1989; Kawata 2010), resulting in the following set of equations:

$$\frac{\partial X}{\partial t} = \upsilon - \beta XY/N - \mu X + D_X \nabla^2 \mathrm{X}, \tag{7.14}$$

$$\frac{\partial Y}{\partial t} = \beta XY/N - \gamma Y - \mu Y + D_Y \nabla^2 \mathrm{Y}, \tag{7.15}$$

$$\frac{\partial Z}{\partial t} = \gamma Y - \mu Z + D_Z \nabla^2 \mathrm{Z}, \tag{7.16}$$

where *X*, *Y*, and *Z* are now functions of both space and time, ∇^2 describes the local diffusion of individuals through space, and D_X, D_Y, and D_Z, are the diffusion rates for the three classes of individuals (Keeling and Rohani 2008; Kawata 2010).

An extension of this approach is to incorporate a transmission kernel that models how transmission risk decreases with distance. This kernel acts to relax the assumption that transmission is solely a localized process, allowing individuals not in the immediate vicinity of an infectious individual to still become infected, albeit at a significantly lower rate (Schofield 2002; Keeling and Rohani 2008). Because reaction-diffusion models provide a very broad-scale view of disease outbreaks, their application has mostly been limited to investigating either epidemics that span large spatial scales (Conner and Miller 2004) or the presence and role of rare but long-range disease transmission (e.g., dispersing animals) in accelerating the spatial diffusion of a newly introduced pathogen (Schofield 2002; Keeling and Rohani 2008).

7.4.2 *Specific Applications*

In 1997, a large-scale and long-term immunization of European foxes against rabies was under consideration to be either terminated or cut back due to diminishing returns and despite lasting sporadic incidences. Tischendorf et al. (1998) used a spatial model to investigate the effectiveness of this program and assess the consequences of either reducing or eliminating it on the persistence of rabies. This approach involved simulating the mating, dispersal, and vaccination of fox social groups across a lattice grid, as well as the contact between and within groups. Unlike the traditional approach, each grid was able to hold one of six possible states—the three “simple” states of infectious, empty, and susceptible and three new “mixed” states of susceptible plus immune, infectious plus immune, and empty plus immune (meaning that only immune foxes are present). The results of this model showed that even in a highly immunized fox population such as the one throughout Europe, rabies can still persist in the form of moving clusters of infection. The probability of disease eradication rises sharply after a mean immunization rate of 70% is reached, although a further 6 years of maintaining this level was required to guarantee rabies eradication (Tischendorf et al. 1998).

In South West England, efforts to eradicate bovine tuberculosis from cattle were complicated by the presence of a wildlife reservoir for the disease in European badgers (*Meles meles*). White and Harris (1995) developed a stochastic lattice-based model that incorporated density-dependent fecundity and cub survival and used badger social groups as the basic unit of measurement. The major result of this model showed that although bovine tuberculosis could persist for a long time in populations with a disease-free equilibrium group size of only four individuals, a group size threshold of six was required for the disease to become endemic. Additionally, increasing intergroup contact rates significantly improved the probability and rate of spread of infection and lowered the group size threshold required for endemicity. As such, perturbations of badger social groups caused by control operations could actually increase the probability of persistence (or spread) of an infection instead of preventing it (White and Harris 1995).

In 1977, a single rabies-positive raccoon was identified in Pennsylvania and determined to be a by-product of a restocking program that transported raccoons from Florida to Virginia and West Virginia in the mid-1970s. By 1996, the rabies virus spread to almost all counties in Pennsylvania, resulting in 3912 confirmed cases of infected raccoons and a further 2137 cases in other animals. Moore (1999) used the timing of first cases in each county to construct a reaction-diffusion model for the entire state and explore what factors influenced the speed and direction of the spread. The resulting contours revealed that the infection first spread northward along the corridors of the Appalachian Mountain and Great Valley sections of the state, before twisting to the west once it reached the high plateau areas in the north. These results differed from the homogeneous point diffusion process originally proposed for the rabies outbreak and were used to inform a strategic oral-bait vaccination strategy (Moore 1999).

Sayers et al. (1985) used a similar approach to investigate how geographic features influenced the diffusion of fox rabies through cities in Germany. The derived pathogen velocity vectors demonstrated that propagation was fastest along a broad range of limestone with elevations up to 600 m and diverse land cover and vegetation. These results were later used by Källen et al. (1985) to design a deterministic model for the spread of this epizootic front westward across Europe that helped identify the width of a control barrier that would prevent the disease from entering the rabies-free United Kingdom. This model predicted a front wave speed of 50 km year^{-1}, consistent with empirical data from the German-Polish border, and resulted in an estimated barrier width of 15 km (Källen et al. 1985). This westward spread was eventually halted through the delivery of a highly effective oral vaccine (Jackson and Wunner 2002).

7.4.3 Future Directions

As illustrated above, the use of spatial models has provided managers with not only the tools to evaluate different control strategies but also with explanations for observed disease distributions. However, there are still many interesting aspects of spatial disease models that warrant additional research.

For example, there is interest in exploring how spatial structure influences basic epidemiological characteristics such as R_0 and the epidemic threshold (Riley et al. 2015). Although calculating these values is relatively simple in models that exhibit homogeneous mixing (i.e., compartmental and metapopulation models), the clumping of individuals in spatial models substantially complicates the calculations. Mollison and Kuulasmaa (1985), for example, demonstrated that the epidemic threshold (i.e., the value of R_0 required for the disease to spread through the population) for a nearest-neighbor lattice-based model is between 2 and 2.4, compared to the usual value of 1. A better understanding of how clumping and other spatial characteristics of a population influence this and other aspects of disease spread can greatly improve the use of spatial models.

Another exciting avenue for future research is to investigate the effects of spatial and temporal resolution on the accuracy of spatial models. Although it may be tempting, for example, to select the finest scale possible based on the available data, it may be more helpful to model the spread of the disease at the same scale as that used for control and management (e.g., counties or townships; Riley et al. 2015). Finer temporal scales may also be tempting, but require substantially more data on all aspects of animal behavior that could contribute to disease spread. Kjaer (2010), for example, used a 2-hour time step to model the spread of chronic wasting disease in a hypothetical population of white-tailed deer (*Odocoileus virginianus*). Berg (2016), on the other hand, modeled a similar system using a 1-week time step. Although this later approach allowed for far richer parameterization of the progress of the disease in individual animals, it came at the expense of being unable to include events that occurred at time intervals shorter than 1 week (e.g., occasional daylong excursions made by deer outside of their home range). Developing generalities about how different spatial and temporal scales influence the modeled spread of a disease can greatly improve the use of spatial models in the management and control of wildlife diseases.

7.5 Network Models

Although many infectious disease models assume random mixing either within the host population as a whole or within each subpopulation or grid cell separately, the number of contacts each individual has in a real population is significantly smaller than the (sub)population size and is highly heterogeneous between individuals. The contact network approach, originally developed for application in statistical physics, provides another means of studying disease dynamics in these heterogeneous populations (Keeling and Eames 2005; Keeling and Rohani 2008; Craft and Caillaud 2011). In this framework, individuals (or groups of individuals) are represented as nodes, and connections between them are referred to as edges. These edges can represent transmission events, contact through which infections could spread, and even movement of animals between groups (Pellis et al. 2015), and can be either undirected (for infections that can pass in both directions with equal probability) or directed (for infections that can pass better in one direction than the other; Keeling and Eames 2005). These connections may also be either binary (representing whether any contact has occurred or not) or weighted (representing the duration or frequency of contact) depending on the host-pathogen system of interest (Godfrey 2013).

The simplest way to represent a contact network is to construct a $N \times N$ adjacency matrix, $\boldsymbol{G}$, where N is the number of individuals in the study population and $\boldsymbol{G}_{ij}$ represents either the presence, duration, or frequency of contact between individual i and individual j. Although this matrix is usually symmetric (i.e., $\boldsymbol{G}_{ij} = \boldsymbol{G}_{ji}$), directed networks yield nonsymmetric matrices (i.e., $\boldsymbol{G}_{ij} \neq \boldsymbol{G}_{ji}$; Keeling and Eames 2005). A number of useful quantities have been derived using adjacency matrices to describe how connections within a network are structured. These include individual-level metrics such as degree (number of edges connected to a single node) and centrality (how important and influential a node is within the network; White et al. 2015), alongside population-level metrics such as clustering coefficient (the extent to which an individual's neighbors are connected to each other; Godfrey 2013) and "small-world-ness" (i.e., the interaction of local clustering and average path length; Humphries and Gurney 2008).

Once the appropriate contact network and associated adjacency matrix are constructed, transitions between disease states are evaluated in discrete time steps (e.g., once every 3 months) on an individual basis using predetermined probabilities (e.g., a susceptible individual who is in contact with an infectious one has a 0.4 probability of becoming infected, while the infectious individual has a 0.2 probability of recovering; Keeling and Rohani 2008). This framework is able to capture complex individual-level structure and heterogeneity in a relatively simple manner and as such has been used not only for animal diseases but also for a variety of human ones such as HIV (Sloot et al. 2008) and SARS (Small and Tse 2005). Network models have also been used in other disciplines, such as sociology, to measure actor prestige (Korfiatis and Sicilia 2007) and to investigate dynamics of rumor spreading (Moreno et al. 2004).

7.5.1 Uses and Limitation

One of the most common uses of network models is to investigate the role of degree heterogeneity and well-connected individuals (i.e., super-spreaders) in driving disease dynamics (Böhm et al. 2009; Hamede et al. 2012; White et al. 2015). For example, because these super-spreaders can have a disproportionately high impact on the basic reproductive ratio and the epidemic threshold, network models are often used to evaluate how either vaccinating or quarantining super-spreaders will influence disease spread (Craft and Caillaud 2011; Tompkins et al. 2011). Normally, such well-connected individuals would be considered to be at highest risk of infection (Godfrey 2013). However, while this relationship does exist for many wildlife populations [e.g., brushtail possums (*Trichosurus vulpecula*); Corner et al. 2003], it appears to be absent in others [e.g., meerkats (*Suricata suricatta*); Drewe 2009]. It is also important to remember that although super-spreaders have been identified in many wildlife populations, they appear to be absent in others [e.g., Tasmanian devils (*Sarcophilus harrisii*); Hamede et al. 2009].

A general trend that has emerged from analyzing network models is that high variation in the degree distribution tends to lower the epidemic threshold and increase the basic reproductive ratio when compared to mean-field compartmental models that use the average degree as a proxy for transmission rates (Bansal et al. 2007). Although several researchers have proposed methods to account for this heterogeneity in mean-field models (e.g., Anderson et al. 1986; Newman 2002), their explicit inclusion via network modeling still provides the best predictor of disease spread (Hamede et al. 2012).

Comparing alternative networks, on the other hand, has been used to gain insights into the importance of different types of animal behavior in disease transmission and to evaluate support for alternative working hypotheses (Godfrey 2013). Drewe (2009), for example, developed separate networks for grooming and aggressive interactions in meerkats and found that not only are the two highly directional (e.g., groomers had a higher risk of tuberculosis infection than groomees), but that grooming was the most important type of social interaction. Fenner et al. (2011) used a similar approach to demonstrate that connectivity to dispersers was more important than connectivity to resident individuals in predicting nematode loads in pygmy blue-tongue lizards (*Tiliqua adelaidensis*).

Networks have also been used to show that group size alone does not necessarily reflect transmission rates. For example, there appears to be no relationship between group size and risk of infection in gidgee skinks (*Egernia stokesii*; Godfrey et al. 2009). Larger primate groups, on the other hand, offset the increased parasite risk associated with group size by imposing within-group substructure (Griffin and Nunn 2011). Network models have been used to show that social interactions are often more important in disease spread than either spatial proximity or home-range overlap (Bull et al. 2012), address the issue of using different spatial scales (Davis et al. 2008), and investigate if density influences contact rates (Ji et al. 2005). Alongside these more specific applications, network models have also been used in much the same way as earlier model types—to evaluate a wide range of disease control measures such as oral vaccines (Delahay et al. 2009) and culling (Ramsey et al. 2002).

As has been illustrated above, one of the main advantages of network models is their versatility, allowing them to be applied to most host-pathogen systems (e.g., both directly and indirectly transmitted viruses, bacteria, and parasites; Godfrey 2013). In fact, they can be parameterized to recreate almost any other model type (e.g., a compartmental model is simply a fully connected network, while a lattice-based model can be replaced by a lattice network with nearest-neighbor connectivity; Craft and Caillaud 2011). The main limitation of these models, however, is that accurate parameterization requires detailed data on the contact and movement patterns of every individual within the study population at a time scale appropriate for the transmission process (Godfrey 2013). Although ongoing advances in GPS and proximity sensor technology are helping to alleviate this problem (e.g., Leu et al. 2010; Hamede et al. 2012; Reynolds et al. 2015), accurate sampling of every individual is still neither financially nor logistically feasible except for the smallest of wildlife populations (Tompkins et al. 2011; Godfrey 2013). Because of this, most studies (e.g., Craft et al. 2010; Hamede et al. 2012; Reynolds et al. 2015) use a "representative" subset of the population to generate contact networks that recreate observed measures of connectivity (e.g., average number of contacts, degree distribution, or amount of clustering).

Another challenge in using contact networks is defining what constitutes a transmissible contact (e.g., what proximity or duration is required to transmit an aerosol pathogen; Craft and Caillaud 2011; Tompkins et al. 2011). Because it is often impossible to determine exactly how transmission occurs without resorting to controlled transmission experiments that are often infeasible, most studies use either spatial proximity (e.g., Fenner et al. 2011) or home-range overlap (e.g., Godfrey et al. 2010) as proxies of contact and transmissible events. Additionally, wildlife networks often, if not always, exhibit temporal variation (e.g., higher contact rates while females are in estrous; Rushmore et al. 2013) that can drastically alter disease dynamics, requiring long-term observations for accurate parameterization (Volz and Meyers 2007, 2009; Hamede et al. 2012).

7.5.2 *Specific Applications*

Although most contact networks are extrapolated from a supposedly representative subset of the study population, certain situations (e.g., endangered or very large animals) may still allow researchers to explicitly capture the full network of contacts. Leu et al. (2010), for example, constructed a directed transmission network for a subpopulation of sleepy lizards (*Tiliqua rugosa*) to determine whether asynchronous use of overnight refuges (a form of indirect and directional transmissible contact) could account for observed ectoparasite loads in individual lizards. Simulating disease spread on this network provided a powerful predictor of ectoparasite loads and revealed that individuals who use many refuges had significantly lower loads as a result of a lower probability of using a previously occupied refuge. This study helped to not only identify increasing the number of available refuges as the optimum strategy for reducing disease in sleepy lizards but also provided one of the first frameworks for using contact networks to investigate the dynamics of indirectly transmitted diseases (Leu et al. 2010).

Network models have been used to demonstrate that heterogeneity in either the connectivity or degree of individuals can have important consequences for the epidemic threshold and rate of spread of a disease. However, it is important to realize that a population's contact structure often operates on multiple hierarchical scales (i.e., individuals form groups that exhibit their own heterogeneous contact network; Craft and Caillaud 2011). Caillaud et al. (2013) constructed an undirected, two-level (i.e., individuals interacting within groups alongside intergroup connectivity) hierarchical model for a hypothetical infectious disease to demonstrate that variation in group size significantly reduces the epidemic thresholds, as well as increases the mean and variance of small outbreak sizes. This suggests that incorporating group size heterogeneity into disease models can greatly improve their predictive capabilities, much like the incorporation of degree heterogeneity described above. The authors also introduced the concept of an epidemiological effective group size, defined as the group size in a homogeneous population that would result in the same epidemic thresholds as the heterogeneous one. By applying this concept to lions (*Panthera leo*) in the Serengeti National Park, Caillaud et al. (2013) demonstrated that the observed network, which exhibited highly variable pride sizes, is comparable to a homogeneous population with approximately 10–20% larger prides.

Recognizing the importance of temporal variation in contact structure, Hamede et al. (2012) used proximity-sensing radio collars to derive seasonal patterns of contact in a distinct population of Tasmanian devils. These patterns were then used to generate sets of contact networks with suitable characteristics (e.g., mean degree and transitivity) that regenerated associations once every 3 months. By stochastically simulating the spread of devil facial tumor disease through these dynamic contact networks, the authors demonstrated that failure to account for seasonal contact structure significantly overestimated the time to host extinction as well as the transmissibility threshold necessary for an epidemic to occur. Interestingly, the differences in time to host extinction between network and compartmental models became negligible for transmission rates close to either zero or one, suggesting that contact heterogeneity has little effect on the dynamics of host extinction from diseases with either very low or very high transmissibility. Unfortunately, because transmission rates are likely to be intermediate for most wildlife diseases, ignoring contact structure can have negative consequences for wildlife conservation (Hamede et al. 2012).

7.5.3 Future Directions

As illustrated above, the use of network models to study the patterns of and driving forces behind disease spread has been growing in the wildlife literature despite many challenges (Craft 2015). As technological advances continue to lessen these challenges, network models will undoubtedly provide the means to investigate many new facets of disease ecology.

For example, Pellis et al. (2015) suggest investigating how specific network characteristics influence the spread of an infectious disease through a system. Previous research has demonstrated the effect of degree and group size heterogeneity on the epidemic threshold and basic reproductive ratio (Hamede et al. 2012; Caillaud et al. 2013). Similar research is needed to investigate other effects such as the impact of clustering and degree correlation on R_0 and the likely size and duration of an outbreak. The role of the distribution and correlation of weights in weighted networks deserves similar attention. A better understanding of the epidemiological consequences of these and other characteristics can greatly improve the use of network models in controlling the spread of infectious diseases (Pellis et al. 2015).

Another exciting avenue for future research is to consider how infections themselves alter the topography and characteristics of a network. Croft et al. (2011), for example, found that the presence of infection in female guppies (*Poecilia reticulata*) decreased the duration of contact between individuals and significantly lowered the clustering of the network. Although this study demonstrated a mechanism by which individuals modified their contact structure in a way that depressed the spread of a disease, other infections may instead alter behavior and the corresponding network in a way that promotes disease spread. Identifying and understanding the consequences of these alterations to the likely spread of a disease is an important step in improving the predictive capabilities of network models.

There is also a need to develop more tractable and realistic methods of incorporating the dynamic nature of real-life contact structures into network models (Pellis et al. 2015). The presence, strength, and direction of contacts between individual animals are neither stagnant nor permanent, but instead change over time due to a variety of intrinsic and extrinsic factors (e.g., resource abundance, migration, or injury). Properly modeling the formation and dissolution of these contacts remains a major challenge in accurately parameterizing dynamic network models of wild animal populations.

Finally, network models can be used to explore how infections have influenced the evolution of social organization in wild animals (Godfrey 2013). One hypothesized cost of this evolution is the increased risk of infection caused by higher contact rates and local population densities (Moller et al. 1993; Altizer et al. 2003). Using network models to investigate how these costs are offset by social behavior (e.g., the use of alternate sleeping or roosting sites) can greatly improve our understanding of social structures and the evolution of certain behaviors in a variety of animals (Godfrey 2013).

7.6 Conclusion

The use of epidemiological models in the study of wildlife diseases began in the early 1980s and has since then grown to be a major component of the wildlife disease literature. Although mean-field compartmental models laid much of the foundation for current epidemiological practices (e.g., the basic reproductive ratio and vaccination threshold), their inability to capture spatial dynamics led to the use of metapopulation and spatial models (Keeling and Rohani 2008; Kawata 2010). More recently, the use of contact networks has emerged as an intuitive and versatile means of studying disease dynamics. As ongoing advances in GPS and proximity sensor technology make it more feasible to capture the level of data required to more accurately parameterize contact structure, network modeling will continue to provide useful insights into infectious diseases of wild animals (Keeling and Eames 2005; Tompkins et al. 2011; Godfrey 2013).

Although much research has evaluated the uses, limitations, and dynamics of different infectious disease models, many potential avenues for future research still exist for each one. For example, although previous research using compartmental models has assessed the role of birth and transmission seasonality in producing annual, semiannual, and even chaotic patterns of disease prevalence (Ireland et al. 2004; Hosseini et al. 2004), little research investigating the role of seasonal patterns of mortality has been undertaken. The development of generalities about what ecological variables (e.g., resource or predator distribution) govern the establishment of contact structures is also of high priority (Craft and Caillaud 2011; Tompkins et al. 2011).

Several more general concerns also exist. For example, there is a substantial need to improve the analytical and computational tools available to infectious disease modelers; emerging methods developed for the analysis of "big data" are particularly intriguing for discerning the contact structure within populations of wild animals (Pellis et al. 2015). These tools include accurate and computationally efficient methods of calculating the basic reproductive ratio, epidemic threshold, and other epidemiologically important quantities from complex model structures (Ball et al. 2015). Identifying ways to combine two or more of the modeling approaches discussed in this chapter (Ball et al. 2015; Pellis et al. 2015), or to combine these models with other tools such as pathogen phylogenies (Frost et al. 2015), to provide a more realistic representation of the complexities involved in the spread of infectious diseases also deserves additional attention.

The ultimate goal of epidemiological models such as those discussed in this chapter is to allow scientists to understand the mechanisms behind and predict the future course of the spread of infectious diseases. As the use and advancement of these models continue, they will undoubtedly continue to aid efforts to minimize risks to human health, decrease loss of livestock, and preserve native and endangered species of wildlife.

Acknowledgments We wish to thank Lauren White, John Fieberg, and Todd Arnold for providing invaluable feedback during the preparation of this chapter.

Compliance with Ethical Standards

Funding This material is based upon work supported by the National Science Foundation Graduate Research Fellowship under Grant No. 00006595 and the University of Minnesota Institute on the Environment.

Conflict of Interest Sergey S. Berg declares that he has no conflict of interest. James D. Forester declares that he has no conflict of interest. Meggan E. Craft declares that she has no conflict of interest.

Ethical Approval This article does not contain any studies with human participants or animals performed by any of the authors.

References

Ajelli M, Gonçalves B, Balcan D et al (2010) Comparing large-scale computational approaches to epidemic modeling: agent-based versus structured metapopulation models. BMC Infect Dis 10:190

Altizer S, Nunn CL, Thrall PH et al (2003) Social organization and parasite risk in mammals: integrating theory and empirical studies. Annu Rev Ecol Evol Syst 34:517–547. https://doi.org/10.1146/annurev.ecolsys.34.030102.151725

Anderson RM, Jackson HC, May RM, Smith AM (1981) Population dynamics of fox rabies in Europe. Nature 289:765–771

Anderson RM, Medley GF, May RM, Johnson AM (1986) A preliminary study of the transmission dynamics of the human immunodeficiency virus (HIV), the causative agent of AIDS. IMA J Math Appl Med Biol 3:229–263

Anneville O, Cury P, Le Page C, Treuil JP (1998) Modelling fish spatial dynamics and local density-dependence relationships: detection of patterns at a global scale. Aquat Living Resour 11:305–314

Antonovics J, Iwasa Y, Hassell MP (1995) A generalized model of parasitoid, venereal, and vector-based transmission processes. Am Nat 145:661–675

Bak P, Chen K, Tang C (1990) A forest-fire model and some thoughts on turbulence. Phys Lett A 147:297–300

Ball F, Britton T, House T et al (2015) Seven challenges for metapopulation models of epidemics, including households models. Epidemics 10:63–67

Bansal S, Grenfell BT, Meyers LA (2007) When individual behaviour matters: homogeneous and network models in epidemiology. J R Soc Interface 4:879–891. https://doi.org/10.1098/rsif.2007.1100

Begon M, Bennett M, Bowers RG et al (2002) A clarification of transmission terms in host-microparasite models: numbers, densities and areas. Epidemiol Infect 129:147–153

Berg SS (2016) Modeling and conservation of wildlife populations in managed landscapes: a trade-off between effort and results. Ph.D. dissertation, University of Minnesota

Beyer HL, Hampson K, Lembo T et al (2012) The implications of metapopulation dynamics on the design of vaccination campaigns. Vaccine 30:1014–1022. https://doi.org/10.1016/j.vaccine.2011.12.052

Birch CPD, Oom SP, Beecham JA (2007) Rectangular and hexagonal grids used for observation, experiment and simulation in ecology. Ecol Model 206:347–359

Böhm M, Hutchings MR, White PCL (2009) Contact networks in a wildlife-livestock host community: identifying high-risk individuals in the transmission of bovine TB among badgers and cattle. PLoS One 4:e5016. https://doi.org/10.1371/journal.pone.0005016

Bradley CA, Altizer S (2005) Parasites hinder monarch butterfly flight: implications for disease spread in migratory hosts: parasites and monarch flight. Ecol Lett 8:290–300. https://doi.org/10.1111/j.1461-0248.2005.00722.x

Broadfoot JD, Rosatte RC, O'Leary DT (2001) Raccoon and skunk population models for urban disease control planning in Ontario, Canada. Ecol Appl 11:295–303

Bull CM, Godfrey SS, Gordon DM (2012) Social networks and the spread of *Salmonella* in a sleepy lizard population. Mol Ecol 21:4386–4392. https://doi.org/10.1111/j.1365-294X.2012.05653.x
Caillaud D, Craft ME, Meyers LA (2013) Epidemiological effects of group size variation in social species. J R Soc Interface 10:20130206–20130206. https://doi.org/10.1098/rsif.2013.0206
Conner MM, Miller MW (2004) Movement patterns and spatial epidemiology of a prion disease in mule deer population units. Ecol Appl 14:1870–1881
Corner LAL, Pfeiffer DU, Morris RS (2003) Social-network analysis of *Mycobacterium bovis* transmission among captive brushtail possums (*Trichosurus vulpecula*). Prev Vet Med 59:147–167. https://doi.org/10.1016/S0167-5877(03)00075-8
Craft ME (2015) Infectious disease transmission and contact networks in wildlife and livestock. Philos Trans R Soc B Biol Sci 370:20140107–20140107. https://doi.org/10.1098/rstb.2014.0107
Craft ME, Caillaud D (2011) Network models: an underutilized tool in wildlife epidemiology? Interdiscip Perspect Infect Dis 2011:1–12. https://doi.org/10.1155/2011/676949
Craft ME, Volz E, Packer C, Meyers LA (2010) Disease transmission in territorial populations: the small-world network of Serengeti lions. J R Soc Interface 8:776–786. https://doi.org/10.1098/rsif.2010.0511
Croft DP, Edenbrow M, Darden SK et al (2011) Effect of gyrodactylid ectoparasites on host behaviour and social network structure in guppies *Poecilia reticulata*. Behav Ecol Sociobiol 65:2219–2227. https://doi.org/10.1007/s00265-011-1230-2
Davis S, Trapman P, Leirs H et al (2008) The abundance threshold for plague as a critical percolation phenomenon. Nature 454:634–637. https://doi.org/10.1038/nature07053
Deal B, Farello C, Lancaster M et al (2004) A dynamic model of the spatial spread of an infectious disease: the case of fox rabies in Illinois. In: Costanza R, Voinov A (eds) Landscape simulation modeling: a spatially explicit, dynamic approach. Springer, New York, NY
Delahay RJ, Smith GC, Hutchings MR (eds) (2009) Management of disease in wild mammals. Springer, New York, NY
Dobson A, Meagher M (1996) The population dynamics of brucellosis in the Yellowstone National Park. Ecology 77:1026–1036
Drewe JA (2009) Who infects whom? Social networks and tuberculosis transmission in wild meerkats. Proc R Soc B Biol Sci 277:633–642. https://doi.org/10.1098/rspb.2009.1775
Erb J, Sampson B, Coy P (2013) Survival and causes of mortality for fisher and marten in Minnesota. Minn Dep Nat Resour Summ Wildl Res Find 2013:112–121
Fenner AL, Godfrey SS, Michael Bull C (2011) Using social networks to deduce whether residents or dispersers spread parasites in a lizard population: social networks and parasite transmission. J Anim Ecol 80:835–843. https://doi.org/10.1111/j.1365-2656.2011.01825.x
Ferguson NM, Donnely CA, Anderson RM (2001) The foot-and-mouth epidemic in Great Britain: pattern of spread and impact of interventions. Science 292:1155–1160. https://doi.org/10.1126/science.1059188
Frost SDW, Pybus OG, Gog JR et al (2015) Eight challenges in phylodynamic inference. Epidemics 10:88–92. https://doi.org/10.1016/j.epidem.2014.09.001
Fulford GR, Roberts MG, Heesterbeek JAP (2002) The metapopulation dynamics of an infectious disease: tuberculosis in possums. Theor Popul Biol 61:15–29. https://doi.org/10.1006/tpbi.2001.1553
Garnett GP (2005) Role of herd immunity in determining the effect of vaccines against sexually transmitted disease. J Infect Dis 191:S97–S106
Gibson GJ (1997) Markov chain Monte Carlo methods for fitting spatiotemporal stochastic models in plant epidemiology. Appl Stat 46:215–233
Gilbert L, Norman RA, Laurenson MK et al (2001) Disease persistence and apparent competition in a three-host community: an empirical and analytical study of large-scale, wild populations. J Anim Ecol 70:1053–1061

Godfrey SS (2013) Networks and the ecology of parasite transmission: a framework for wildlife parasitology. Int J Parasitol Parasites Wildl 2:235–245. https://doi.org/10.1016/j.ijppaw.2013.09.001

Godfrey SS, Bull CM, James R, Murray K (2009) Network structure and parasite transmission in a group living lizard, the gidgee skink, *Egernia stokesii*. Behav Ecol Sociobiol 63:1045–1056. https://doi.org/10.1007/s00265-009-0730-9

Godfrey SS, Moore JA, Nelson NJ, Bull CM (2010) Social network structure and parasite infection patterns in a territorial reptile, the tuatara (*Sphenodon punctatus*). Int J Parasitol 40:1575–1585. https://doi.org/10.1016/j.ijpara.2010.06.002

Gremillion-Smith C, Woolf A (1988) Epizootiology of skunk rabies in North America. J Wildl Dis 24:620–626

Griffin RH, Nunn CL (2011) Community structure and the spread of infectious disease in primate social networks. Evol Ecol 26:779–800. https://doi.org/10.1007/s10682-011-9526-2

Hamede RK, Bashford J, McCallum H, Jones M (2009) Contact networks in a wild Tasmanian devil (*Sarcophilus harrisii*) population: using social network analysis to reveal seasonal variability in social behaviour and its implications for transmission of devil facial tumour disease. Ecol Lett 12:1147–1157. https://doi.org/10.1111/j.1461-0248.2009.01370.x

Hamede R, Bashford J, Jones M, McCallum H (2012) Simulating devil facial tumour disease outbreaks across empirically derived contact networks: simulating DFTD outbreaks in contact networks. J Appl Ecol 49:447–456. https://doi.org/10.1111/j.1365-2664.2011.02103.x

Hanski I (1998) Metapopulation dynamics. Nature 396:41–49

Hanski I (1999) Metapopulation ecology. Oxford University Press, Oxford

Hess GR (1994) Conservation corridors and contagious disease: a cautionary note. Conserv Biol 8:256–262

Hess GR (1996) Disease in metapopulation models: implications for conservation. Ecology 77:1617–1632

Hosseini PR, Dhondt AA, Dobson A (2004) Seasonality and wildlife disease: how seasonal birth, aggregation and variation in immunity affect the dynamics of *Mycoplasma gallisepticum* in house finches. Proc R Soc B Biol Sci 271:2569–2577. https://doi.org/10.1098/rspb.2004.2938

Humphries MD, Gurney K (2008) Network "small-world-ness": a quantitative method for determining canonical network equivalence. PLoS One 3:e0002051. https://doi.org/10.1371/journal.pone.0002051

Ireland JM, Norman RA, Greenman JV (2004) The effect of seasonal host birth rates on population dynamics: the importance of resonance. J Theor Biol 231:229–238. https://doi.org/10.1016/j.jtbi.2004.06.017

Jackson A, Wunner W (eds) (2002) Rabies. Academic, London

Jesse M, Ezanno P, Davis S, Heesterbeek JAP (2008) A fully coupled, mechanistic model for infectious disease dynamics in a metapopulation: movement and epidemic duration. J Theor Biol 254:331–338. https://doi.org/10.1016/j.jtbi.2008.05.038

Ji W, White PCL, Clout NM (2005) Contact rates between possums revealed by proximity data loggers. J Appl Ecol 42:595–604

Källen A, Arcuri P, Murray JD (1985) A simple model for the spatial spread and control of rabies. J Theor Biol 116:377–393

Kao RR (2003) The impact of local heterogeneity on alternative control strategies for foot-and-mouth disease. Proc R Soc B Biol Sci 270:2557–2564. https://doi.org/10.1098/rspb.2003.2546

Kawata Y (2010) Susceptible-infective-recovered type epidemiological models in wild animal management: a literature review. Res Bull Obhiro Univ 31:1–12

Keeling MJ (2001) Dynamics of the 2001 UK foot and mouth epidemic: stochastic dispersal in a heterogeneous landscape. Science 294:813–817. https://doi.org/10.1126/science.1065973

Keeling MJ, Eames KTD (2005) Networks and epidemic models. J R Soc Interface 2:295–307. https://doi.org/10.1098/rsif.2005.0051

Keeling MJ, Rohani P (2008) Modeling infectious diseases in humans and animals. Princeton University Press, Princeton, NJ

Keeling MJ, Woolhouse MEJ, May RM et al (2003) Modelling vaccination strategies against foot-and-mouth disease. Nature 421:136–142

Keeling MJ, Danon L, Vernon MC, House TA (2010) Individual identity and movement networks for disease metapopulations. Proc Natl Acad Sci 107:8866–8870

Keller SM, Gabriel M, Terio KA et al (2012) Canine distemper in an isolated population of fishers (*Martes pennanti*) from California. J Wildl Dis 48:1035–1041. https://doi.org/10.7589/2011-12-350

Kermack WO, McKendrick AG (1927) A contribution to the mathematical theory of epidemics. Proc R Soc Lond Math Phys Eng Sci 115:700–721

Kjaer LJ (2010) Individual-based modeling of white-tailed deer (*Odocoileus virginianus*) movements and epizootiology. Ph.D. Dissertation, Southern Illinois University

Korfiatis N, Sicilia M-A (2007) Social measures and flexible navigation on online contact networks. In: Proceedings of the IEEE international conference on fuzzy systems (FUZZ-IEEE). IEEE, London, pp 1–6

Langlois JP, Fahrig L, Merriam G, Artsob H (2001) Landscape structure influences continental distribution of hantavirus in deer mice. Landsc Ecol 16:255–266

Laurenson MK, Norman RA, Gilbert L et al (2003) Identifying disease reservoirs in complex systems: mountain hares as reservoirs of ticks and louping-ill virus, pathogens of red grouse. J Anim Ecol 72:177–185

Leu ST, Kappeler PM, Bull CM (2010) Refuge sharing network predicts ectoparasite load in a lizard. Behav Ecol Sociobiol 64:1495–1503. https://doi.org/10.1007/s00265-010-0964-6

Levins R (1969) Some demographic and genetic consequences of environmental heterogeneity for biological control. Bull Entomol Soc Am 15:237–240

Lloyd-Smith JO, Getz WM, Westerhoff HV (2004) Frequency-dependent incidence in models of sexually transmitted diseases: portrayal of pair-based transmission and effects of illness on contact behaviour. Proc R Soc B Biol Sci 271:625–634. https://doi.org/10.1098/rspb.2003.2632

López G, López-Parra M, Fernández L et al (2009) Management measures to control a feline leukemia virus outbreak in the endangered Iberian lynx. Anim Conserv 12:173–182. https://doi.org/10.1111/j.1469-1795.2009.00241.x

McCallum H (2016) Models for managing wildlife disease. Parasitology 143:805–820

McCallum H, Barlow N, Hone J (2001) How should pathogen transmission be modelled? Trends Ecol Evol 16:295–300

McNeil WH (1998) Plagues and peoples. Anchor, New York

Miksch F, Haim C, Schneckenreither G, Breite-necker F (2013) Comparison of differential equations and cellular automata for epidemic simulation. ERK 2013 Proc 22:137–140

Moller AP, Dufva R, Allander K (1993) Parasites and the evolution of host social behavior. Adv Study Behav 22:65–102

Mollison D, Kuulasmaa K (1985) Spatial epidemic models: theory and simulations. In: Bacon PJ (ed) Population dynamics of rabies in wildlife. Academic, London

Moore DA (1999) Spatial diffusion of raccoon rabies in Pennsylvania, USA. Prev Vet Med 40:19–32

Moreno Y, Nekovee M, Pacheco AF (2004) Dynamics of rumor spreading in complex networks. Phys Rev E 69:66130

Munoz-Zanzi C, Mason M, Encina C et al (2014) Household characteristics associated with rodent presence and Leptospira infection in rural and urban communities from southern Chile. Am J Trop Med Hyg 90:497–506. https://doi.org/10.4269/ajtmh.13-0334

Murray JD, Stanley EA, Brown DL (1986) On the spatial spread of rabies among foxes. Proc R Soc B Biol Sci 229:111–150. https://doi.org/10.1098/rspb.1986.0078

Newman ME (2002) Spread of epidemic disease on networks. Phys Rev E 66:16128

Nishiura H, Klinkenberg D, Roberts M, Heesterbeek JAP (2009) Early epidemiological assessment of the virulence of emerging infectious diseases: a case study of an influenza pandemic. PLoS One 4:e6852

Packer C, Holt RD, Hudson PJ et al (2003) Keeping the herds healthy and alert: implications of predator control for infectious disease. Ecol Lett 6:797–802. https://doi.org/10.1046/j.1461-0248.2003.00500.x

Pellis L, Ball F, Bansal S et al (2015) Eight challenges for network epidemic models. Epidemics 10:58–62. https://doi.org/10.1016/j.epidem.2014.07.003

Plowright RK, Foley P, Field HE et al (2011) Urban habituation, ecological connectivity and epidemic dampening: the emergence of Hendra virus from flying foxes (*Pteropus* spp.). Proc R Soc B Biol Sci 278:3703–3712. https://doi.org/10.1098/rspb.2011.0522

Poggi S, Neri FM, Deytieux V et al (2013) Percolation-based risk index for pathogen invasion: application to soilborne disease in propagation systems. Phytopathology 103:1012–1019

Pons-Salort M, Serra-Cobo J, Jay F et al (2014) Insights into persistence mechanisms of a zoonotic virus in bat colonies using a multispecies metapopulation model. PLoS One 9:e95610. https://doi.org/10.1371/journal.pone.0095610

Ramsey D, Spencer N, Caley P et al (2002) The effects of reducing population density on contact rates between brushtail possums: implications for transmission of bovine tuberculosis. J Appl Ecol 39:806–818

Ramsey DSL, O'brien DJ, Cosgrove MK et al (2014) Forecasting eradication of bovine tuberculosis in Michigan white-tailed deer: forecasting bTB in white-tailed deer. J Wildl Manag 78:240–254. https://doi.org/10.1002/jwmg.656

Reynolds JJH, Hirsch BT, Gehrt SD, Craft ME (2015) Raccoon contact networks predict seasonal susceptibility to rabies outbreaks and limitations of vaccination. J Anim Ecol 84:1720–1731. https://doi.org/10.1111/1365-2656.12422

Riley S, Eames K, Isham V et al (2015) Five challenges for spatial epidemic models. Epidemics 10:68–71. https://doi.org/10.1016/j.epidem.2014.07.001

Rushmore J, Caillaud D, Matamba L et al (2013) Social network analysis of wild chimpanzees provides insights for predicting infectious disease risk. J Anim Ecol 82:976–986. https://doi.org/10.1111/1365-2656.12088

Ryder JJ, Miller MR, White A et al (2007) Host-parasite population dynamics under combined frequency- and density-dependent transmission. Oikos 116:2017–2026. https://doi.org/10.1111/j.2007.0030-1299.15863.x

Sayers BM, Ross AJ, Saengcharoenrat P, Mancourian BG (1985) Pattern analysis of the case occurrences of fox rabies in Europe. In: Bacon PJ (ed) Population dynamics of rabies in wildlife. Academic, London

Schofield P (2002) Spatially explicit models of Turelli-Hoffmann Wolbachia invasive wave fronts. J Theor Biol 215:121–131. https://doi.org/10.1006/jtbi.2001.2493

Sloot P, Ivanov S, Boukhanovsky A et al (2008) Stochastic simulation of HIV population dynamics through complex network modelling. Int J Comput Math 85:1175–1187

Small M, Tse CK (2005) Small world and scale free model of transmission of SARS. Int J Bifurc Chaos 15:1745–1755

Smith GC, Cheeseman CL (2002) A mathematical model for the control of diseases in wildlife populations: culling, vaccination and fertility control. Ecol Model 150:45–53

Smith DL, Lucey B, Waller LA et al (2002) Predicting the spatial dynamics of rabies epidemics on heterogeneous landscapes. Proc Natl Acad Sci 99:3668–3672

Swinton J, Harwood J, Grenfell BT, Gilligan CA (1998) Persistence thresholds for phocine distemper virus infection in harbour seal *Phoca vitulina* metapopulations. J Anim Ecol 67:54–68

Thompson D, Muriel P, Russell D et al (2002) Economic costs of the foot and mouth disease outbreak in the United Kingdom in 2001. Rev Sci Tech-Off Int Epizoot 21:675–685

Thorne ET, William ES (1988) Disease and endangered species: the black-footed ferret as a recent example. Conserv Biol 2:66–74

Tischendorf L, Thulke H-H, Staubach C et al (1998) Chance and risk of controlling rabies in large-scale and long-term immunized fox populations. Proc R Soc B Biol Sci 265:839–846. https://doi.org/10.1098/rspb.1998.0368

Tompkins DM, Dunn AM, Smith MJ, Telfer S (2011) Wildlife diseases: from individuals to ecosystems: ecology of wildlife diseases. J Anim Ecol 80:19–38. https://doi.org/10.1111/j.1365-2656.2010.01742.x

Van Baalen M, Rand DA (1998) The unit of selection in viscous populations and the evolution of altruism. J Theor Biol 193:631–648

Volz E, Meyers LA (2007) Susceptible-infected-recovered epidemics in dynamic contact networks. Proc R Soc B Biol Sci 274:2925–2934. https://doi.org/10.1098/rspb.2007.1159

Volz E, Meyers LA (2009) Epidemic thresholds in dynamic contact networks. J R Soc Interface 6:233–241. https://doi.org/10.1098/rsif.2008.0218

Webb CT, Brooks CP, Gage KL, Antolin MF (2006) Classic flea-borne transmission does not drive plague epizootics in prairie dogs. Proc Natl Acad Sci U S A 103:6236–6241

White PCL, Harris S (1995) Bovine tuberculosis in badger (*Meles meles*) populations in southwest England: the use of a spatial stochastic simulation model to understand the dynamics of the disease. Philos Trans R Soc B Biol Sci 349:391–413. https://doi.org/10.1098/rstb.1995.0126

White LA, Forester JD, Craft ME (2015) Using contact networks to explore mechanisms of parasite transmission in wildlife: contact networks: wildlife parasite transmission. Biol Rev. https://doi.org/10.1111/brv.12236

Xie F, Horan RD (2009) Disease and behavioral dynamics for brucellosis control in elk and cattle in the Greater Yellowstone Area. J Agric Resour Econ 34:11–33

Yachi S, Kawasaki K, Shigesada N, Teramoto E (1989) Spatial patterns of propagating waves of fox rabies. Forma 4:3–12

Yip PSF, Watson R, Chen Q (2007) Estimation of vaccine efficacy and the vaccination threshold. Stat Med 26:4475–4488. https://doi.org/10.1002/sim.2874

Chapter 8
The Ecology of Pathogen Spillover and Disease Emergence at the Human-Wildlife-Environment Interface

Kathleen A. Alexander, Colin J. Carlson, Bryan L. Lewis, Wayne M. Getz, Madhav V. Marathe, Stephen G. Eubank, Claire E. Sanderson, and Jason K. Blackburn

Abstract Novel diseases are increasingly emerging into human populations through the complex—and often, unseen—stepwise process of spillover from a combination of wildlife, livestock, vectors, and the abiotic environment. Characterizing and modeling the spillover interface are a key part of how eco-epidemiologists respond to the growing global burden of emerging infectious diseases; but the diversity of pathogen life cycles and transmission modes poses a complex challenge for ecologists and clinicians alike. We review our current understanding of the spillover process and present a framework that relates spillover rates and human-to-human transmissibility to the basic reproduction number (R_0). Using pathogens that exemplify important transmission pathways (anthrax, Ebola, influenza, and Zika), we illustrate key aspects of the spillover interface and discuss implications to public health and management of emerging infectious disease.

K. A. Alexander (✉) · C. E. Sanderson
Department of Fish and Wildlife Conservation, Virginia Tech, Blacksburg, VA, USA
e-mail: kathyalx@vt.edu; clairees@vt.edu

C. J. Carlson · W. M. Getz
Department Environmental Science, Policy and Management, UC Berkeley, Berkeley, CA, USA
e-mail: cjcarlson@berkeley.edu; wgetz@berkeley.edu

B. L. Lewis
Department of Population Health Sciences, Department of Computer Science, Biocomplexity Institute, Virginia Tech, Blacksburg, VA, USA
e-mail: blewis@vt.edu

M. V. Marathe · S. G. Eubank
Network Dynamics and Simulation Sciences Laboratory, Biocomplexity Institute, Virginia Tech, Blacksburg, VA, USA
e-mail: mmarathe@vbi.vt.edu; seubank@vbi.vt.edu

J. K. Blackburn
Department of Geography, University of Florida, Gainesville, FL, USA
e-mail: jkblackburn@ufl.edu

C. J. Hurst (ed.), *The Connections Between Ecology and Infectious Disease*, Advances in Environmental Microbiology 5,
https://doi.org/10.1007/978-3-319-92373-4_8

8.1 Introduction

The spread of environmental or animal pathogens into human populations is a natural and perpetual process that dates back as far as human records extend. One of the earliest recorded epidemics, the plague of Athens (also called the Thucydides syndrome), took place between 430 and 425 B.C.E. and is believed to have been an outbreak of Ebola virus (Olson et al. 1996) or typhoid fever (Papagrigorakis et al. 2006). Measles emerged in human populations after evolving from rinderpest, a disease of cattle, sometime around the eleventh or twelfth century, and perhaps the most famous disease outbreak in recorded history, the Black Death, was caused by plague (*Yersinia pestis*), a vector-borne zoonosis. In this sense, the emergence of diseases is a natural process, the by-product of pathogen "chatter" between susceptible host populations of different species, with the emergence of diseases into human populations only a small subset of the broader cross talk among mammals, and animals writ large.

This process of spillover is stochastic and, from the perspective of pathogen success, fairly imperfect. The majority of mammalian viruses lack the capacity to even infect humans, and the majority of spillover events that do occur fail to establish circulating infection in the new host. But despite the long and complex history of peoples and plagues, scientific consensus indicates that the rate of infectious disease emergence has been accelerating in more recent history. Differentiating the next pandemic from background noise is a nearly insurmountable challenge and one that is ongoing as new pathogens evolve or are discovered. As participants in clinically relevant research settings, the task of ecologists studying spillover revolves around three key issues:

1. Describing the process of spillover and interactions and state changes between hosts, pathogens, and the environment that lead to disease emergence
2. Developing predictive conceptual and quantitative models that can explain spillover patterns and forecast disease emergence, epidemic behavior, and/or intervention impact
3. Applying ecological and clinical knowledge to develop interventions that mitigate, control, and potentially prevent epidemics and pandemics

Each of these independently poses a substantial challenge. The diversity of pathogens, transmission modes, and eco-epidemiological interfaces further complicates the development of a conceptual framework that will adequately address any one of these challenges, let alone all three.

In this chapter, we review existing research on the spillover interface and the relevant ecological advances that have been made especially in the last two decades. We begin by reviewing the most recent work characterizing the process of pathogen emergence. Building on this, we present a conceptual framework that engages the stages of pathogen emergence as a dynamic continuum rather than a binary, or even linear, set of sequential steps. We use this framework to evaluate a number of pathogens that illustrate key features important to the spillover process. We conclude by discussing how this approach lends itself to a more quantitative and focused assessment of spillover itself—the key element shaping pathogen emergence potential at the human-animal-environmental interface.

8.2 The Usual Suspects: What Makes an Emerging Zoonosis?

Zoonotic disease emergence arises as a consequence of transmission of a multi-host pathogen from an animal host (direct or indirect transmission) to a human, with or without onward human-to-human transmission and establishment (Antia et al. 2003). We define a **zoonotic reservoir** as one or more epidemiologically connected animal populations and/or environments where a pathogen can be maintained and transmitted to humans (Haydon et al. 2002) and, more broadly, domestic animals as well. Regardless of transmission mode, the process by which a pathogen moves from one host population (or environmental reservoir) to another host population is referred to as **spillover** and arises from complex bidirectional interactions among people, animals, pathogen communities, and environments. This is a key step in zoonotic disease emergence but remains poorly understood and often not explicitly quantified in transmission models (Lloyd-Smith et al. 2009; Iacono et al. 2016), which tend to focus on dynamics within a single population.

Some of the most significant quantitative work at the spillover interface has focused not on predicting rates of zoonotic spillover but on identifying and predicting the common features of spillover diseases and the pathways they tend to exploit. Viruses have been best studied in this context; basic rules of thumb are well established, such that single-stranded RNA viruses are predisposed to become zoonoses, likely due to a predisposition toward evolvability and host switching. Other rules have emerged from studying the networks of viral transmission between different host groups and humans. For example, a study by Johnson et al. (2015) demonstrated a number of key rules, perhaps most importantly that viruses with greater host plasticity in animal reservoirs are more likely to demonstrate human-to-human transmissibility. Some virus families, such as arenaviruses and filoviruses, are predisposed to human-to-human transmission. Correspondingly, some groups of mammals were especially common reservoirs of zoonotic viruses, including rodents (especially for arena- and bunyaviruses), primates (retroviruses), and bats (paramyxoviruses and rhabdoviruses). Pathways of spillover were also noted to differ, with bush meat hunting being a notable source of exposure to nonhuman primate viruses, while proximity of homes to fields was a significant driver for exposure to rodent-borne viruses (Dearing and Dizney 2010). Although sometimes not considered a "pathway of spillover," vectors can also act as a critical component in spillover dynamics with, for example, Johnson et al. showing that virus spillover from bird reservoirs was disproportionately vector borne.

A recent study by Olival et al. (2017) refined and elaborated these findings, confirming that RNA viruses are more commonly zoonotic and that the phylogenetic breadth of viral hosts is the strongest predictor of zoonotic potential (analogous to

Johnson's findings about viral plasticity). Olival et al. also confirmed that bat viruses were disproportionately zoonotic, an unsurprising finding given that many well-known viruses have originated from bat reservoirs [e.g., Ebola and Marburg viruses, Hendra and Nipah viruses, and even severe acute respiratory syndrome (SARS) and, likely, Middle East respiratory syndrome (MERS)]. Olival et al. confirmed that viruses with arthropod vectors have a greater plasticity in their mammalian hosts and showed that viral replication in the cytoplasm also predicts zoonotic potential. Finally, their study shows that human population density in host ranges and the phylogenetic relatedness of wildlife reservoirs to humans were important predictors of zoonotic emergence. These aspects are likely related to the evolutionary processes that occur at the spillover interface, where viruses coevolve with new hosts and incrementally develop greater emergence potential over time (which we discuss below).

Studies like Johnson's and Olival's are a key part of how disease ecologists consolidate information about the emergence of zoonotic diseases and reveal important information about the patterns—and, ideally, processes—that transpire at the spillover interface. Given the recent preponderance of devastating viral outbreaks, such as Ebola and Zika, it is unsurprising that so much attention has been paid to the factors that drive viral emergence. Comparatively less work has been done to consolidate these types of results for other major pathogen groups, like bacteria, protozoans, and fungi. While broad studies like these elucidate general patterns and help develop predictive tools for identifying the zoonotic potential of recently discovered pathogens, these studies do little to explain outbreak process and spillover dynamics for known zoonoses.

8.3 Characterizing Pathogen Spillover

The diversity of existing emerging pathogens can be overwhelming, with over 300 documented emerging infectious diseases occurring between 1940 and 2004 (Jones et al. 2008), and several more, which have since emerged. Greater still, it must be expected, is the diversity of potential zoonotic threats that have not yet emerged into human populations but could do so in the coming years as viruses, hosts, and landscapes experience accelerated change. Frameworks provide an opportunity to generalize patterns across pathogens, supporting model development and interventions that can be especially valuable for diagnosing epidemic profiles and responding during the early days of a pathogen's emergence. Several frameworks have been developed that organize and describe pathogens based on stages of emergence, but the spillover interface is still poorly captured for many diseases, and most frameworks have limited direct applicability to quantitative methods. In this section, we review existing frameworks and, building on this, present our own framework that is explicitly directed at modeling the unique mathematical behavior of pathogen spillover.

8.3.1 Existing Frameworks

In contemporary disease ecology literature, two major frameworks have been proposed that describe the stages of pathogen emergence. The first, and most commonly referenced, was proposed by Wolfe et al. (2007) and divides pathogen emergence into five progressive stages through which a pathogen may pass sequentially:

Wolfe's Stages of Spillover

I. Agent only in animals
II. Primary infection in humans
III. Limited outbreak in humans
IV. Long outbreak in humans
V. Exclusive human agent

In this classification scheme, zoonotic emergence occurs between Stages II and IV. In the most constrained case (Stage II), spillover only ever leads to primary infection in humans with no onward transmission within human populations. Whereas Stage III is characterized by short chains of transmission ("stuttering chains"), Stage IV features human-to-human sustained epidemic transmission. The line between the two is obviously not only subjective but may create difficulties when characterizing expected pathogen behavior. For example, outbreaks of Ebola virus beginning in 1976 were typically small and situated in rural areas where outbreaks could have been classified as Stage III. However, since 2014, Ebola has been considered a Stage IV zoonosis, and modeling work undertaken during the 2014–2015 Ebola epidemic in West Africa showed that "the same epidemiological conditions that were present in 1976 could have generated a large outbreak purely by chance" (Camacho et al. 2014). The line, therefore, between short-chain and epidemic transmission for a pathogen is sometimes unclear.

A second framework, proposed by Lloyd-Smith et al. (2009), refines the Wolfe classification scheme by more formally considering the mathematical definitions of the three intermediate categories:

II. Spillover into humans (no human-to-human transmission; $R_0 = 0$)
III. Stuttering transmission in humans ($R_0 \leq 1$)
IV. Long outbreak in humans ($R_0 > 1$)

Here, R_0 is defined as the number of secondary infections caused by a typical single infective individual in a wholly susceptible population during its period of infectiousness (Diekmann et al. 1990). Incorporating an explicitly quantitative aspect to the definition of the stages improves the framework in obvious ways and facilitates the more direct integration of epidemiological modeling and zoonosis research. However, the same data limitations pose a problem for this framework, as a fairly newly emerged zoonosis is unlikely to have enough accompanying surveillance data to characterize its basic R_0, let alone elucidate situational variance due to socioeconomic and ecological factors. Later, we will discuss modeling approaches that can resolve some of these problems.

What do these frameworks contribute to disease ecology? At the most basic level, frameworks for classifying pathogens help us abstract and generalize patterns across hosts, pathogens, and landscapes. Furthermore, developing a more standardized language for describing spillover and disease emergence helps bridge the gap between ecology, infectious disease research, and clinical and public health. With the increasing burden of emerging infectious diseases and the accelerating rate of emergence of novel zoonoses, frameworks for classifying zoonotic disease emergence can help develop and refine prioritization schemes.

8.3.2 Advancing the Spillover Framework

The past few decades of ecological research have shown that pathogen life history is far more flexible and dynamic than any linear set of steps characterizing emergence might capture. A framework oriented on a sequential progression can be useful for conceptualizing the emergence of some pathogens like measles or human immunodeficiency virus (HIV); but the diversity of emerging zoonoses contains many more life histories. Moreover, the development of a framework with an inherent directionality risks "adaptationism," misconstruing the random evolutionary trajectories of pathogens that include humans as a (non-special) part of a broader ecosystem including wildlife, vectors, livestock, and even the soil and water. To improve our understanding of spillover into human populations, we must first better conceptualize the complicated dynamics of interspecies spillover (human or not; Fig. 8.1).

As discussed previously, spillover and emergence of zoonotic pathogens have generally been represented within a framework that focuses on the objects of systems (hosts, pathogens, entities in the environment) (Daszak et al. 2000; Childs et al. 2007) rather than the interacting and cascading systems' processes themselves. It is the active steps of emergence (exposure, contact, invasion, and onward transmission), influenced by host-pathogen evolutionary processes, that ultimately determine the outcome of interactions among entities. In short, spillover and emergence are stochastic processes with outcomes that depend on the probabilities of the occurrence of underlying events. As we will make clear, the variability associated with spillover and emergence is large because the size of groups involved (vis-à-vis number of infected individuals) is small and hence subjects to the vagaries of sampling: it is only after sufficiently many of these events have been observed that we can begin to meaningfully characterize spillover and emergence in terms of expected rates of pathogen transmission. Using our framework, three key processes are delineated that determine (1) the occurrence of pathogen spillover from reservoirs either to an intermediary or directly to a focal host; (2) the occurrence of primary transmission from an intermediary to a focal host, when intermediary hosts

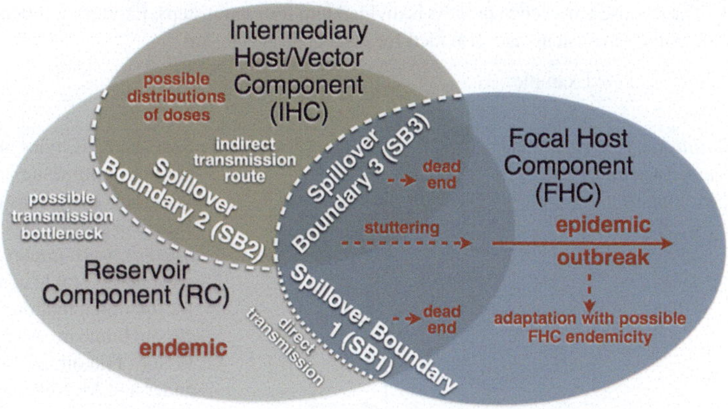

Fig. 8.1 Conceptual model of the process of pathogen spillover and emergence. The ellipses represent three conceptual components: reservoir (RC, either environmental or wildlife populations serving as an ecological system in which the pathogen is endemically maintained), intermediary host/vector (IHC) population that serves to transfer the pathogen from the reservoir to the focal host, and focal host (FHC), generally human or species of special concern (e.g., domestic or highly threatened species). The dotted lines represent three spillover boundaries between RC-FHC (SB1 direct transmission route), RC-IHC (SB2), and IHC-FHC (SB3) (the concatenation of these latter two constitutes the indirect transmission route)

play a role (which is in most cases); and (3) the occurrence of secondary focal host-to-host transmission required for true disease emergence in the focal host. We use the term true emergence to contrast with the observation of multiple focal host cases of an infectious disease arising exclusively from spillover (Antia et al. 2003). Using emerging zoonotic disease examples, we explore this framework and provide recommendations for evaluating zoonotic outbreaks, managing interventions during outbreaks, and planning future research.

8.3.2.1 Structural Elements

The model has three main compartments: a reservoir component (RC, self-sustaining or endemic source of pathogen that may be purely environmental or involve at least one host species as well), an intermediary host/vector component (IHC) where prevalence would drop to zero if the reservoir were removed (if this component is self-sustaining, then it just becomes part of the reservoir), and a focal host component (FHC) where epidemic outbreaks may occur and the potential for endemic establishment of the pathogen exists through evolutionary processes. Each of these compartments in the model is influenced by various factors and interactions (e.g., factors which influence the state of the host, pathogen, and environment and the resultant interactions; see Table 8.1 for examples), as well as the processes of source-host contact, intensity and duration of exposure, pathogen invasion potential for individual hosts (i.e., interactions between pathogens and the immune system), and onward transmission and adaptation within host populations. We find it insightful to

Table 8.1 The disease emergence process is divided into four main steps. Examples of both factors and zoonotic pathogen systems are provided for each emergence step

Process	Example		
	Factor	Disease	Mechanism
Host exposure (spatial overlap between host and pathogen)	Host behavior	Chagas disease	Sedentary lifestyles and rearing of domestic animals in particular, such as the Brazilian guinea pigs (*Cavia aperea*), were important in creating favorable conditions for the domiciliation (successful home invasion) of the vector (Triatominae) (Alexander and McNutt 2010)
	Habitat alteration and fragmentation	*Borrelia burgdorferi*, Lyme disease	Deforestation and habitat fragmentation results in decreasing mammalian species diversity and increasing population densities of the reservoir with concomitant increased human exposure (Allan et al. 2003)
	Agricultural processes	*Cryptosporidium parvum*	Fecal contamination of water sources in agriculture production systems and food processing (Orlandi et al. 2002)
	Commercial resource use	Ebola	Logging of forests and exposure to virus, commercial sale of bush meat
	Climate	*Yersinia pestis*, plague	El Niño Southern Oscillation events increase the number of *Yersinia pestis* flea vectors and rodent host populations and are linked to prairie dog colony extinction events due to this pathogen (Stapp et al. 2004)
	Movement of infected hosts	*Trypanosoma brucei rhodesiense*, sleeping sickness	Movement of infected cattle is linked to an outbreak of human sleeping sickness in a previously unaffected area of Uganda (Fèvre et al. 2005)
	Seasonal dynamics	Lassa virus	Dry season shifts increases in rodent host densities in houses may increase the risk of contact with rodent host and excreta (Fichet-Calvet et al. 2007)
	Host density	Hantavirus	Increases in rodent populations and density are

(continued)

Table 8.1 (continued)

Process	Example		
	Factor	Disease	Mechanism
			associated with increased contact and pathogen spill-over to humans (Wells et al. 1997)
	Disruption of reservoir host social structure	*Mycobacterium bovis*, bovine tuberculosis	Culling of badgers to control TB exposure induced badger long-distance movement and dispersal (Donnelly et al. 2003)
Contact (nature of contact with the pathogen will determine if pathogen can invade host)	Predation of infected hosts	Bluetongue virus	Bluetongue virus, a vector-borne pathogen, infects ruminants, shrews, and some rodent species. Ingestion of infected prey allows virus transmission to African predators without vector involvement (Alexander et al. 1994)
	Predation of infected hosts	African horse sickness virus	African horse sickness is a vector-borne disease that principle affects equids. Domestic dog and African predators are thought to be infected from ingestion of infected equid species (Alexander et al. 1995)
	Diet	Ebola	Hosts that eat certain fruits are brought together where transmission of the virus can occur from the bat reservoir through saliva contamination of fruit and ingestion by other species such as gorillas (Dobson 2005)
	Socioeconomic status of human communities	Leptospirosis	Poverty and compromised sanitation infrastructure increase contact with environmental sources of *Leptospira* (Reis et al. 2008)
	Pathogen release and change in immune protection in the human host	Monkeypox virus	Cross-immunity between smallpox and monkeypox virus and cessation of vaccination have allowed host immunologic release and emergence of monkeypox virus in the human host (Rimoin et al. 2010)

(continued)

Table 8.1 (continued)

	Example		
Process	Factor	Disease	Mechanism
Pathogen invasion and adaptation (pathogen enters and replicates in the susceptible host)	Virus evolution and strain variability	Influenza viruses	Influenza interspecies transmission is influenced by strain variability. New strains can evolve when two different viruses infect individual cells. Segments derived from each of the infecting "parents" may reassort and create a new strain with a changed host invasion potential (Webster et al. 1997)
	Host physiology, coinfection	*Cryptococcus neoformans*	Immunosuppression of the human host with HIV/AIDS can increase pathogen invasion risk (Chuck and Sande 1989)
	Coinfection can decrease macroparasite infestations	Nematode infections	The gastrointestinal nematode *Heligmosomoides polygyrus* decreases infestation of the tick *Ixodes ricinus* in free-living yellow-necked mice, *Apodemus flavicollis* (Ferrari, Cattadori et al. 2009)
	Tissue tropism	Hendra and Nipah virus	Viral transmission and spillover is influenced by the tissue tropism of the virus and access to the exterior of the host (Hooper et al. 2001; Childs 2004
	Human behavior	Sudden acute respiratory syndrome (SARS)	Wet markets allow different hosts and viruses to have concentrated contact with each other and human, supporting viral change and host adaptation (Brown 2004)
	Mutation frequency, genetic diversity	Venezuelan equine encephalitis virus	RNA mutations allow production of amplification-competent (high equine viremia) viral strains, which can then invade the human host through vector-mediated transmission (Anishchenko et al. 2006)
	Virus evolution and strain variability	SARS	Genetic variations in critical genes, such as the *Spike* gene

(continued)

Table 8.1 (continued)

Process	Example Factor	Disease	Mechanism
			(protein responsible for host cell receptor binding) influence pathogen invasion potential (Song et al. 2005)
	Human sexual behavior	HIV	Persons at an increased risk of transmitting or being infected by the HIV virus were more likely to practice unprotected sex (Halperin 1999)
Sustained onward transmission in new host(s) (secondary cases of human-human transmission)	Culture and human behavior	Ebola virus	Culturally driven practices such as ritual handwashing and sharing of communal meals at funerals of Ebola-infected individuals have been significantly associated with Ebola viral transmission (Griffin et al. 2003)
	Globalized travel and contact	SARS	Air travel and global traffic facilitate the spread of the pathogen (Ali and Keil 2006)
	Within human host adaptive mutation potential	SARS	Non-synonymous changes in the spike gene are found in viruses with sustained human-to-human transmission. These genetic changes are not found in viruses circulating in the reservoir host (palm civets) or spillover viruses that did not successfully move to the host epidemic space where sustained human-to-human transmission occurs (Pepin et al. 2010)

consider pathogen spillover as typically involving four conceptually distinct, transmission processes: spillover from the reservoir component (RC) directly to the focal host component (FHC) (i.e., spillover boundary 1, SB1), spillover from the RC to the intermediary host/vector component (IHC) (i.e., spillover boundary 2, SB2), spillover from the IHC to the FHC (i.e., spillover boundary 3, SB3), and host-to-host transmission within the FHC itself.

Direct transmission across SB1 occurs between focal susceptible individuals and either individuals from reservoir species, the environment, or fomites (e.g., infected intermediary host fecal material or carcasses). Transmission across SB2 often presents an important bottleneck to pathogens ultimately entering FHC. Transmission across SB3 is generally the focal spillover process, particularly those vectored by mosquitoes and parasites. Transmission across these spillover boundaries may either result in a dead-end infection of a focal host, a stuttering transmission chain in the focal host ($R_0 < 1$ in FHC), or an epidemic outbreak (R_0 necessarily > 1 in FHS). One or more such outbreaks may lead to adaptation and FHC endemicity.

Situations may exist where some focal hosts have been infected across the SB1 or SB3, resulting in mixed transmission. Emergence within the FHC (i.e., a zoonotic outbreak) requires that novel host(s) then transmit the pathogen to other members of the same species, with the degree of onward transmission determining the nature of the epidemic in the FHC. In some cases, pathogens may routinely spillover without resulting in any onward transmission in the FHC. In this case, the novel host is a dead end, and the epidemic is purely driven by spillover events. In humans, we see examples of this with anthrax and rabies. Once a spillover proceeds to include direct focal host-to-host transmission, if $R_0 < 1$, the process ends in a transmission chain that has a geometric distribution with an expected length of $(1 - 1/R_0)$ (distribution of chains that arise from several such transmission events is referred to as stuttering chains, Antia et al. 2003). On the other hand, if $R_0 > 1$, then the probability of an epidemic outbreak in FHC rises steeply with the value of R_0 (Antia et al. 2003).

Finally, it is worth stressing that the reservoir (RC) may itself be a vector-host system with another vector acting as the intermediary (i.e., IHC) between RC and the focal host (FHC). This is the case for West Nile virus (WNV) where RC is maintained by a passerine bird and *Culex* mosquitoes host-vector system but where *Aedes* mosquitoes are an intermediary host/vector component (Kilpatrick 2011; Petersen et al. 2013). Thus, when humans get bitten by WNV-infected *Culex* individuals, transmission can be regarded as direct (i.e., occurring across SB1); but when humans get bitten by WNV-infected *Aedes* individuals, transmission can be regarded as indirect with the pathogen needing to cross both SB2 and SB3 to be transferred from RC to FHC. The reason for this discussion relates to how we ultimately use our model to assess probabilities of outbreaks of zoonotic disease in humans and other focal species of interest.

8.3.2.2 IHC Computational Elements

It is important to note that all models are abstractions. Deciding on the appropriate level of detail to include in a model depends on the questions to be addressed and an understanding of the processes needed in the model to adequately address these questions. In our framework, we assume that little is known about the structure and dynamics of the reservoir. Thus, we simply represent the reservoir in terms of a spatiotemporal risk-of-infection function $R\ (x, t)$, where x is a point in space (typically 2D) and t is a point in time. More ambitiously, we endeavor to represent the intermediary host in terms of both its population density $Y\ (x, t)$ and the prevalence of infection $IY\ (x, t)$ over space and time. Calculation of $PY\ (x, t)$ in terms of $R\ (x, t)$ and $Y\ (x, t)$ represents the real challenge, particularly if the impacts of various critical factors, as elaborated in the sections below, are to be incorporated. The primary focus is on the IHC where most of the interesting dynamics take place prior to an outbreak within the FHC. Once an outbreak begins in the focal host population, then the emphasis switches to the dynamics of infection within the FHC. This will only progress beyond the dead-end or stuttering phases if the pathogen is sufficiently virulent within FHC to have an associated $R_0 > 1$.

As will become clear from our discussion, many different types of zoonotic systems exist, each requiring its own approach to the development of an appropriate model. Here, we focus on general concepts, as well as some novel elements that have not been introduced elsewhere in the literature. The essence of an IHC model is the intermediary population. Without going into detail regarding age, sex, or spatial structure, the dynamics of this population should minimally be described by a model that includes a demographic recruitment process (e.g., births and immigration), a natural mortality process, a pathogen exposure process, a disease class structure (e.g., susceptible S, infected but not yet infectious E, infectious I, recovered with some level of immunity R that may wane over time), a disease-induced mortality process, and a disease progression transfer matrix. The latter may be rather simple when considering transfers from E to I to R and possibly back to S, but the real challenge lies in modeling the transfer of individuals from class S to class E—i.e., the disease transmission process.

Disease transmission can be unpacked as a concatenation of (1) a source/susceptible host contact process that possibly includes notions of the intensity and duration of exposure to a dose of pathogen and (2) the probability of succumbing to infection given the characteristics of the contact. Contact itself requires an understanding of host behavioral and movement ecology, while dose-exposure computations require explicit characterization of the pathogen encounter risk distribution function $R(x, t)$ introduced above. Note that the movement of susceptible individuals in IHC may alter their movement behavior in response to the distribution of $R(x, t)$, as is the case for anthrax. Finally the probability of succumbing to infection given a contact (C) of intensity w and duration τ will depend on the immunological state of the susceptible in IHC. If a susceptible is immunologically naive, then we would expect the probability of succumbing to be given by a function $C(w, \tau)$ that has the following properties: $C(0, \tau) = 0$ for all $\tau \geq 0$ and $C(w, 0) = 0$ for all $w \geq 0$ imply no dose and no transfer; $C(w, \tau) \geq 0$ is an increasing function of w for fixed τ and of τ for fixed w but is constrained to satisfy $C(w, \tau) \leq 1$ for all $w \geq 0$ and $\tau \geq 0$; and, most importantly, $C(w, k)$ is an increasing function of w when $w\tau = k$. The latter implies that for a constant dose k, it is more contagious to be exposed to a higher dose rate (dose per unit time) for a shorter time than a lower dose rate for a longer time. Perhaps the simplest two-parameter function that satisfies these properties is

$$C(w, \tau) = \left(\frac{c_1 w\tau}{1 + c_1 w\tau}\right)\left(\frac{c_2 w}{1 + c_2 w}\right) \tag{8.1}$$

where $c1 > 0$ and $c2 > 0$ are the parameters in question. If a susceptible is not naive, then we can reduce the value of $C(w, \tau)$ accordingly. This function for C is of course a model, and therefore an oversimplification (and should not be taken to suggest that low doses imply negligible exposure), but offers a useful conceptual framework and is a starting point for more complicated epidemiological models.

8.3.2.3 FHC Empirical Elements

Susceptible individuals within FHC have three possible sources of infection: from RC across spillover boundary SB1, from IHC across spillover boundary SB3, and from host-to-host transmission within FHC. With recent advances in RNA and DNA sequencing technology, we now have the ability to trace the lineage of pathogens that have high mutational rates, in particular single-stranded RNA viruses (Duffy et al. 2008) within families such as the Coronaviridae (including SARS-CoV and MERS-CoV), Flaviviridae (e.g., yellow fever, West Nile, dengue, Zika, and hepatitis C viruses), and Filoviridae (Ebola virus and Marburg virus). Monitoring molecular data over time allows an opportunity to assess the spillover rates of viruses across SB1 and SB3. Genetic sequences from pathogens originating in RC and IHC are necessary here but not always available. Contact tracing can also be used in human populations to infer whether transmission is primary (i.e., across SB1 or SB3) or secondary (i.e., human-to-human within FHC). Finally, once secondary transmission has been identified in the FHC, then contact tracing can also be used to construct next-generation distributions where estimates of R_0 can be developed. Only if $R_0 > 1$ can an outbreak occur, and even then if $R_0 > 1$ is close to 1, epidemic fade-out may be more likely than an epidemic outbreak (Parrish et al. 2008; Lloyd-Smith et al. 2005).

The quantity R_0 is often considered the most important element studied in epidemiology, providing critical insight into epidemic behavior (Heesterbeek 2002). The value of R_0 can be reduced by shortening the period of infectiousness, decreasing the rate of new infections, or by increasing pathogen mortality (Hudson et al. 2008). In a full FHC epidemic with eventual fade-out, host factors themselves may drive the process through changes in contact behavior, treatment, and/or other interventions that reduce R_0, leading to fade-out and cessation of the epidemic. For example, interventions such as quarantining infected individuals limited the SARS outbreak in China in 2003 to fewer than 10,000 cases, even though the international travel of infected individuals puts hundreds of millions of individuals at risk (Smith 2006). Alternatively, the pathogen may establish itself in the new host population, sustained by secondary transmission, leading ultimately to endemic infection (e.g., HIV/AIDS; see below) or eventual epidemic fade-out. In the latter case, the pathogen cannot persist in the population as might happen, for example, when the proportion of susceptible individuals has been greatly reduced by the action of the epidemic itself [e.g., 1918 influenza pandemic, "the Spanish flu" (Taubenberger and Morens 2006)]. Ultimately, it is the convergence and interplay of specific factors from the involved compartments and host-pathogen interactions that will determine if pathogen transmission will successfully occur and the nature of the resultant epidemic(s).

8.3.2.4 Factors Influencing the Disease Emergence Process

A unique suite of factors will influence each step of the pathogen invasion process, determining the potential for disease emergence (Table 8.1). As alluded above, pathogen exposure is the fundamental requirement for pathogen spillover in a new host and is influenced by the spatial overlap and density of infected and susceptible hosts. Here, various factors can influence the outcome of exposure. External forcings such as climate or extreme weather events are able to drive host and pathogen distributions and disease occurrence (Alexander et al. 2012b). For example, El Niño Southern Oscillation events have been linked to increases in the number of *Yersinia pestis* flea vectors and rodent host populations, leading to increased pathogen invasion and mortality in prairie dog (*Cynomys ludovicianus*) colonies and colony extinction events (Stapp et al. 2004). Likewise, human behavior can influence domestic animal pathogen exposure to wildlife. For example, cattle herding behavior of dog owners influenced contact between African wild dogs (*Lycaon pictus*) and domestic dogs, exposing the wild dogs to canine pathogens and causing catastrophic declines in wild dog populations (Alexander and McNutt 2010).

While spatiotemporal overlap between a potential host and a pathogen reservoir may occur, the nature of the contact between host and pathogen must be appropriate to support host invasion (Morris et al. 2016). For example, socioeconomic status can influence human exposure to leptospirosis where poverty and compromised sanitation infrastructure increase contact with environmental sources of *Leptospira* (Reis et al. 2008). Similarly, bluetongue virus was historically thought to infect only ruminants, shrews, and some rodent species with pathogen transmission being vector dependent. Bluetongue virus in African predators, however, appears to be related to the ingestion of virus-infected prey with prevalence levels associated with feeding behavior and organ access (Alexander et al. 1994).

Pathogen spillover and replication in the new host can be influenced by the presence of other pathogens (Rigaud et al. 2010). Cross-immunity between pathogens can influence disease outcomes, as seen with smallpox and monkeypox virus, where cessation of smallpox vaccination allowed host immunologic release and emergence of monkeypox virus in the human host (Rimoin et al. 2010). Evolutionary change can also influence pathogenicity of an infectious disease organism or modify host resistance and pathogen invasion potential (Tack et al. 2012). In this case, coevolutionary selection occurs in response to variation in a myriad of processes acting on both the host and pathogen affecting their interactions across space and time. These effects make it difficult to generalize pathogen invasion behavior and predict host-pathogen interaction outcomes.

Pathogen factors pertaining primarily to evolutionary change [neutral drift, coevolution with the host, or adaptive evolution (Antia et al. 2003)] often become critical in determining the final outcome of the invasion in situations where regular outbreaks of the pathogen lead to a strain endemic to the FHC. This is the case for all human diseases that have clear origins in wildlife populations. For example, non-synonymous changes in the Spike gene were found only in SARS viruses where sustained human-to-human transmission occurred. These genetic changes were not found in viruses circulating in the reservoir host (palm civets; *Paradoxurus hermaphroditus*) or in those spillover viruses that did not successfully move to the FHC where sustained human-to-human transmission occurred (Pepin et al. 2010). A more dramatic case is HIV-AIDS that has two different simian sources, resulting in two distinct groups of HIV pathogens: one from gorillas (*Gorilla gorilla*; HIV-1) and the other from sooty mangabeys (*Cercocebus atys*; HIV-2) (Lemey et al. 2003; Keele et al. 2006).

8.3.2.5 Zoonotic Disease Emergence and R_0

Many studies of multi-host pathogen systems have mathematically incorporated the process of spillover (McCormack and Allen 2007; Kilpatrick et al. 2006; Dobson 2004), but failure to monitor spillover rates long term has limited our ability to assess the magnitude of spillover as isolated phenomena from within host species transmission. For example, in diseases such as rabies (Zinsstag et al. 2009) and brucellosis (Zinsstag et al. 2005), where R_0 in humans is zero, we often find calculations of R_0 developed for reservoir populations, without any explicit quantitative consideration of the magnitude of spillover into the FHC. This is also seen in a recent paper that explicitly models dog-to-human transmission for rabies (Zinsstag et al. 2009), including the development of clinical disease in both human and dog populations. R_0 is calculated for dog-to-dog transmission in IHC, but no specification of spillover rates across SB3 is made. This tells us that there is sustained transmission in the dog population, prior to any introduction of interventions; but the estimated dog-to-human contact rate is never used to provide an explicit corresponding estimation of the size of the spillover event. In the context of vector-borne diseases, a similarly approached assessment can be found with *Plasmodium knowlesi* transmission from monkeys to humans through *Anopheles leucosphyrus* mosquitoes, an interesting example of spillover (Cox-Singh and Singh 2008). A recent model of this system (Yakob et al. 2010) could be usefully extended to include the elaboration of spillover values: monkeys to humans and humans back to monkeys, in both cases through mosquitoes. Incorporation of such measures would be useful in understanding and characterizing the dynamic process of disease emergence.

8.3.2.6 R_0, H_T, and Spillover

When transmission has a significant density-dependent component, R_0 is related to the establishment threshold H_T (Diekmann et al. 1990) (but see Lloyd-Smith et al. 2005), defined as the minimum density of susceptible hosts necessary for establishment of the pathogen in a new host population. This threshold may disappear when frequency-dependent transmission predominates (Getz and Pickering 1983), as is the case for sexually transmitted diseases. Establishment of the pathogen will only occur if the necessary threshold density of susceptible individuals is identified and is required to ensure that $R_0 \geq 1$; otherwise, failing to have the necessary density of susceptibles, the pathogen will fade out with $R_0 < 1$. Disease control or eradication efforts are then focused on using this threshold principle to manipulate pathogen fade-out in the new host population through the reduction of susceptibles as, for example, with H1N1 and the use of vaccination and school closures to control the outbreak. However, R_0 and H_T do not apply to transmission across spillover boundaries SB1 and SB3 where primary transmission of zoonotic disease occurs. These concepts are, by definition, only applicable to the transmission within the reservoir or spillover host populations themselves. Further, spillover and onward transmission differ in an important respect: spillover is a series of onetime events, while onward transmission can either lead to stuttering chains, involving small groups of individuals from the animal and human populations interacting in the FHC, or full-blown outbreaks. Consequently, relative to time, the accumulation of spillover events is a linear process coupled to the nonlinear exponential processes of transmission in the reservoir and spillover host populations (Fig. 8.2). To assess the potential for outbreaks in the FHC, we need estimates of both spillover rates across SB1 and SB3, as well as estimates of the R_0 that will ensue once host-to-host transmission is established in the FHC.

By way of illustration, we can examine how public health systems consider zoonotic diseases. Human health objectives are focused on minimizing and preventing human morbidity and mortality associated with zoonotic disease transmission, and control efforts are directed at reducing or even eliminating initial spillover events. This public health perspective highlights the importance of characterizing spillover both qualitatively and quantitatively. Many emerging zoonotic pathogens present as a spillover epidemic only in the human population with $R_0 = 0$ (e.g., anthrax) or a mixed epidemic type where there is spillover with highly inefficient human-to-human transmission ($R_0 < 1$ in the human host, e.g.,

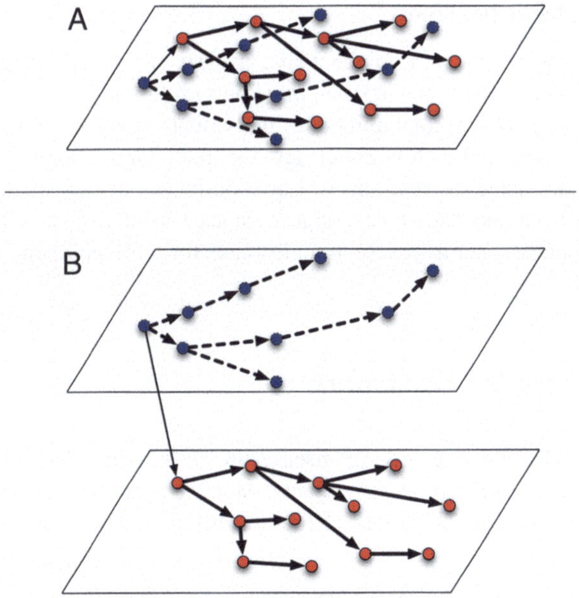

Fig. 8.2 Mixed epidemic dynamics (**a**) shows spillover and onward transmission in a mixed epidemic (spillover epidemic and host epidemic). Blue circles are members of the reservoir population; red, the new host. Edges represent transmission. (**b**) shows exactly the same graph but explicitly separating the two populations. In (**b**), it is obvious that there are two separate processes, linked by a single event. The single spillover event does not lead to a geometrically growing number of spillovers, whereas onward transmission in the new host may. The appropriate interventions and surveillance methods for the two processes are different

monkeypox virus). The immediate focus of control for many emerging zoonoses is not the reduction of infectious animals in the reservoir populations (the reservoir species may not even be known) but minimizing human behaviors and other factors that are thought to contribute to pathogen exposure and invasion risk in humans (e.g., Ebola; see below). In these instances, R_0 in the human or in the reservoir population is not the central factor of interest. Rather, we need to focus on the complex processes of zoonotic pathogen spillover that must occur in the first place. Quantifying spillover rates provides information on the number of cases that can be expected in the FHC over time, a process that is not captured by the traditional R_0 in either the IHC or FHC.

Applying R_0 to our framework suggests that emergence outcomes can therefore be characterized in one of four ways (Lloyd-Smith et al. 2005):

1. The infected host is a dead end—dying and not passing the pathogen on to another human because human-to-human transmission is not possible ($R_0 = 0$ in the FHC).
2. The epidemic stutters along and inevitably fades out because $R_0 \leq 1$ in the FHC.
3. Though $R_0 > 1$ in the FHC, the infected host may or may not pass it on to another human, but the chain of transmission dies out by chance with probability $p_{\text{fadeout}} = 1 - 1/R_0$ (**epidemic fade-out**).
4. $R_0 > 1$ in the FHC, and the transmission chain takes off with probability $p_{\text{breakout}} = 1/R_0$ (**epidemic breakout**).

Fortunately, when $R_0 > 1$ in the FHC, a clear statistical demarcation exists between the epidemic fade-outs (#3) and breakouts (#4), with indicated probabilities. The distribution of sizes of total number of individuals infected is bimodal, with the low-end mode associated with fade-outs and the much larger, upper-end mode with breakouts [cf. simulated distributions of Ebola outbreaks in Getz et al. (2015)]. The emergence of a zoonotic disease thus requires at least a stuttering process in the FHC to occur sufficiently often to ensure an index case for a full epidemic in the FHC.

8.3.3 Applying the Framework

Effective management of emerging disease threats requires that we differentiate among spillover and FHC processes when designing interventions. Our framework lends itself to its own classification scheme for differentiating zoonoses:

Our Framework for Classifying Epidemics

A. Spillover epidemic, either fading or breaking out after initial spillover
B. Mixed epidemic (outbreak in both RC and FHC)
C. Human host epidemic of zoonotic origin (initial spillover, with subsequent adaptation of pathogen to become endemic in the FHC)

(continued)

D. Animal host epidemic of human origin (as in previous case with roles of human and animal hosts reversed)
E. Human host epidemic with zoonotic genetic source contribution
F. Animal host epidemic with human genetic source contribution

To illustrate these points, we describe four systems of pathogen emergence that represent the epidemic types along the continuum (Table 8.2).

Table 8.2 Epidemic types, examples, and their description

#	Name	Examples	Description
1	Spillover epidemic	Zoonoses—rabies, anthrax (humans) Anthropozoonoses	Only spillover transmission occurs, as the spillover host is a dead-end host for the pathogen, i.e., never enters the host epidemic space. Three kinds of transmission are identified: (a) Zoonoses—animal-to-human pathogen transmission. Also known as obligate zoonotic when referring to pathogens of animal origin (b) Anthropozoonses—human-to-animal pathogen transmission
2	Mixed epidemic	Ebola virus (human) Paramyxoviruses (great apes) Human tuberculosis (elephants; persistent mixed transmission can occur bidirectionally)	Spillover (zoonotic and anthropozoonotic) may be persistent (with varying quiescent periods) but is accompanied by limited secondary human-to-human or animal-to-animal transmission in the host epidemic space
3	Human host epidemic of zoonotic origin	SARS, HIV	A zoonotic pathogen makes a species jump from the zoonotic host departing from the spillover epidemic space, becoming human host adapted with sustained human-to-human transmission
4	Animal host epidemic of human origin	*Mycobacterium bovis*	A human source pathogen makes a species jump from the human host departing from the spillover epidemic space, becoming animal host adapted with sustained animal-to-animal transmission
5	Human host epidemic with zoonotic genetic source contribution	Swine-origin influenza A H1N1	A pathogen co-circulates in human or animal reservoirs, where genetic reassortment and transmission occur
6	Animal host epidemic with human genetic source contribution	Human H1N2 and human-swine reassortant H1N2 and H1N1 influenza A viruses in pigs	

8.3.3.1 Anthrax (Spillover and Epidemic Fade-Out, Obligate Spillover Pathogen: Almost Exclusively RC)

Anthrax is a zoonosis caused by the spore-forming, gram-positive bacterium, *Bacillus anthracis*. This zoonosis affects livestock and wildlife nearly worldwide (Hugh-Jones and Blackburn 2009), with recent reemergence in humans in several agricultural areas (Kracalik et al. 2015). Under certain environmental conditions, spores can persist for long periods of time in the soil and cause subsequent outbreaks (Blackburn et al. 2007; Cherkasskiy 1999; Dragon and Rennie 1995). Across its known geographic distribution, epizootics can range from a few cases (sporadic) to massive outbreaks (Blackburn 2006).

Spillover As a soilborne pathogen, spillover begins at the host-environment interface, and epizootics can persist for weeks to months (Hugh-Jones and Blackburn 2009). Infection may be from either direct ingestion of spores from soil or contaminated vegetation (primarily in grazing herbivores), ingestion of leaves contaminated with blowfly emesis (primarily for browsers) (Blackburn et al. 2010; Braack and De Vos 1990), or direct inoculation from biting flies (Blackburn 2010; Krishna et al. 1958; Blackburn et al. 2014b). Though poorly studied, animal inhalation of spores is not implausible (Turnbull et al. 1998). It has been suggested that during large epizootics, high case numbers of a primary species (e.g., American bison; *Bison bison*) may cause enough environmental contamination that secondary host species (e.g., moose, *Alces alces*, or elk, *Cervus canadensis*) may succumb later in the outbreak (Fig. 8.3a) (Hugh-Jones and Blackburn 2009; Dragon et al. 1999). Specific mechanisms for transmission remain poorly understood and require further research. Human infection is considered secondary and often linked with handling contaminated carcasses or slaughtering infected animals (Woods et al. 2004) and therefore constrained to the spillover. The spatial boundaries of the spillover are limited by the distribution of environmental conditions that support the pathogen; though when livestock control is limited, contaminated meat movement can increase spillover into urban areas (Kracalik et al. 2013, 2015). Limited mechanical transmission may occur within or between herbivorous hosts if tabanid flies do in fact play a role in the transmission cycle (Blackburn et al. 2014a). However, this is likely limited by the seasonality of fly life cycles and conditions that promote pathogen survival. Onward transmission in humans is unlikely. Fade-out results as a consequence of outbreak control interventions [e.g., carcass burning or burial, livestock vaccination campaigns (Kracalik et al. 2014)] or seasonal limits on the pathogen or mechanical vectors, though these conditions are not fully understood.

FHC Limited onward transmission may occur within an ungulate host species if tabanid flies do in fact play a role in the transmission cycle. Onward transmission in humans is highly unlikely. Fade-out results as a consequence of either outbreak control interventions (e.g., carcass burning or burial and sustained preventative vaccination when tenable) or seasonal limits on either the pathogen or mechanical vectors.

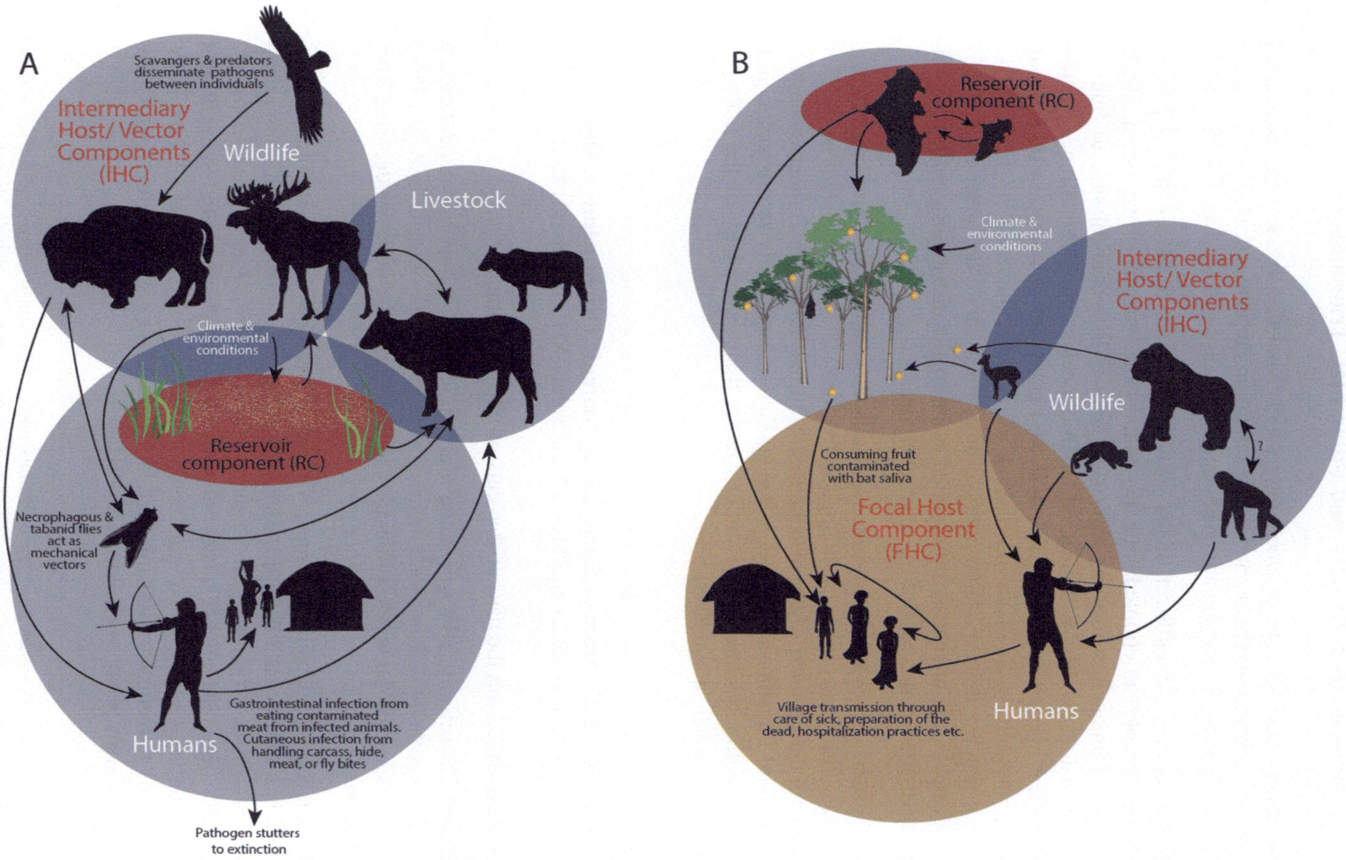

Fig. 8.3 Anthrax disease (**a**) is caused by the spore-forming soil bacterium *Bacillus anthracis* with spillover occurring directly in the environmental reservoir. Host behavior and environmental conditions affect epizootics, with onward transmission limited to nonhuman hosts affected by biting flies. Human cases are

8.3.3.2 Ebola (Mixed Epidemic Type, Mixed Pathogen Type: Spillover Pathogen from RC/IHC and within FHC Transmission)

Ebola hemorrhagic fever is an emerging zoonotic viral disease in West and Central Africa causing severe morbidity and high mortality in humans and wildlife (Alexander et al. 2015). Outbreaks are sporadic with viral quiescent periods upward of 20 years.

Spillover Successful spillover of the pathogen appears to be a complex process involving a number of coupled networks and seasonal drivers (Pinzon et al. 2004), linking the human host-to-virus reservoirs (Fig. 8.3b). Transmission to humans results from direct contact with infected wildlife species through handling and eating of bush meat [duiker; primates, SB3; bats, SB1 (Leroy et al. 2009)] or ingestion of fruit contaminated with Ebola-infected bat saliva [(Alexander et al. 2015), SB1]. Three bat species are considered putative virus reservoirs (Alexander et al. 2015). Spillover epidemics are necessarily limited to the spatial distribution of the reservoir host or distribution network of infected bush meat.

FHC Sustained onward transmission in humans results from close contact with blood, secretions, or tissues of infected individuals. The spatial extent of the host epidemic is then limited to the distribution of human-to-human contacts necessary for successful transmission of the pathogen. Fade-out generally results as a consequence of outbreak control interventions rather than the biology of the pathogen (barrier techniques, quarantine).

8.3.3.3 HIV/AIDS (Human Epidemic of Zoonotic Origin, Human Host Adapted: Initially Spillover, Now Solely FHC Transmission)

HIV-1 and HIV-2 are the causative agents of AIDS in humans.

Spillover Historically, HIV originated from spillover of simian immunodeficiency virus (SIV) pathogens from nonhuman primate species to humans in Africa [SB3 (Heeney et al. 2006)]. On adaptation to the human host, the virus has departed from the spillover.

FHC Successful adaptation of these viruses to sustained human-to-human transmission is a rare event. Of 35 different primate species infected with lentiviruses, only a few viruses from two primate species (chimpanzees, *Pan troglodytes troglodytes*, and sooty mangabeys) successfully invaded and have persisted in the human host population causing global pandemics (Heeney et al. 2006). Here, human-to-human virus transmission dynamics are largely driven by sociocultural factors that influence human behavior (Halperin 1999).

Fig. 8.3 (continued) secondary to animal infections, and onward transmission is unlikely. Ebola (**b**) is an emerging zoonotic disease where spillover dynamics appear to include up to four coupled systems. Population dynamics and seasonal influences appear to be primary drivers of the process of pathogen invasion (Alexander et al. 2012b)

8.3.3.4 Influenza (Epidemic with Reservoir Genetic Source Contribution: Spillover with Potential Bidirectional Transmission Within and Between IHC and FHC)

Spillover Influenza viruses circulate in a wide array of domestic animal and wildlife species with frequent spillover to humans (SB3). Spillback from humans to animals can also occur.

FHC Reassortment of genetic material from these animal reservoirs has been associated with changes in virulence, invasion potential, and adaptation in the human host (Smith et al. 2009; Olsen 2002). Some strains (e.g., H5N1, "bird flu") demonstrate limited human-to-human transmission potential while being highly virulent (Yang et al. 2007). Other strains (e.g., H1N1, "swine flu") are highly adapted to the human host, have relatively low virulence, and cause pandemic disease (Lagace-Wiens et al. 2010).

8.3.4 Evaluating the Framework

Characterization of epidemic types within our framework allows us to more clearly identify direct implications for outbreak control. For example, zoonotic pathogen transmission at SB1 or SB3 will be spatially restricted in occurrence to where the RC or IHC species and humans intersect. In contrast, once a pathogen has evolved and adapted to the human host niche (FHC), the pathogen is freed from this spatial restriction. This moves the outbreak from being described principally by ecological variables such as host range or environment (e.g., Ebola, anthrax) to an outbreak described principally by socioculturally shaped transmission dynamics (e.g., HIV). A caveat here is the movement of infected animal products in the form of bush meat, which can expand the spatial restrictions in spillover also driven by sociocultural influences (Alexander et al. 2012a). The movement of bush meat is a growing concern in the public health and agricultural sectors.

Our framework identifies important gaps in the quantitative evaluation of the zoonotic outbreaks by identifying the need to link together the concatenated stochastic processes of spillover (SB1-3) and onward transmission in the FHC. The real problem, however, of characterizing transmission spillover boundaries is that sample sizes are rather small and observations are only partial: we may have some observations of successful spillovers (e.g., Ebola outbreaks over the past 40 years), but we have no idea how many infected animal-susceptible human contact events actually occurred. Further, during the early stages of a zoonotic outbreak, we do not have data that provides what proportion of transmission events took place across SB1 and SB3 versus that occurring within the FHC. For example, a hunter may be infected with Ebola virus from eating infected gorilla meat (Fig. 8.4). This infected individual may then go on to infect four other people in his family that care for him. Alternatively, five people in this family may handle and eat the meat of an infected gorilla directly and become infected, with no human-to-human transmission occurring. In both cases, five

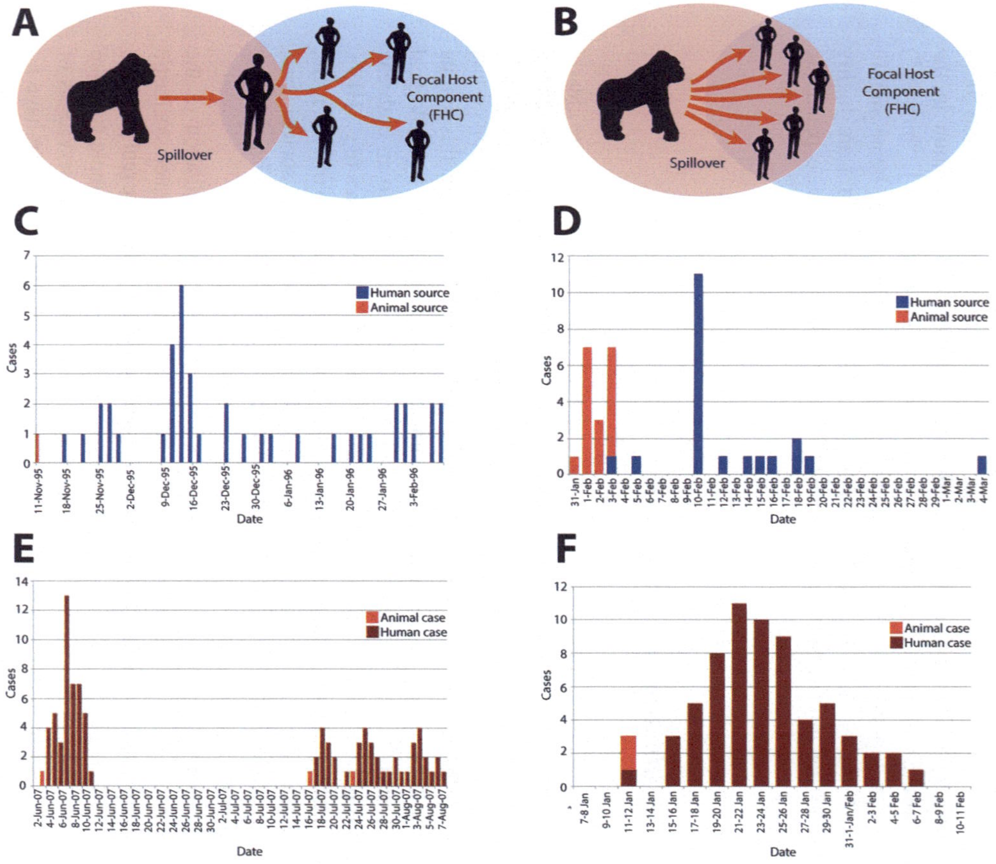

Fig. 8.4 Scenarios of Ebola spillover. In (**a**), the spillover is limited with the majority of transmission occurring in the host-epidemic focal host component (FHC). In (**b**), there is only a spillover epidemic with no FHC ($R_0 = 0$). Insets **c** and **d** illustrate Ebola outbreaks in Gabon (Georges et al. 1999) where both spillover types

individuals are infected, but in the first case, two separate parameters must be estimated to predict the pathogen emergence process (transmission rates in both the spillover and the FHC), while in the second case, only the transmission rate in the FHC is needed, given that a spillover event has occurred. It is possible that genetic data may allow us to differentiate between these two cases, though we will need to know something about the rates of Ebola virus strain evolution within animal and humans hosts, respectively. In the second case, however, estimating the probability that isolated spillover events occur over time requires long-term samples of outbreak rates and the assumption of stationary conditions (i.e., the environmental and human population factors involved remain unchanged over time). While the temporal dynamics of these two cases would be different, in a mixed ongoing epidemic, characterization and application of how the two stochastic processes, one at the spillover boundary and one in the FHC, are intertwined are essential to gaining the necessary insight into the outbreak etiology and efficacy of different management actions.

Explicitly integrating the separate stochastic processes of spillover and onward transmission facilitates consideration of new questions that are not routinely discussed in emerging infectious disease literature. Increasing our research focus on this integration from both the mathematical and conceptual perspective will be important to our ability to model the dynamics of coupled spillover and FHC transmission processes. Emphasis is given to the possibility that some spillovers will stutter to extinction even though we may have $R_0 >> 1$ in the human population. Conversely, some spillovers with $R_0 < 1$ in the human population will, nevertheless, lead to outbreaks sustained by continuous transmission from animal to human spillover. When the public health focus is on the pathogen, not the strain (as in influenza), we can then consider the case of multiple spillovers of different strains with a distribution of R_0 in the human population. The public health question is then not so much will an HkNℓ ($k = 1, ..., 18$; $\ell = 1, ..., 11$) variant of influenza spillover and seed an epidemic as it is how often will a reassorted influenza virus spillover and seed an epidemic. The spillover component is then the critical ingredient in an analysis.

8.4 Ongoing Quantitative Challenges at the Spillover Interface

The spillover interface is, epidemiologically, unique and complex; and, as we have highlighted here, unique quantitative approaches are correspondingly needed to describe the rate and dynamics of spillover. The most basic concepts like R_0 can

Fig. 8.4 (continued) (**a**, **b**) are illustrated. Red indicates human cases directly from animals; blue indicates human-to-human transmission. (**e**) and (**f**) illustrate livestock anthrax outbreaks with human spillover in Bengal (Ray et al. 2009) and Zimbabwe (Gombe et al. 2010). In both, human cases (dark red) were directly from animals (light red), illustrating anthrax is limited to the intermediary host/vector component (IHC)

be less meaningful and may need special formulations that more accurately summarize the different classes of epidemics we describe above. At the actual spillover boundary (Fig. 8.1), basic concepts like the force of infection break down. In normal epidemiology, for a pathogen circulating within a population, the force of infection λ is conventionally expressed as

$$\lambda = \frac{I}{N}\beta \tag{8.2}$$

where β is the transmission rate. Within a single species, contact rates can be defined based on assumptions of density or frequency dependence, using the number or proportion of infected individuals, respectively, to parameterize λ. Between two species, contact rates are harder to define based on I alone, and so the force of infection is a problematic concept at the spillover interface. Lloyd-Smith et al. (2015) suggested a customized definition of the **spillover force of infection**, adapted to the language we use here:

$$\lambda_{\mathrm{s}} = \left(\frac{I_{\mathrm{RC}}}{N_{\mathrm{RC}}}\right) \times (\mathrm{RC} : \mathrm{FHC}\ \text{contact rate}) \times \beta_{\mathrm{s}} \tag{8.3}$$

where

$$\beta_{\mathrm{s}} = P\,(\text{infection}|\text{contact}) \tag{8.4}$$

Parameterizing the RC:FHC contact rate—especially as a single model variable—poses a serious quantitative challenge. Increasingly complex models will be needed that break down the different processes at the RC:FHC interface—as well as those that separate the role of the IHC—helping researchers more readily conceptualize and parameterize models.

Recent work on Lassa fever has especially highlighted just how far models at the spillover interface have come in the last 5 years. A model recently published by Iacono et al. (2016) establishes a "unified framework" for modeling zoonotic spillover with horizontal transmission, at the most fundamental level based on the notion that spillovers are a Poisson point process with rate λ, such that the probability of a given number k of spillover events during interval τ is

$$P(k) = \frac{e^{-\lambda\tau}(\lambda\tau)^k}{k!} \tag{8.5}$$

They suggest that spillovers are a **self-exciting** process, as new human infections generate more human infections, but are also a **self-correcting** process, as acquired immunity in human populations depletes susceptible populations and decreases future transmission rates. Different framings of the relative importance of those processes lend themselves to different corresponding mathematical formulations, such as self-correcting Poisson processes or—if random variation in λ is stochastic enough to be important to model accuracy—a Poisson-gamma mixture with feedback. In their model, the force of infection is expressed as a combination of zoonotic and human-to-human components:

$$\widehat{\lambda}(t) = S_{\mathrm{H}}(t)\eta_{\mathrm{R}}(N_{\mathrm{R}})Pr_{\mathrm{R}}(N_{\mathrm{R}})\chi R + S_{\mathrm{H}}(t)\eta_{\mathrm{H}}(N_{\mathrm{H}})Pr_{\mathrm{H}}(N_{\mathrm{H}}, t)\chi_{\mathrm{H}} \quad (8.6)$$

where S_{H} is the susceptible human population size; Pr_{R} and Pr_{H} are the prevalence of infected rodents and humans, respectively; N_{H} and N_{R} are population sizes of humans and rodents, respectively; and χ_{R} is a parameter that represents "the ability of the reservoir to excrete a suitable dosage of the virus and the human response to it" (with a corresponding χ_{H} for human-to-human transmission). In their application to Lassa fever, the authors show that the flexibility of this model produces dramatic responses to subtle changes. For example, moving from a constant rate of spillover to a seasonally peaking process (as a mode of hypothesis testing) reduces human-to-human contributions to outbreaks from roughly 90% to 20%.

With expanding accuracy in models of the spillover process, simulation methods have now begun to be developed that appropriately predict spatial risk patterns for Lassa fever. A study by Redding et al. (2016) combined a (simpler) model of RC: FHC transmission (λ, which they term the "force of zoonotic infection") with habitat suitability modeling for reservoir hosts to develop an "environmental-mechanistic zoonotic spillover model." Models like these, which consider spatial variation in human-reservoir contact and spillover risk, represent a tremendous advancement in the quantification of the spillover interface. Most spatial patterns in zoonotic disease emergence are currently studied with more correlative methods like **ecological niche models**, which are tremendously powerful but subject to a number of user-end decisions, only relate patterns of occurrence to environmental variables (transmission risk rather than transmission rates), and suffer from problems relating to lack of consensus among conflicting models. Both modeling frameworks offer the possibility of extending models for predictions under global change; but models that explicitly and mechanistically account for human case burdens arising from spillover (rates, not risks) have a clear advantage when the data exists to parameterize them.

8.5 Conclusion

Articulation of zoonotic outbreaks as a set of processes that take place dynamically (Fig. 8.1) allows a deeper understanding of how to computationally categorize different classes of zoonotic diseases, with important implications to both public health policy and management of zoonotic outbreaks. Using this approach it becomes clearer why influenza is so different from many other zoonotic pathogens because the spillover-FHC concatenation can be bidirectional between humans and animals as well as animals and the environment. While R_0 is often ubiquitously applied to analysis of all infectious disease, this framework helps us understand why this concept is only part of the story for zoonotic diseases where key stochastic events and interdependent processes influence spillover probability at any or all interfaces. This emphasis on the spillover component of zoonotic disease emergence points to a need for greater computational and mechanistic focus on spillover itself.

Compliance with Ethical Standards

Funding This work has been partially supported by NSF CNH 114953, NSF CNH 1518486, and NSF EEID 1518663 (KAA), NSF HSD Grant SES-0729441, NIH MIDAS project 2U01GM070694-7, NSF PetaApps Grant OCI-0904844, DTRA15R&D Grant HDTRA1-0901-0017, DTRA CNIMS Grant HDTRA1-07-C-0113 (MM,SE), NSF EEID 1617982 (WMG), and NIH 1R01GM117617-01 (JKB). We thank B. Singer for comments on this chapter.

Conflict of Interest All authors declare they have no conflict of interest.

Ethics Approval This article does not contain any studies with human participants or animals performed by any of the authors.

References

Alexander KA, McNutt JW (2010) Human behavior influences infectious disease emergence at the human–animal interface. Front Ecol Environ 8(10):522–526

Alexander KA, MacLachlan NJ, Kat PW et al (1994) Evidence of natural blue-tongue virus infection among African carnivores. Am J Trop Med Hyg 51(5):568–576

Alexander K, Kat P, House J et al (1995) African horse sickness and African carnivores. Vet Microbiol 47(1–2):133–140

Alexander KA, Blackburn JK, Vandewalle ME et al (2012a) Buffalo, bush meat, and the zoonotic threat of brucellosis in Botswana. PLoS One 7(3):e32, 842

Alexander KA, Lewis BL, Marathe M et al (2012b) Modeling of wildlife-associated zoonoses: applications and caveats. Vector Borne Zoonotic Dis 12(12):1005–1018

Alexander KA, Sanderson CE, Marathe M et al (2015) What factors might have led to the emergence of Ebola in West Africa? PLoS Negl Trop Dis 9(6):e0003, 652

Ali S, Keil R (2006) Global cities and the spread of infectious disease: the case of severe acute respiratory syndrome (SARS) in Toronto, Canada. Urban Stud 43(3):491

Allan B, Keesing F, Ostfeld R (2003) Effect of forest fragmentation on Lyme disease risk. Conserv Biol 17(1):267–272

Anishchenko M, Bowen R, Paessler S et al (2006) Venezuelan encephalitis emergence mediated by a phylogenetically predicted viral mutation. Proc Natl Acad Sci U S A 103(13):4994

Antia R, Regoes RR, Koella JC et al (2003) The role of evolution in the emergence of infectious diseases. Nature 426(6967):658

Blackburn J (2006) Evaluating the spatial ecology of anthrax in North America: Examining epidemiological components across multiple geographic scales using a GIS-based approach [Doctoral dissertation]. Louisiana State University, Baton Rouge

Blackburn JK (2010) Integrating geographic information systems and ecological niche modeling into disease ecology: a case study of *bacillus anthracis* in the United States and Mexico. In: O'Connell KP, Skowronski EW, Bakanidze L, Sulakvelidze A (eds) Emerging and endemic pathogens. Springer, Dordrecht, pp 59–88

Blackburn JK, McNyset KM, Curtis A et al (2007) Modeling the geographic distribution of *bacillus anthracis*, the causative agent of anthrax disease, for the contiguous united states using predictive ecologic niche modeling. Am J Trop Med Hyg 77(6):1103–1110

Blackburn JK, Curtis A, Hadfield TL et al (2010) Confirmation of *Bacillus anthracis* from flesh-eating flies collected during a West Texas anthrax season. J Wildl Dis 46(3):918–922

Blackburn JK, Hadfield TL, Curtis AJ et al (2014a) Spatial and temporal patterns of anthrax in white-tailed deer, *Odocoileus virginianus*, and hematophagous flies in West Texas during the summertime anthrax risk period. Ann Am Assoc Geogr 104(5):939–958

Blackburn JK, Van Ert M, Mullins JC et al (2014b) The necrophagous fly anthrax transmission pathway: empirical and genetic evidence from wildlife epizootics. Vector Borne Zoonotic Dis 14(8):576–583

Braack L, De Vos V (1990) Feeding habits and flight range of blow-flies (*Chrysomyia* spp.) in relation to anthrax transmission in the Kruger National Park, South Africa. Onderstepoort J Vet Res 57(2):141–142

Brown C (2004) Emerging zoonoses and pathogens of public health significance – an overview. Revue Scientifique et Technique-Office International des Epizooties 23(2):435–442

Camacho A, Kucharski A, Funk S et al (2014) Potential for large outbreaks of Ebola virus disease. Epidemics 9:70–78

Cherkasskiy B (1999) A national register of historic and contemporary anthrax foci. J Appl Microbiol 87(2):192–195

Childs J (2004) Zoonotic viruses of wildlife: hither from yon. In: Emergence and control of zoonotic viral encephalitides. Springer, Berlin

Childs JE, Richt JA, Mackenzie JS (2007) Introduction: conceptualizing and partitioning the emergence process of zoonotic viruses from wildlife to humans. In: Childs JE, Mackenzie JS, Richt JA (eds) Wildlife and emerging zoonotic diseases: the biology, circumstances and con sequences of cross-species transmission. Springer, Heidelberg, pp 1–31

Chuck S, Sande M (1989) Infections with *Cryptococcus neoformans* in the acquired immuno-deficiency syndrome. N Engl J Med 321(12):794–799

Cox-Singh J, Singh B (2008) Knowlesi malaria: newly emergent and of public health importance? Trends Parasitol 24(9):406–410

Daszak P, Cunningham AA, Hyatt AD (2000) Emerging infectious diseases of wildlife–threats to biodiversity and human health. Science 287(5452):443–449

Dearing MD, Dizney L (2010) Ecology of hantavirus in a changing world. Ann N Y Acad Sci 1195(1):99–112

Diekmann O, Heesterbeek JAP, Metz JA (1990) On the definition and the computation of the basic reproduction ratio r0 in models for infectious diseases in heterogeneous populations. J Math Biol 28(4):365–382

Dobson A (2004) Population dynamics of pathogens with multiple host species. Am Nat 164(S5): S64–S78

Dobson AP (2005) What links bats to emerging infectious diseases? Science 310(5748):628

Donnelly C, Woodroffe R, Cox D et al (2003) Impact of localized badger culling on tuberculosis incidence in British cattle. Nature 426(6968):834–837

Dragon DC, Rennie RP (1995) The ecology of anthrax spores: tough but not invincible. Can Vet J 36(5):295

Dragon D, Elkin B, Nishi J et al (1999) A review of anthrax in Canada and implications for research on the disease in northern bison. J Appl Microbiol 87(2):208–213

Duffy S, Shackelton LA, Holmes EC (2008) Rates of evolutionary change in viruses: patterns and determinants. Nat Rev Genet 9(4):267

Ferrari N, Cattadori I, Rizzoli A et al (2009) Heligmosomoides polygyrus reduces infestation of Ixodes ricinus in free-living yellow-necked mice, *Apodemus flavicollis*. Parasitology 136: 305–316

Fèvre E, Picozzi K, Fyfe J et al (2005) A burgeoning epidemic of sleeping sickness in Uganda. Lancet 366(9487):745–747

Fichet-Calvet E, Lecompte E, Koivogui L, Soropogui B, Doré A, Kourouma F, Sylla O, Daffis S, Koulémou K, Ter Meulen J (2007) Fluctuation of abundance and Lassa virus prevalence in *Mastomys natalensis* in Guinea, West Africa. Vector-Borne Zoonotic Dis 7(2):119–128

Getz WM, Pickering J (1983) Epidemic models: thresholds and population regulation. Am Nat 121:892–898

Getz WM, Gonzalez J-P, Salter R et al (2015) Tactics and strategies for managing Ebola outbreaks and the salience of immunization. Comput Math Methods Med 2015:736507

Georges AJ, Leroy EM, Renaut AA et al (1999) Ebola hemorrhagic fever outbreaks in Gabon, 1994–1997: epidemiologic and health control issues. J Infect Dis 179:S65–S75

Gombe N, Nkomo B, Chadambuka A et al (2010) Risk factors for contracting anthrax in Kuwirirana ward, Gokwe North, Zimbabwe. Afr Health Sci 10:159–164

Griffin D, Donaldson K, Paul J et al (2003) Pathogenic human viruses in coastal waters. Clin Microbiol Rev 16(1):129

Halperin DT (1999) Heterosexual anal intercourse: prevalence, cultural factors, and HIV infection and other health risks, Part I. AIDS Patient Care STDs 13(12):717–730

Haydon DT, Cleaveland S, Taylor LH et al (2002) Identifying reservoirs of infection: a conceptual and practical challenge. Emerg Infect Dis 8:1468–1473

Heesterbeek J (2002) A brief history of R0 and a recipe for its calculation. Acta Biotheor 50: 189–204

Heeney JL, Dalgleish AG, Weiss RA (2006) Origins of HIV and the evolution of resistance to AIDS. Science 313:462

Hooper P, Zaki S, Daniels P et al (2001) Comparative pathology of the diseases caused by Hendra and Nipah viruses. Microbes Infect 3(4):315–322

Hudson P, Perkins S, Cattadori I et al (2008) The emergence of wildlife disease and the application of ecology. Princeton University Press, Princeton, NJ, pp 347–367

Hugh-Jones M, Blackburn J (2009) The ecology of *Bacillus anthracis*. Mol Asp Med 30(6): 356–367

Iacono GL, Cunningham AA, Fichet-Calvet E et al (2016) A unified framework for the infection dynamics of zoonotic spillover and spread. PLoS Negl Trop Dis 10:e0004957

Johnson CK, Hitchens PL, Evans TS et al (2015) Spillover and pandemic properties of zoonotic viruses with high host plasticity. Sci Rep 5:14830

Jones K, Patel N, Levy M et al (2008) Global trends in emerging infectious diseases. Nature 451:990

Keele BF, Van Heuverswyn F, Li Y et al (2006) Chimpanzee reservoirs of pandemic and nonpandemic HIV-1. Science 313:523–526

Kilpatrick AM (2011) Globalization, land use, and the invasion of West Nile virus. Science 334:323–327

Kilpatrick AM, Kramer LD, Jones MJ et al (2006) West Nile virus epidemics in North America are driven by shifts in mosquito feeding behavior. PLoS Biol 4:e82

Kracalik IT, Malania L, Tsertsvadze N et al (2013) Evidence of local persistence of human anthrax in the country of Georgia associated with environmental and anthropogenic factors. PLoS Negl Trop Dis 7(9):e2388

Kracalik I, Abdullayev R, Asadov K et al (2014) Changing patterns of human anthrax in Azerbaijan during the post-soviet and preemptive livestock vaccination eras. PLoS Negl Trop Dis 8:e2985

Kracalik I, Malania L, Imnadze P et al (2015) Human Anthrax transmission at the urban–rural Interface, Georgia. Am J Trop Med Hyg 93:1156–1159

Krishna R, Mohiyuden S et al (1958) Tabanus flies as transmitters of anthrax-a field experience. Indian Vet J 38(7):348–353

Lagace-Wiens PR, Rubinstein E, Gumel A (2010) Influenza epidemiology past, present, and future. Crit Care Med 38:e1–e9

Lemey P, Pybus OG, Wang B et al (2003) Tracing the origin and history of the HIV-2 epidemic. Proc Natl Acad Sci U S A 100(11):6588–6592

Leroy EM, Epelboin A, Mondonge V et al (2009) Human Ebola outbreak resulting from direct exposure to fruit bats in Luebo, Democratic Republic of Congo, 2007. Vector Borne Zoonotic Dis 9(6):723–728
Lloyd-Smith JO, Cross PC, Briggs CJ et al (2005) Should we expect population thresholds for wildlife disease? Trends Ecol Evol 20(9):511–519
Lloyd-Smith JO, George D, Pepin KM et al (2009) Epidemic dynamics at the human-animal interface. Science 326(5958):1362–1367
Lloyd-Smith JO, Funk S, McLean AR et al (2015) Nine challenges in modelling the emergence of novel pathogens. Epidemics 10:35–39
McCormack RK, Allen LJ (2007) Disease emergence in multi-host epidemic models. Math Med Biol 24(1):17–34
Morris LR, Proffitt KM, Asher V et al (2016) Elk resource selection and implications for anthrax management in Montana. J Wildl Manag 80(2):235–244
Olival KJ, Hosseini PR, Zambrana-Torrelio C et al (2017) Host and viral traits predict zoonotic spillover from mammals. Nature 546(7660):646–650
Olsen CW (2002) The emergence of novel swine influenza viruses in North America. Virus Res 85(2):199–210
Olson PE, Hames C, Benenson A et al (1996) The thucydides syndrome: Ebola deja vu? (or Ebola reemergent?). Emerg Infect Dis 2(2):155
Orlandi P, Chu D, Bier J, Jackson G (2002) Parasites and the food supply. Food Technol Champaign Then Chicago 56(4):72–79
Papagrigorakis MJ, Yapijakis C, Synodinos PN et al (2006) DNA examination of ancient dental pulp incriminates typhoid fever as a probable cause of the plague of Athens. Int J Infect Dis 10(3):206–214
Parrish CR, Holmes EC, Morens DM et al (2008) Cross-species virus transmission and the emergence of new epidemic diseases. Microbiol Mol Biol Rev 72(3):457–470
Pepin KM, Lass S, Pulliam JRC et al (2010) Identifying genetic markers of adaptation for surveillance of viral host jumps. Nat Rev Microbiol 8(11):802–813
Petersen LR, Brault AC, Nasci RS (2013) West Nile virus: review of the literature. JAMA 310(3): 308–315
Pinzon JE, Wilson JM, Tucker CJ et al (2004) Trigger events: enviroclimatic coupling of Ebola hemorrhagic fever outbreaks. Am J Trop Med Hyg 71(5):664–674
Ray TK, Hutin YJ, Murhekar MV (2009) Cutaneous anthrax, West Bengal, India, 2007. Emerg Infect Dis 15(3):497
Redding DW, Moses LM, Cunningham AA et al (2016) Environmental-mechanistic modelling of the impact of global change on human zoonotic disease emergence: a case study of lassa fever. Methods Ecol Evol 7(6):646–655
Reis RB, Ribeiro GS, Felzemburgh RD et al (2008) Impact of environment and social gradient on leptospira infection in urban slums. PLoS Negl Trop Dis 2(4):e228
Rigaud T, Perrot-Minnot MJ, Brown MJ (2010) Parasite and host assemblages: embracing the reality will improve our knowledge of parasite transmission and virulence. Philos Trans R Soc Lond B Biol Sci 277(1701):3693–3702
Rimoin AW, Mulembakani PM, Johnston SC et al (2010) Major increase in human monkeypox incidence 30 years after smallpox vaccination campaigns cease in the democratic republic of congo. Proc Natl Acad Sci U S A 107(37):16,262–16,267
Smith RD (2006) Responding to global infectious disease outbreaks: lessons from sars on the role of risk perception, communication and management. Soc Sci Med 63(12):3113–3123
Smith GJ, Bahl J, Vijaykrishna D, Zhang J et al (2009) Dating the emergence of pandemic influenza viruses. Proc Natl Acad Sci U S A 106(28):11,709–11,712
Song H, Tu C, Zhang G et al (2005) Cross-host evolution of severe acute respiratory syndrome coronavirus in palm civet and human. Proc Natl Acad Sci U S A 102(7):2430
Stapp P, Antolin MF, Ball M (2004) Patterns of extinction in prairie dog metapopulations: plague outbreaks follow el nino events. Front Ecol Environ 2(5):235–240

Tack AJ, Thrall PH, Barrett LG et al (2012) Variation in infectivity and aggressiveness in space and time in wild host–pathogen systems: causes and consequences. J Evol Biol 25(10):1918–1936

Taubenberger JK, Morens DM (2006) Influenza en 1918: La madre de todas las pandemias. Revista Biomedica 17(1):69–79

Turnbull PC, World Health Organization et al (1998) Guidelines for the surveillance and control of anthrax in humans and animals. World Health Organization, Department of Communicable Diseases Surveillance and Response, Geneva

Webster R, Shortridge K, Kawaoka Y (1997) Influenza: interspecies transmission and emergence of new pandemics. FEMS Immunol Med Microbiol 18(4):275–279

Wells R, Young J, Williams R et al (1997) Hantavirus transmission in the United States. Emerg Infect Dis 3(3):361

Wolfe ND, Dunavan CP, Diamond J (2007) Origins of major human infectious diseases. Nature 447(7142):279

Woods CW, Ospanov K, Myrzabekov A et al (2004) Risk factors for human anthrax among contacts of anthrax-infected livestock in Kazakhstan. Am J Trop Med Hyg 71(1):48–52

Yakob L, Bonsall MB, Yan G (2010) Modelling knowlesi malaria transmission in humans: vector preference and host competence. Malaria J 9(1):329

Yang Y, Halloran ME, Sugimoto JD et al (2007) Detecting human-to-human transmission of avian influenza a (h5n1). Emerg Infect Dis 13(9):1348

Zinsstag J, Roth F, Orkhon D, Chimed-Ochir G et al (2005) A model of animal–human brucellosis transmission in mongolia. Prev Vet Med 69(1–2):77–95

Zinsstag J, Dürr S, Penny M et al (2009) Transmission dynamics and economics of rabies control in dogs and humans in an African city. Proc Natl Acad Sci U S A 106(35):14996–15001

Chapter 9
Integrating Landscape Hierarchies in the Discovery and Modeling of Ecological Drivers of Zoonotically Transmitted Disease from Wildlife

Douglas G. Goodin, Colleen B. Jonsson, Linda J. S. Allen, and Robert D. Owen

Abstract Changes in landscape and land use can drive the emergence of zoonoses, and hence, there has been great interest in understanding how land cover change and the cascade of ecological effect associated with it are associated with emerging infectious diseases. In this chapter, we review how a spatially hierarchical approach can be used to guide research into the links between landscape properties and zoonotic diseases. Methodological advances have played a role in the revival of landscape epidemiology and we introduce the role of methodologies such as geospatial analysis and mathematical modeling. Importantly, we discuss cross-scale analysis and how this would provide a richer perspective of the ecology of zoonotic diseases. Finally, we will provide an overview of how hierarchical research strategies and modeling might be generally used in analyses of infectious zoonoses originating in wildlife.

D. G. Goodin
Department of Geography, Kansas State University, Manhattan, KS, USA

C. B. Jonsson (✉)
Department of Microbiology, Immunology and Biochemistry, University of Tennessee Health Science Center, Memphis, TN, USA
e-mail: cjonsson@utk.eduof

L. J. S. Allen
Department of Mathematics and Statistics, Texas Tech University, Lubbock, TX, USA

R. D. Owen
Department of Biological Sciences, Texas Tech University, Lubbock, TX, USA

C. J. Hurst (ed.), *The Connections Between Ecology and Infectious Disease*, Advances in Environmental Microbiology 5,
https://doi.org/10.1007/978-3-319-92373-4_9

9.1 Introduction

Zoonotic diseases, or zoonoses, are infectious diseases transmitted to humans from animals and may be bacterial, viral, or parasitic in origin. Approximately 58% of the pathogens associated with infectious diseases in humans have originated through spillover from wildlife—e.g., ebolaviruses, hantaviruses, coronaviruses, henipaviruses (Jones et al. 2008; Lloyd-Smith et al. 2009; Smith et al. 2014; Woolhouse and Gowtage-Sequeria 2005). Since 1980, zoonotic pathogens represent the bulk of the outbreaks in human populations in both number (87% versus 13% by vector-borne pathogens) and diversity (Smith et al. 2014). The recently reported increase in zoonoses has been attributed to a variety of reasons, although a major driver is changes in land use resulting from the increased demands of human populations on the natural environment through agriculture intensification and deforestation (Jones et al. 2013; Woolhouse and Gowtage-Sequeria 2005). Land alteration for food production, broadly defined to include both agriculture and pastoral activities, has produced profound changes in the type and structure of the earth's vegetation cover. It has also altered the way humans interact with their environment, for example, in some geographical areas suppressing wildlife-human interaction (e.g., contact with large predators common among hunter-gathers) and instead favoring contact between humans and peridomestic species such as rodents, which are common reservoirs for many zoonotic pathogens. Numerous examples of emergent zoonoses have often accompanied land clearance, and hence, there has been great interest in understanding how land cover change and the cascade of ecological effects associated with it are ecologically associated with emerging infectious diseases (McFarlane et al. 2013; Woolhouse and Gowtage-Sequeria 2005).

A persistent question in modeling of pathogen emergence is what is the spatial scale at which these processes occur? How do we measure and integrate the impact of processes that occur at varying and nested spatial scales that result in the observed disease distribution (Watts et al. 2005)? For example, climate (which can vary in scale from local to continental) and landscape are each associated with disease patterns and have been cited as factors in zoonotic disease outbreaks. Land cover change and land-use intensification often occurs at finer scales than climate and can thus be thought of as nested within the climate system. At still finer scales, population dynamics and habitat interactions of the pathogen and its reservoir communities (and, indeed, with the human populations vulnerable to disease transmissions from these communities) occur within the climate and landscape scales and are influenced, but not necessarily determined, by them. This hierarchical nesting (Fig. 9.1) of these processes complicates the overall study and modeling of the spillover and transmission dynamics of wildlife pathogens to human populations. However, there are existing bodies of theory which can help shed light on these processes. Hence, we review some of these conceptual ideas, focusing especially on landscape epidemiology, a body of theory first developed in the 1930s and more recently updated to include modern tools and techniques for studying both disease and environmental process (Ostfeld et al. 2005), and hierarchy theory, a framework which incorporates the idea of multiple, nested spatial scales (Allen and Starr 1982).

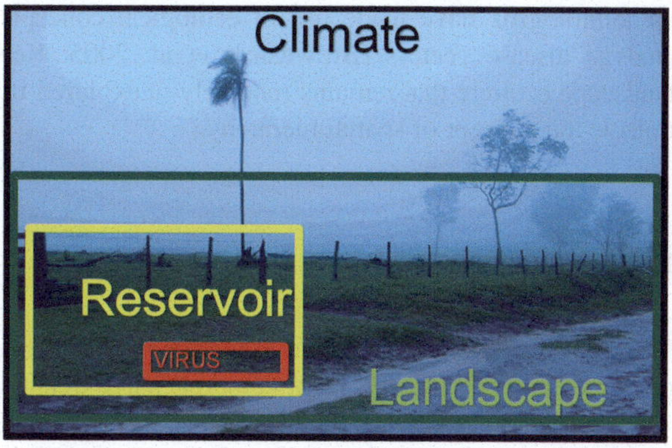

Fig. 9.1 Illustration of hierarchical nesting of possible key components in pathogen emergence in wildlife. Each process outlined with a "box" represents a distinct spatial scale to consider in developing models

In the following, we will review how a spatially hierarchical approach can be used to guide research into the links between landscape properties and zoonotic diseases. This will show how landscape concepts have been used to analyze the occurrence and spatial patterning of zoonotic diseases and how many of these studies are being conducted at particular spatial scales. This review will be followed by an illustration of how an integrated, cross-scale analysis might be employed to gain a deeper understanding of the ecology of zoonotic diseases, using an example from our own research. As earlier noted, methodological advances have played a role in the revival of landscape epidemiology. We will therefore discuss the role of methodologies such as geospatial analysis and mathematical modeling—again using examples drawn from our current research. We will show how cross-scale analysis might be employed to gain a deeper understanding of the ecology of zoonotic diseases. Finally, we will provide an overview of how hierarchical research strategies and modeling might be generally used in analyses of infectious zoonoses originating in wildlife.

9.2 Landscape Epidemiology and Ecology

Landscape epidemiology is the study of spatial patterns of disease and disease risk arising from underlying environmental causes. The fundamental concepts of landscape epidemiology, first proposed by Evegeniĭ Pavlovskiĭ, stem from the idea that the spatial occurrence of disease could be understood by studying landscape and environmental factors associated with the disease (Pavlovskiĭ 1966). Pavlovksiĭ's ideas have undergone a revival, stimulated in part by widespread availability of geospatial data, analysis tools, and models. In particular, satellite remote sensing data analyzed within a geographic information systems (GIS) framework has equipped landscape epidemiologists with a powerful suite of tools for analyzing environmental patterns associated with disease occurrence. Landscape epidemiology has also benefited from the theoretical perspective of landscape ecology, another relatively new discipline that attempts to understand the relationship between spatial pattern and ecological process (Meentemeyer et al. 2012). For example, landscape epidemiologists have made use of ecological concepts such as fragmentation to analyze disease vectors (Brownstein et al. 2005; Reisen 2010). One aspect of landscape ecology that remains relatively unexplored in infectious disease applications is the concept of spatial hierarchy.

9.2.1 Hierarchy Theory

Hierarchy in ecology is a multifaceted theory incorporating elements of nonlinear dynamics and complexity; for a comprehensive treatment of hierarchy theory, see Allen and Starr (1982). The fundamental concept of hierarchy theory is that processes occurring at finer scales (i.e., "lower" in the spatial hierarchy) are constrained by processes at higher levels. Hierarchical levels can also be distinguished by the rates at which ecological processes occur—faster at finer scales, slower at coarser ones.

Hierarchy theory in ecology arose as a response to the need for a rigorous method of handling middle-number systems, that is, systems whose components are too few to treat statistically but too many to address with classical Newtonian mathematics. Hierarchy provides a framework by which these middle number systems can be decomposed into a series of manageable units, whose environmental drivers can be characterized by the scale (and thus the rate) at which they occur. Such a framework is amenable to the study of landscape epidemiology, since the linkages between environmental factors and disease are often multivariate, nonlinear, and not confined to a specific spatial scale.

Spatial scale is a crucial aspect of hierarchy theory. Any discussion of the relationship between ecological process and scale therefore must first define how the term is used and to what characteristic dimensions these terms apply. These definitions are complicated by the number of ways that the term "scale" is used. Scale is frequently referred to by descriptive adjectives such as "large" or "small," which Meentemeyer (1989) noted can have opposing meaning depending on whether one is referring to cartographic scale (where small scale refers to less spatial detail) or ecological scale, where smaller scale equates to smaller spatial area and greater detail. Ecologists generally distinguish between the grain and extent of a process, where extent refers to the area over which a process occurs, and grain denotes the smallest resolvable component of the process. Typically, grain can be described as either "fine," indicating small resolvable elements, or "coarse," indicating larger elements. Thus, we might characterize the scale process like tropical deforestation as occurring at relatively large extent, because a large area is affected, but at relatively fine grain, because the individual deforested units can be quite small. Csillag et al. (2000) argue that for ecological uses of scale, it is preferable to adopt specific terminology to distinguish between extent and grain. In hierarchy theory, spatial extent is probably the more commonly used sense of scale (Jenerette and Wu 2000). In describing spatial hierarchies, it is often useful to apply descriptive names to realms of scale. Terms such as "global," "continental," "regional," and "local" are often used, although the precise areas referred to often vary. For our review of hierarchy in infectious disease analysis, we will operationally define continental scale as areas exceeding 10^6 km^2, regional scales as ranging from 10^3 km^2–10^6 km^2, local scale from 10^1–10^3 km^2, and microscale <10^1 km^2. The lower end of the microscale represents the general size range of a small rodent's world.

Landscape analyses across spatial scale frequently use remote sensing data (Kitron et al. 2006; Wu 1999). Currently, imagery is available from an array of orbital sensors with widely varying spatial, temporal, and spectral resolutions (Table 9.1). Grain size (or resolution, as it is more commonly termed in remote sensing) is an engineered property of these sensors, dependent on the optical characteristics of the sensor and orbital characteristics of the platform. Theoretically,

Table 9.1 Partial list of current of current and past orbital remote-sensing instruments, frequently used in landscape epidemiological studies

Instrument	Platform	Spatial resolution (m)	Notes
Pleiades 1A	Pleiades	2.0 m (VNIR) 0.5 m (PAN)	Available 2011–Present
Worldview	Worldview 1–4	1.41–1.84 (VNIR) 0.31m–0.46 m (PAN)	Resolution is at nadir. Finer resolution sensor on WV-3 and -4. Available 2007–present
GeoEye-1	GeoEye	1.84 m (VNIR) 0.46 (PAN)	Available 2008–present Resolution is at nadir value
QuickBird	QuickBird	2.63 (VNIR) 0.73 (PAN)	Available 2001–2015
IKONOS	IKONOS	3.2 (VNIR) 0.82 (PAN)	Available 2007–2105
HRV	SPOT 1–3	20 m (VNIR) 10 m (PAN)	Available 1986–1997
HRVIR	SPOT 4–5	20m (VNIR, SWIR) 10 m (Pan)	Available 1998–2015
Vegetation	SPOT 4–5	1000 m	Produces a vegetation index (NDVI) product
Azersky	SPOT 6–7	6 m (VNIR, SWIR) 1.5 m (PAN)	Available 2012–present
Multispectral scanner (MSS)	Landsat 1–5	79 m	Available 1972–1999
Thematic Mapper/ Enhanced Thematic Mapper + (TM/ETM+)	Landsat 4,5,7	30 m (VNIR, SWIR) 120 m (TIR) 20 m (Pan)	Available 1984–2013 (ETM+ experienced partial failure in 1999)
OLI/TIRS	Landsat 8	30 m (VNIR, SWIR) 100 m (TIR) 15 m (Pan)	Available 2013–present
LISS	IRS 1C/1D	23.5 m (VNIR, SWIR) 7.5 PAN	Available 2003–present
Sentinel-2	Sentinel-2	10 m (VNIR) 20 m (SWIR)	Available 2015–present
Terra/Aqua	MODIS	250 m (Red, NIR 500 m (VNIR, SWIR) 1000 m (VNIR, SWIR, TIR)	Available 1999–present
AVHRR	NOAA polar orbiters	1000 m (VNIR, TIR)	Available 1979–present

there are no limitations to the extent of any of these systems, although practical limitations (particularly cost of data acquisition) dictate a rough correspondence between extent and resolution. Thus, the extents listed in Table 9.1 are based on these practical limitations. The variety of remote sensor data available has certainly facilitated hierarchical landscape analysis, but in some sense, it has also imposed limitations. Each remote sensor represents a "window" through which ecological process at some combination of grain size and extent can be observed. These limitations also extend to the range of spectral wavelengths each sensor can detect and the number and width of bands in which these wavelengths are detected. Combining these discrete views into an integrated picture is a central challenge for hierarchical analysis of infectious disease processes.

9.3 Landscape and Zoonotic Disease: A Selected Review of Literature

There is a large body of work relating infectious disease to environmental factors. Most of these studies have concentrated on a single class of causative factor operating over a characteristic spatial scale. In the following, we will briefly review several examples which represent different classes of landscape factors that have been shown to impact zoonotic viral emergence. Since our intent is to show how landscape spatial hierarchies influence disease processes, we will concentrate on examples that show the influence of landscape and land cover processes at various spatial scales, using the concept and terminology of scale developed in the previous section.

For convenience, we will group the literature reviewed here into two categories. First, we will review the relationship between land cover disturbance (including anthropogenic and natural disturbance) and zoonotic disease emergence. Included in this category are causes due to landscape structure, including natural landscape barriers, land cover change and disturbance, and fragmentation. This category also includes disturbance due to agricultural practices. The second class of landscape impacts that we will review are climate-driven landscape changes, which include climate-influenced changes in vegetation phenology patterns as well as persistent or transient superficial changes such as flooding. Processes contained within these two categories are not exclusive; however we will review them based on their predominant process.

9.3.1 Disturbance-Driven Landscape Change

Landscape structure refers to the pattern and arrangement of habitats on the Earth's surface. Like all landscape variables, structure is complex. It varies with scale and per organism. The same landscape might be structurally very different for birds, small mammals such as rodents, and larger animals, and since zoonotic disease reservoirs vary in their body size and habitat, structural effects on disease also vary. Structure can arise naturally due to topographic, edaphic meaning related to the soil, or climatic conditions, as well as through human landscape alteration. Both types of structural changes can be relevant to disease processes. Landscape structure has been related to zoonotic disease at a variety of spatial scales. At regional scale, Russell et al. (2006) show how natural landscape barriers such as rivers and preferred habitat can limit rabies spread by raccoons and be used to manage wildlife through vaccination efforts. Smith et al. (2002) used landscape heterogeneity as a predictor of the spread of rabies. Langlois et al. (2001) showed that the distribution of hantavirus-bearing rodents in North America was influenced by landscape fragmentation. Their analysis was unusual in that the extent of the study was continental, but the grain size was local ($\approx$1 km). Estrada-Peña and Oteo (1991) showed that landscape structure, particularly landscape connectivity, showed a strong influence on the abundance of Lyme disease vectors in Spain.

Land cover changes are known to affect zoonotic diseases through controls on the population dynamics of reservoir species (especially wild mammals) as well as disease vectors (Patz et al. 2008). Giraudoux et al. (2003) used regional-scale land cover changes in France and China to show how host mammal communities affect transmission dynamics of the endoparasite *Echinococcus multilocularis.* Using the ROMPA (ratio of optimal to marginal patch area) hypothesis of Lidicker and others, it has been shown that regional-scale landscape dynamics of intermediate host species can in turn affect parasite egg survival and transmission (Lidicker 1995). The Giraudoux et al. (2003) study also considered the role of landscape change on establishing minimum thresholds, which they termed "filters" or "screens" of suitability for disease transfer.

Agriculture practices led to the Nipah virus outbreak in Malaysia (Epstein et al. 2006; Pulliam et al. 2012). Using a combination of field, laboratory, and modeling approaches, these efforts have supported the hypothesis that emergences of viruses such as Nipah are due to ecological and not evolutionary drivers. These findings underscore the importance of having multidisciplinary teams work together to build predictive models for discovery of the relationships between anthropogenic environmental change and the transmission or spillover of infectious agents.

9.3.2 Climate-Driven Landscape Change

Climate-driven landscape change in this context refers to the effect of atmospheric processes (most notably precipitation) on the habitat of zoonotic host organisms. Variations in rainfall magnitude and frequency have notable effects on vegetation phenology, causing variations in surface greenness that can be tracked using remote-sensing instruments (De Beurs and Henebry 2004; Reed et al. 2009). The emergence of several zoonotic diseases including hantavirus pulmonary syndrome (HPS) (Yates et al. 2002), Argentine hemorrhagic fever (Simone et al. 2010), and Bolivian hemorrhagic fever (Kilgore et al. 1995) can be clearly linked to landscape.

Several studies have linked climate-driven changes to patterns of disease occurrence at different spatial scales. Two groups have evaluated the relationship between temporal patterns of Normalized Difference Vegetation Index (NDVI) and occurrence of Ebola virus in West Africa (Pinzon et al. 2004; Tucker et al. 2002). They found that NDVI trajectories showed distinctive "trigger events" prior to occurrences of the disease in humans and apes, which they hypothesized might be used to forecast conditions conducive to outbreaks of Ebola hemorrhagic fever. The remotely sensed data for this analysis came from the NOAA-AVHRR sensor, with continental spatial extent and observation grain size (pixel resolution) of 1 km^2. The observed NDVI trajectories were related to precipitation patterns, reinforcing the link between climate and disease occurrence. Estrada-Peña and Oteo (1991) and Estrada-Peña et al. (2006) used coarse resolution vegetation index data to model and predict the continental-scale relationship between climate-driven landscape change and Lyme disease. Again, the resolution and extent of this study were consistent with the idea that climate constrains disease processes at higher levels in the spatial hierarchy. At finer spatial scales, Glass et al. (2000) used patterns of reflectance in Landsat Thematic Mapper (TM) data to statistically model the presence of HPS in the Southwestern United States. At this spatial scale, the spectral response of the surface incorporates climatic factors (especially antecedent precipitation) but also integrates structural and compositional factors of the vegetation canopy itself. Models based on these techniques have shown some utility for predicting hantavirus cases (Glass et al. 2002).

9.4 Hierarchical Analysis of an Emergent Zoonosis: An Example

One of the important advantages of a hierarchical approach is that it allows a multifactorial explanation for the occurrence of the infection and disease. The environmental, landscape, and climatic processes each contribute to processes that may alter species interactions within their habitat. These extrinsic factors can alter the reservoir population dynamics, drive extinction, and affect maintenance (persistence) of the microorganism. Essentially, these factors can create constraints within their spatial scale and across scales. In this section, we explore an example of hierarchical observation that can be used in conjunction with modeling to test hypotheses regarding the effects of environment, landscape, and climate upon zoonotic pathogen distribution. For this we draw upon our work and others in the study of *Hantavirus* with a focus on South America (Jonsson et al. 2010; Palma et al. 2012).

9.4.1 Continental Scale: Phylogeography

Hierarchy theory suggests that processes at coarser grain also occur over longer time frames and can often be assumed to be static with respect to finer-scale processes. The phylogeographical patterns of South American hantaviruses within the Southern Cone ("El Cono Sur," a subcontinental region roughly defined as consisting of the country Argentina, often including Chile, plus sometimes considered to include Uruguay and Paraguay) mapped with environment at the coarsest spatial scales (Chu et al. 2006). In other words, the phylogenetic clades of hantaviruses from the Southern Cone of South America appear tied to coherent spatial patterns consistent with subcontinental-scale biogeographic features such as the major biomes (Chu et al. 2006). In Fig. 9.2, the locations of strains from three major subclades of South American hantaviruses are shown in the context of the major biomes based on the World Wildlife Fund terrestrial ecoregions data (Olson et al. 2001). For example, we find members from one subclade (i.e., Laguna Negra, Rio Mamoré, and Alto Paraná viruses) carried by rodent reservoirs that span the tropical grass and shrubland and dry broadleaf forest areas along the western, central regions of South America (Chu et al. 2006; Johnson et al. 1997; Richter et al. 2010; Yahnke et al. 2001). In contrast, a second subclade of rodents that harbor Jabora, Maporal, and Necocli hantaviruses inhabit mainly the moist broadleaf forest biome stretching from Venezuela and Colombia into Paraguay (Chu et al. 2009; de Oliveira et al. 2011; Fulhorst et al. 2004; Londoño et al. 2011). Knowing the extent of prevalence of these closely related viruses over this vast region would certainly be fascinating from a phylogeographical point of view. It is interesting that these viruses are not yet associated with cases of HPS. There are numerous strains of hantaviruses that have been identified in the Atlantic Forest (extends along the eastern coast of Brazil and into eastern Paraguay) and in the temperate grass and shrubland of Argentina. The third clade of rodents, those that harbor the Juquitiba, Oran, and Lechiguanas viruses, resides in the more humid Lower Chaco (a region that encompasses the flooded savannas of Southern Paraguay and forms a transitional environment between the arid Gran Chaco) and temperate coniferous forests in Argentina, Brazil, and Uruguay (De Araujo et al. 2015; Delfraro et al. 2008). Andes, Maciel, and Pergamino viruses reside in rodents in the temperate grass and shrubland biome in Argentina (Bohlman et al. 2002; González-Ittig et al. 2014). Andes virus is also found in Chile in Mediterranean woodland and shrub biome (Torres-Perez et al. 2004). Araraquara viruses are associated with rodents that largely reside within agriculturally transformed areas of Brazil (i.e., sugarcane production fields) that were formerly areas of moist broadleaf forest biome (De Araujo et al. 2015; de Sousa et al. 2008; Suzuki et al. 2004).

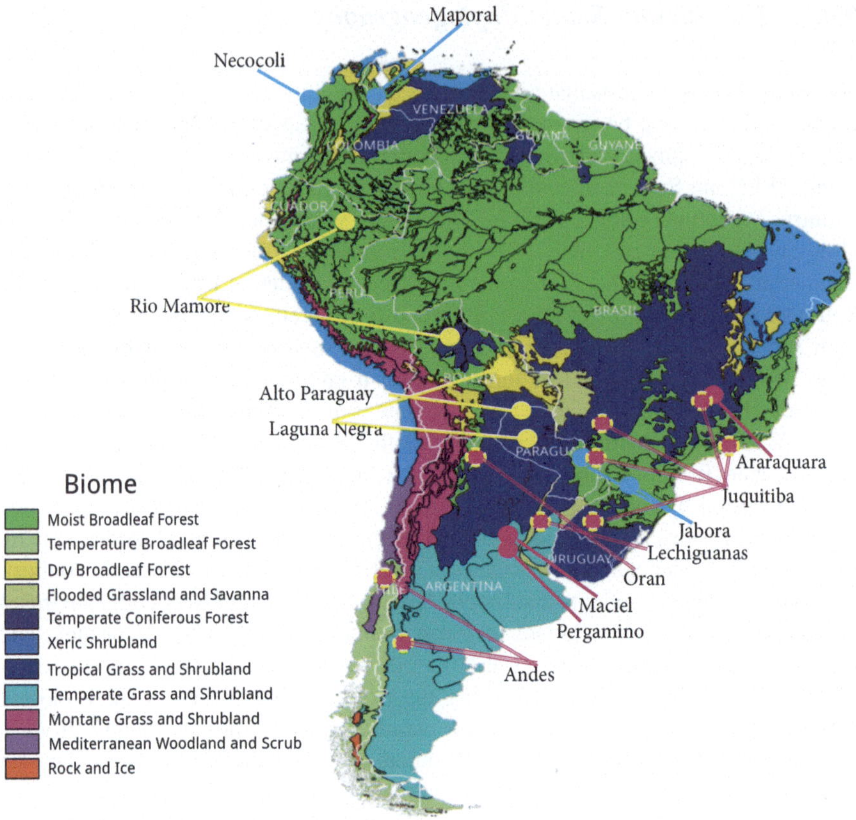

Fig. 9.2 Illustration of the major biomes of South America from the World Wildlife Fund terrestrial ecoregions data (Olson et al. 2001). Selected strains of closely related hantaviruses are presented from three distinct subclades as indicated by the color of the yellow or pink dot at their location. The viruses represented by the pink dot and pink with yellow represent two distinct lineages from one subclade

9.4.2 *Regional-Scale Land-Cover Association with Pathogen Prevalence*

Coarse-scale analysis has shown a relationship between hantavirus genetics and broad ecological pattern. Narrowing this view to a more regional scale begins to reveal how land cover (and land cover disturbance) affects the spatial variability of hantaviruses. Like all zoonotic disease, the ecology of each species of *Hantavirus* is closely related to that of its host organism; thus, generalization of virus-landscape relationships cannot be made without considering the habitat characteristics of the reservoir host. We do not have abundant data on the microhabitat characteristics of many host species (Lozada and Guthmann 1998). However, general habitat types defined at grains sizes of ~1 km might be useful to study preferences among various host species~~,~~ and be relevant to the landscape epidemiology of hantaviruses~~,~~ regardless of whether certain land cover types are more closely associated with the presence of the virus.

A regional-scale analysis of rodent reservoirs of hantaviruses in Paraguay (Goodin et al. 2006) showed that the host species do indeed show patterns of land cover preference, even when land cover is mapped into very general categories. The most common hantavirus host rodents in Paraguay, *Akodon montensis* and *Oligoryzomys* spp., both showed disproportionately high probabilities of occurring in areas subjected to large-scale agricultural disturbance (Goodin et al. 2006). Even more significant, however, was the fact that rodents found to be antibody positive for hantavirus (indicating exposure to the virus at some point) were more likely to be associated with the human-disturbed land cover types. This relationship held even when the underlying differences in habitat preference were controlled. This finding suggests that some aspect of land cover disturbance or change increases the likelihood that a member of a host species will be infected with the virus. While analysis at the regional scale shows that human land cover alteration is associated with hantavirus presence, discovering the specific nature and causes for this observation cannot be addressed at this scale. Questions of causation must be addressed at finer spatial scales, in large part because that is the scale to which most host populations and individuals are interacting with their environment (Lozada and Guthmann 1998; Owen et al. 2010).

9.4.3 Local Scale: Host and Habitat Associations

Simply presumed, because rodent~~s~~ species have specific requirements with respect to habitat, and because each hantavirus species seemingly persists only in certain rodent~~s~~ species, one might assume that a predictive map can be made for ~~of~~ the prevalence of a particular hantavirus based on knowledge of the associated rodent species. Unfortunately, for most rodents, we know very little about their general life cycles and other important biological information that is critical for modeling (e.g., age at sexual maturity, birth rates, litter sizes). Further, we have only a minimal amount of information on their habitat preferences, although some recent studies have indicated that fine-grained evaluations are necessary to understand rodent species distributions and community composition (Goodin et al. 2009; Lozada and Guthmann 1998; Poindexter et al. 2012; Schnell et al. 2010).

9.4.4 Microscale: Virus and Host Interactions

At the microscale level, pathogen survival and reproduction depend on the dynamics within the reservoir host, which in turn depend on the habitat or the reservoir. A rodent may live within an approximate 1 km range for most of its life barring fire, flooding, and other natural disasters. In the case of a viral pathogen, the virus must overcome physical barriers of the host and be compatible with cell receptors to gain entry into a target cell (Allen et al. 2012). Viral replication depends on host and viral genetics and other host factors such as prior pathogen or other immunogenic exposure, nutritional status, coinfection, age, sex, reproductive status, and the host immune response (Allen et al. 2012). These and other factors determine outcome when a pathogen enters a host, with the possibilities including severity of any associated disease and whether the pathogen either is to be cleared by the host immune system~~,~~ or the pathogen will persist~~s~~ in the host. For example, the survival of hantaviruses in nature depends on maintenance of persistent infections within their specific rodent reservoir. Hantaviruses infect and persist only in the rodent reservoir with which the virus has coevolved, and the infection is believed to last the life of the animal (Meyer and Schmaljohn 2000). Notably, persistent infection of rodent reservoirs by hantaviruses shows continuous virus replication, without complete clearance by the immune system, and no pathological changes (Jonsson et al. 2010; Vaheri et al. 2013). Humans are not a natural reservoir for these viruses, and as such humans typically become infected only upon contact with aerosolized excreta from the rodent reservoir. In humans, hantavirus infection can result in severe disease although outcomes vary with different hantaviral species. The molecular basis for different disease outcomes in humans has been attributed to difference in receptor preferences of nonpathogenic and pathogenic hantaviruses. Models that connect data on the outcomes associated with immune response in reservoir versus human hosts are just beginning to be developed.

9.5 Role of Mathematical Modeling in Spatial Ecology of Infectious Diseases

Mathematical models are valuable tools for synthesizing information and testing hypotheses to provide insight into how and why disease outbreaks or spatial patterns of infections might arise in wildlife populations. Several books and review articles summarize some of the modeling efforts on zoonotic infectious diseases (Alexander et al. 2012; Allen et al. 2012; Grenfell and Dobson 1995; Heesterbeek et al. 2015; Hudson et al. 2002; Lloyd-Smith et al. 2009). A variety of modeling formats, deterministic and stochastic, have been applied to the study of zoonotic diseases. These models include compartmental, agent-based, individual-based, metapopulation, network, and ecological niche, but these classifications overlap. Agent-based, individual-based, and metapopulation models may be classified under network models, where the nodes are infectious agents, individuals or populations, connected via an underlying network (Riley et al. 2015). Simple compartment models (SEIR—susceptible, exposed, infectious, and recovered) are connected via a dispersal network in what is often called a patch model or metapopulation model (Allen et al. 2009; Arino et al. 2005; McCormack and Allen 2007b). Ecological niche models contain less detail about individual dynamics, and instead they are closely related to landscape, presence/absence data, and GIS-based climatic and environmental data (Alexander et al. 2012; Peterson 2014).

Mathematical tractability and the complexity of interactions among pathogens, reservoirs, and human hosts and the environment have often restricted the model formulation to the reservoir host and to a single spatial scale—landscape, population, or within-host. Coupling temporal scales (hours to years or longer), biological complexity levels (genes to cells to ecosystems), and spatial scales (local to global) have been a continuous challenge to modelers (Heesterbeek et al. 2015). Recent theoretical investigations on coupling within-host and between-host models are advancing (Feng et al. 2013; Gilchrist and Coombs 2006; Mideo et al. 2008).

Hantaviruses and rabies are two examples where spatial patterns of infection involving multiple species have been observed (Chu et al. 2009, 2006; Haydon et al. 2002; Rhodes et al. 1998; Smith et al. 2002). Spatially explicit computer simulations that incorporate landscape heterogeneity and spatial genetic structure are being applied to study control of rabies and other zoonotic diseases (Alexander et al. 2012; Parratt et al. 2016; Real and Biek 2007; Rees et al. 2013). Models with multi-host species and multi-pathogens are being investigated. For example, mathematical models for hantavirus infection in rodents have been studied in the context of multiple host species, spatial spread, and environmental variability (Abramson and Kenkre 2002; Abramson et al. 2003; Allen et al. 2006a, b, 2009; McCormack and Allen 2007a, b). These models have shown that those random or seasonal variations which impact an ecosystems carrying capacity for a particular rodent species can trigger outbreaks when rodent densities and contacts rates are high (Allen et al. 2006b). In addition, theoretical analyses have shown that when multiple species, as opposed to a single species, are involved in the transmission process, there may be a dilution or amplification effect that impacts disease persistence (Dobson 2004; McCormack and Allen 2007a). Models for spatial spread among discrete patches have shown the importance of there being at least one patch where the disease persists (Arino et al. 2005; Allen et al. 2009; McCormack and Allen 2007b).

9.6 Conclusions

Wu and Loucks (1995) have suggested that the most significant contribution of hierarchy theory is as a framework for explicitly incorporating heterogeneity and scale into ecological analysis. In this chapter, we have tried to show some ways in which hierarchical consideration of spatial (and to some extent, temporal) scale can be incorporated into ecological analyses of zoonotic diseases. Our review of the literature also suggests some of the opportunities in infectious disease research resulting from the hierarchical consideration of scale but also some of the challenges. Spatial patterns of zoonotic hosts, the pathogens they harbor, and the host-zoonosis relationships may appear quite different depending on the scale at which we view them. Although patterns derived from different size scales are often referred to as "emergent" patterns, it is unclear whether the newly recognized large-scale or the small-scale patterns might not be emergent from scale levels either above or below them. Although we could view these scales to be nested subsets of one another, it may be more useful both conceptually and in practice, to view the patterns derived from different scales as complementary information requiring integration, rather than as contradictory results requiring amelioration.

One area of opportunity for the application of hierarchical concepts in zoonotic disease ecology corresponds to scale-related questions facing the global change research community in general; how does global environmental change, especially global warming, manifest itself spatially? It has long been recognized that the spatial distribution of both individuals and communities of species is linked to climate and that these are two-way linkages; however much uncertainty remains about where, when, and how species will respond to climate change (Potter et al. 2013). Fundamentally, the issue reduces to one of scale; the potential impacts of climate change are most often conceptualized at the macroscale (i.e., regional or global) but operate across a range of scales including those more proximate to the host or reservoir organisms (Ashcroft et al. 2009; Diffenbaugh et al. 2005; Suggitt et al. 2011). Adoption of the complementary, hierarchical view of scale provides a framework to address these and similar questions in disease ecology.

Model selection among the wide array of potential formats depends on the virus-host ecological system being investigated, data availability, and the questions to be addressed. Many challenges remain in model formulation, analysis, and simulation of zoonotic disease dynamics that relate to landscape and climate and the wide range of temporal and spatial scales (Allen et al. 2012; Buhnerkempe et al. 2015; Heesterbeek et al. 2015; Lloyd-Smith et al. 2009; Pellis et al. 2015). Addressing the challenges of scale can be met through mathematical and computational approaches and methods being developed in a variety of fields including computer science, ecology, geography, immunology, genetics, mathematics, statistics, and virology. This will require the continued close collaboration across numerous disciplines to converge toward models that reflect the immense biological diversity of pathogen-host ecology.

Acknowledgments CBJ and LJSA acknowledge the support of the National Science Foundation Grants (NSF) DMS-1516011 and DMS-1517719. CBJ and RDO acknowledge the support of the National Institutes of Health Grant (NIH) I103053. CBJ, LJSA, RDO, and DG acknowledge support from NIH R01 TW006986-01 through the NIH-NSF Ecology of Infectious Disease Program. RDO was partially supported by the Programa Nacional de Incentivo a los Investigadores (CONACYT, Paraguay).

Compliance with Ethical Standards

Funding This study was funded by National Science Foundation grants (DMS-1516011 and DMS-1517719) and National Institutes of Health grant (R010AI103053). This work was supported by a grant from the Fogarty International Center (R01 TW006986-01) under the NIH-NSF Ecology of Infectious Diseases initiative.

Conflict of Interest Douglas G. Goodin declares that he has no conflict of interest. Colleen B. Jonsson declares that she has no conflict of interest. Linda J. S. Allen declares that she has no conflict of interest. Robert D. Owen declares that he has no conflict of interest.

Ethical Approval This article does not contain any studies with human participants or animals performed by any of the authors.

References

Abramson G, Kenkre VM (2002) Spatiotemporal patterns in the hantavirus infection. Phys Rev E Stat Nonlin Soft Matter Phys 66:011912

Abramson G, Kenkre VM, Yates TL, Parmenter RR (2003) Traveling waves of infection in the hantavirus epidemics. Bull Math Biol 65:519–534

Alexander KA, Lewis BL, Marathe M, Eubank S, Blackburn JK (2012) Modeling of wildlife-associated zoonoses: applications and caveats. Vector Borne Zoonotic Dis 12:1005–1018

Allen TFH, Starr TB (eds) (1982) Hierarchy: perspectives for ecological complexity. University of Chicago Press, Chicago

Allen LJ, McCormack RK, Jonsson CB (2006a) Mathematical models for hantavirus infection in rodents. Bull Math Biol 68:511–524

Allen LJS, Allen EJ, Jonsson CB (2006b) The impact of environmental variation on hantavirus infection in rodents. In: Gumel AB, Castillo-Chavez C, Mickens RE, Clemence DP (eds) Modeling the dynamics of human diseases: emerging paradigms and challenges. AMS, Providence, RI, pp 1–15

Allen LJS, Wesley CL, Owen RD, Goodin DG, Koch D, Jonsson CB, Chu Y-K, Hutchinson S, Paige R (2009) A habitat-based model for the spread of hantavirus between reservoir and spillover species. J Theor Biol 260:510–522

Allen LJ, Brown VL, Jonsson CB, Klein SL, Laverty SM, Magwedere K, Owen JC, van den Driessche P (2012) Mathematical modeling of viral zoonoses in wildlife. Nat Resour Model 25:5–51

Arino J, Davis JR, Hartley D, Jordan R, Miller JM, van den Driessche P (2005) A multi-species epidemic model with spatial dynamics. Math Med Biol 22:129–142

Ashcroft MB, Chisholm LA, French KO (2009) Climate change at the landscape scale: predicting fine-grained spatial heterogeneity in warming and potential refugia for vegetation. Global Change Biol 15:656–667

Bohlman MC, Morzunov SP, Meissner J, Taylor MB, Ishibashi K, Rowe J, Levis S, Enria D, St Jeor SC (2002) Analysis of hantavirus genetic diversity in Argentina: S segment-derived phylogeny. J Virol 76:3765–3773

Brownstein JS, Skelly DK, Holford TR, Fish D (2005) Forest fragmentation predicts local scale heterogeneity of Lyme disease risk. Oecologia 146:469–475

Buhnerkempe MG, Roberts MG, Dobson AP, Heesterbeek H, Hudson PJ, Lloyd-Smith JO (2015) Eight challenges in modelling disease ecology in multi-host, multi-agent systems. Epidemics 10:26–30
Chu YK, Milligan B, Owen RD, Goodin DG, Jonsson CB (2006) Phylogenetic and geographical relationships of hantavirus strains in Eastern and Western Paraguay. Am J Trop Med Hyg 75:1127–1134
Chu YK, Goodin D, Owen RD, Koch D, Jonsson CB (2009) Sympatry of two hantavirus strains, Paraguay, 2003-2007. Emerg Infect Dis 15:1977–1980
Csillag F, Fortin M-J, Dungan JL (2000) On the limits and extensions of the definition of scale. Bull Ecol Soc Am 81:230–232
De Araujo J, Duré AIL, Negrão R, Ometto T, Thomazelli LM, Durigon EL (2015) Co-circulation in a single biome of the Juquitiba and Araraquara hantavirus detected in human sera in a sub-tropical region of Brazil. J Med Virol 87:725–732
De Beurs KM, Henebry GM (2004) Land surface phenology, climatic variation, and institutional change: analyzing agricultural land cover change in Kazakhstan. Remote Sens Environ 8:497–509
de Oliveira RC, Padula PJ, Gomes R, Martinez VP, Bellomo C, Bonvicino CR, Freire E Lima DI, Bragagnolo C, Caldas ACS, D'Andrea PS, de Lemos ERS (2011) Genetic characterization of hantaviruses associated with sigmodontine rodents in an endemic area for hantavirus pulmonary syndrome in southern Brazil. Vector Borne Zoonotic Dis 11:301
de Sousa RL, Moreli ML, Borges AA, Campos GM, Livonesi MC, Figueiredo LT, Pinto AA (2008) Natural host relationships and genetic diversity of rodent-associated hantaviruses in Southeastern Brazil. Intervirology 51:299–310
Delfraro A, Tomé L, D'Elía G, Clara M, Achával F, Russi JC, Rodonz JRA (2008) Juquitiba-like hantavirus from 2 nonrelated rodent species, Uruguay. Emerg Infect Dis 14:1447–1451
Diffenbaugh NS, Pal JS, Trapp RJ, Giorgi F (2005) Fine-scale processes regulate the response of extreme events to global climate change. Proc Natl Acad Sci U S A 102:15774–15778
Dobson A (2004) Population dynamics of pathogens with multiple host species. Am Nat 164(Suppl 5):S64–S78
Epstein JH, Field HE, Luby S, Pulliam JR, Daszak P (2006) Nipah virus: impact, origins, and causes of emergence. Curr Infect Dis Rep 8:59–65
Estrada-Peña A, Oteo JA (1991) Vectors of Lyme disease in Spain. Res Rev Parasitol 51:101–102
Estrada-Peña A, Venzal JM, Acebedo Sanchez C (2006) The tick *Ixodes ricinus*: distribution and climate preferences in the Western Palaearctic. Med Vet Entomol 20:350–359
Feng Z, Velasco-Hernandez J, Tapia-Santos B (2013) A mathematical model for coupling within-host and between-host dynamics in an environmentally-driven infectious disease. Math Biosci 241:49–55
Fulhorst CF, Cajimat MNB, Utrera A, Milazzo ML, Duno GM (2004) Maporal virus, a hantavirus associated with the fulvous pygmy rice rat (*Oligoryzomys fulvescens*) in Western Venezuela. Virus Res 104:139–144
Gilchrist MA, Coombs D (2006) Evolution of virulence: interdependence, constraints, and selection using nested models. Theor Popul Biol 69:145–153
Giraudoux P, Craig PS, Delattre P, Bartholomot B, Harraga S, Quere JP, Raoul F, Wang Y, Shi D, Vuitton DA (2003) Interactions between landscape changes and host communities can regulate *Echinococcus multilocularis* transmission. Parasitology 127:121–131
Glass GE, Cheek JE, Patz JA, Shields TM, Doyle TJ, Thoroughman DA, Hunt DK, Enscore RE, Gage KL, Irland C, Peters CJ, Bryan R (2000) Using remotely sensed data to identify areas at risk for hantavirus pulmonary syndrome. Emerg Infect Dis 6:238–247
Glass GE, Yates TL, Fine JB, Shields TM, Kendall JB, Hope AG, Parmenter CA, Peters CJ, Ksiazek TG, Li CS, Patz JA, Mills JN (2002) Satellite imagery characterizes local animal reservoir populations of Sin Nombre virus in the Southwestern United States. Proc Natl Acad Sci U S A 99:16817–16822

González-Ittig RE, Rivera PC, Levis SC, Calderón GE, Gardenal CN (2014) The molecular phylogenetics of the genus *Oligoryzomys* (Rodentia: C ricetidae) clarifies rodent host–hantavirus associations. Zool J Linnean Soc 171:457–474

Goodin DG, Koch DE, Owen RD, Chu Y-K, Hutchinson JMS, Jonsson CB (2006) Land cover associated with hantavirus presence in Paraguay. Global Ecol Biogeogr 15:519

Goodin DG, Paige R, Owen RD, Ghimire K, Koch DE, Chu YK, Jonsson CB (2009) Microhabitat characteristics of *Akodon montensis*, a reservoir for hantavirus, and hantaviral seroprevalence in an Atlantic forest site in Eastern Paraguay. J Vector Ecol 34:104–113

Grenfell BT, Dobson AP (1995) Ecology of disease in natural populations. Cambridge University Press, Cambridge, UK

Haydon DT, Cleaveland S, Taylor LH, Laurenson MK (2002) Identifying reservoirs of infection: a conceptual and practical challenge. Emerg Infect Dis 8:1468–1473

Heesterbeek H, Anderson RM, Andreasen V, Bansal S, De Angelis D, Dye C, Eames KT, Edmunds WJ, Frost SD, Funk S, Hollingsworth TD, House T, Isham V, Klepac P, Lessler J, Lloyd-Smith JO, Metcalf CJ, Mollison D, Pellis L, Pulliam JR, Roberts MG, Viboud C, Isaac Newton Institute, I.D.D.C. (2015) Modeling infectious disease dynamics in the complex landscape of global health. Science 347:4339

Hudson PJ, Rizzoli A, Grenfell BT, Heesterbeek H, Dobson AP (2002) The ecology of wildlife diseases. Oxford University Press, Oxford

Jenerette GD, Wu J (2000) On the definitions of scale. Bull Ecol Soc Am 81:104–105

Johnson AM, Bowen MD, Ksiazek TG, Williams RJ, Bryan RT, Mills JN, Peters CJ, Nichol ST (1997) Laguna Negra Virus associated with HPS in Western Paraguay and Bolivia. Virology 238:115–127

Jones KE, Patel NG, Levy MA, Storeygard A, Balk D, Gittleman JL, Daszak P (2008) Global trends in emerging infectious diseases. Nature 451:990

Jones BA, Grace D, Kock R, Alonso S, Rushton J, Said MY, McKeever D, Mutua F, Young J, McDermott J, Pfeiffer DU (2013) Zoonosis emergence linked to agricultural intensification and environmental change. Proc Natl Acad Sci USA 110:8399

Jonsson CB, Figueiredo LT, Vapalahti O (2010) A global perspective on hantavirus ecology, epidemiology, and disease. Clin Microbiol Rev 23:412–441

Kilgore PE, Peters CJ, Mills JN, Rollin PE, Armstrong L, Khan AS, Ksiazek TG (1995) Prospects for the control of Bolivian hemorrhagic fever. Emerg Infect Dis 1:97–100

Kitron U, Clennon JA, Cecere MC, Gürtler RE, King CH, Vazquez-Prokopec G (2006) Upscale or downscale: applications of fine scale remotely sensed data to Chagas disease in Argentina and schistosomiasis in Kenya. Geospatial Health 1:49–58

Langlois JP, Fahrig L, Merriam G, Artsob H (2001) Landscape structure influences continental distribution of hantavirus in deer mice. Landsc Ecol 16:255–266

Lidicker WZ (1995) Landscape approaches in mammalian ecology and conservation. University of Minnesota Press, Minneapolis

Lloyd-Smith JO, George D, Pepin KM, Pitzer VE, Pulliam JR, Dobson AP, Hudson PJ, Grenfell BT (2009) Epidemic dynamics at the human-animal interface. Science 326:1362–1367

Londoño AF, Díaz FJ, Agudelo-Flórez P, Levis S, Rodas JD (2011) Genetic evidence of hantavirus infections in wild rodents from northwestern Colombia. Vector Borne Zoonotic Dis 11:701

Lozada M, Guthmann N (1998) Microhabitat selection under experimental conditions of three sigmodontine rodents. Ecoscience 5:51–55

McCormack RK, Allen LJS (2007a) Disease emergence in multi-host epidemic models. Math Med Biol 24:17–34

McCormack RK, Allen LJS (2007b) Multi-patch deterministic and stochastic models for wildlife diseases. J Biol Dyn 1:63–85

McFarlane RA, Sleigh AC, McMichael AJ (2013) Land-use change and emerging infectious disease on an island continent. Int J Environ Res Public Health 10:2699–2719

Meentemeyer V (1989) Geographical perspectives of space, time, and scale. Landsc Ecol 3:163–173

Meentemeyer RK, Haas SE, Vaclavik T (2012) Landscape epidemiology of emerging infectious diseases in natural and human-altered ecosystems. Annu Rev Phytopathol 50:379–402

Meyer BJ, Schmaljohn CS (2000) Persistent hantavirus infections: characteristics and mechanisms. Trends Microbiol 8:61–67

Mideo N, Alizon S, Day T (2008) Linking within- and between-host dynamics in the evolutionary epidemiology of infectious diseases. Trends Ecol Evol 23:511–517

Olson DM, Dinerstein E, Wikramanayake ED, Burgess ND, Powell GVN, Underwood EC, D'Amico JA, Itoua I, Strand HE, Morrison JC, Loucks CJ, Allnutt TF, Ricketts TH, Kura Y, Lamoreux JF, Wettengel WW, Hedao P, Kassem KR (2001) Terrestrial ecoregions of the world: a new map of life on Earth. Bioscience 51:933–938

Ostfeld RS, Glass GE, Keesing F (2005) Spatial epidemiology: an emerging (or re-emerging) discipline. Trends Ecol Evol 20:328–336

Owen RD, Goodin DG, Koch DE, Chu Y-K, Jonsson CB (2010) Spatiotemporal variation in *Akodon montensis* (Cricetidae: Sigmodontinae) and hantaviral seroprevalence in a subtropical forest ecosystem. J Mammal 91:467

Palma RE, Polop JJ, Owen RD, Mills JN (2012) Ecology of rodent-associated hantaviruses in the Southern Cone of South America: Argentina, Chile, Paraguay, and Uruguay. J Wildl Dis 48:267–281

Parratt SR, Numminen E, Laine A-L (2016) Infectious disease dynamics in heterogeneous landscapes. Annu Rev Ecol Evol Syst 47:281–306

Patz JA, Olson SH, Uejio CK, Gibbs HK (2008) Disease emergence from global climate and land use change. New Emerg Infect Dis 92:1473–1491

Pavlovskiĭ EN (1966) Natural nidality of transmissible diseases, with special referece to the landscape epidemiology of zooanthroponoses. University of Illinois Press, Urbana, IL

Pellis L, Ball F, Bansal S, Eames K, House T, Isham V, Trapman P (2015) Eight challenges for network epidemic models. Epidemics 10:58–62

Peterson AT (2014) Mapping disease transmission risk in geographic and ecological contexts. John Hopkins University Press, Baltimore, MD

Pinzon JE, Wilson JM, Tucker CJ, Arthur R, Jahrling PB, Formenty P (2004) Trigger events: enviroclimatic coupling of Ebola hemorrhagic fever outbreaks. Am J Trop Med Hyg 71:664–674

Poindexter CJ, Schnell GD, Sánchez-Hernández C, Romero-Almaraz MdL, Kennedy ML, Best TL, Wooten MC, Owen RD (2012) Variation in habitat use of coexisting rodent species in a tropical dry deciduous forest. Mammal Biol 77:249–257

Potter KA, Woods A, Pincebourde S (2013) Microclimatic challenges in global change biology. Global Change Biol 19:2932–2939

Pulliam JR, Epstein JH, Dushoff J, Rahman SA, Bunning M, Jamaluddin AA, Hyatt AD, Field HE, Dobson AP, Daszak P, Henipavirus Ecology Research G (2012) Agricultural intensification, priming for persistence and the emergence of Nipah virus: a lethal bat-borne zoonosis. J R Soc Interface 9:89–101

Real LA, Biek R (2007) Spatial dynamics and genetics of infectious diseases on heterogeneous landscapes. J R Soc 4:935–948

Reed BC, Schwartz MD, Xiao X (2009) Remote sensing phenology. In: Noormets A (ed) Phenology of ecosystem processes: applications in global change research. Springer, New York, NY, pp 231–246

Rees EE, Pond BA, Tinline RR, Bélanger D (2013) Modelling the effect of landscape heterogeneity on the efficacy of vaccination for wildlife infectious disease control. J Appl Ecol 50:881–891

Reisen WK (2010) Landscape epidemiology of vector-borne diseases. Annu Rev Entomol 55:461–483

Rhodes CJ, Atkinson RPD, Anderson RM, Macdonald DW (1998) Rabies in Zimbabwe: reservoir dogs and the implications for disease control. Philos Trans Soc Lond B Biol Sci 353:999–1010

Richter MH, Hanson JD, Cajimat MN, Milazzo ML, Fulhorst CF (2010) Geographical range of Rio Mamoré virus (family Bunyaviridae, genus Hantavirus) in association with the small-eared pygmy rice rat (*Oligoryzomys microtis*). Vector Borne Zoonotic Dis 10:613

Riley S, Eames K, Isham V, Mollison D, Trapman P (2015) Five challenges for spatial epidemic models. Epidemics 10:68–71

Russell CA, Real LA, Smith DL (2006) Spatial control of rabies on heterogeneous landscapes. PLoS One 1:e27

Schnell GD, De Lourdes Romero-almaraz M, Martínez-chapital ST, Sánchez-hernández C, Kennedy ML, Best TL, Wooten MC, Owen RD (2010) Habitat use and demographic characteristics of the west Mexican cotton rat (*Sigmodon mascotensis*). Mammalia 74:379–393

Simone I, Cagnacci F, Provensal C, Polop J (2010) Environmental determinants of the small mammal assemblage in an agroecosystem of central Argentina: the role of *Calomys musculinus*. Mammal Biol 75:496–509

Smith DL, Lucey B, Waller LA, Childs JE, Real LA (2002) Predicting the spatial dynamics of rabies epidemics on heterogeneous landscapes. Proc Natl Acad Sci U S A 99:3668–3672

Smith KF, Goldberg M, Rosenthal S, Carlson L, Chen J, Chen C, Ramachandran S (2014) Global rise in human infectious disease outbreaks. J Royal Soc 11:20140950

Suggitt AJ, Gillingham PK, Hill JK, Huntley B, Kunin WE, Roy DB, Thomas CD (2011) Habitat microclimates drive fine-scale variation in extreme temperatures. Oikos 120:1–8

Suzuki A, Bisordi I, Levis S, Garcia J, Pereira LE, Souza RP, Sugahara TK, Pini N, Enria D, Souza LT (2004) Identifying rodent hantavirus reservoirs, Brazil. Emerg Infect Dis 10:2127–2134

Torres-Perez F, Navarrete-Droguett J, Aldunate R, Yates TL, Mertz GJ, Vial PA, Ferres M, Marquet PA, Palma RE (2004) Peridomestic small mammals associated with confirmed cases of human hantavirus disease in southcentral Chile. Am J Trop Med Hyg 70:305–309

Tucker CJ, Wilson JM, Mahoney R, Anyamba A, Linthicum KJ, Myers M (2002) Climatic and ecological context of the 1994-1996 ebola outbreaks. Photogr Eng Rem Sens 68:147–152

Vaheri A, Strandin T, Hepojoki J, Sironen T, Henttonen H, Makela S, Mustonen J (2013) Uncovering the mysteries of hantavirus infections (Disease/Disorder overview). Nat Rev Microbiol 11:539

Watts DJ, Muhamad R, Medina DC, Dodds PS (2005) Multiscale, resurgent epidemics in a hierarchical metapopulation model. Proc Natl Acad Sci U S A 102:11157

Woolhouse MEJ, Gowtage-Sequeria S (2005) Host range and emerging and reemerging pathogens. Emerg Infect Dis 11:1842

Wu J (1999) Hierarchy and scaling: extrapolating information along a scaling ladder. Can J Rem Sens 25:367–380

Wu J, Loucks OL (1995) From balance of nature to hierarchical patch dynamics: a paradigm shift in ecology. Q Rev Biol 70:439–466

Yahnke CJ, Meserve PL, Ksiazek TG, Mills JN (2001) Patterns of infection with Laguna Negra virus in wild populations of *Calomys laucha* in the central Paraguayan chaco. Am J Trop Med Hyg 65:768

Yates TL, Mills JN, Parmenter CA, Ksiazek TG, Parmenter RR, Vande Castle JR, Calisher CH, Nichol ST, Abbott KD, Young JC, Morrison ML, Beaty BJ, Dunnum JL, Baker RJ, Salazar-Bravo J, Peters CJ (2002) The ecology and evolutionary history of an emergent disease: hantavirus pulmonary syndrome. BioScience 52:989–998

The manufacturer's authorised representative in the EU is Springer Nature Customer Service Centre GmbH, Europaplatz 3, 69115 Heidelberg, Germany. If you have any concerns regarding our products, please contact ProductSafety@springernature.com

Printed and bound by CPI Group (UK) Ltd, Croydon, CR0 4YY
30/09/2026
02225434-0013